'A World of New Ideas'

VOLUME I

SCIENTISTS OF WALES

Series Editors
Gareth Ffowc Roberts
Bangor University

John V. Tucker
Swansea University

Iwan Rhys Morus
Aberystwyth University

SCIENTISTS OF WALES

'A World of New Ideas'

1650–1820

VOLUME I: THE ISLES

PAUL FRAME

UNIVERSITY OF WALES PRESS
2025

www.uwp.co.uk

British Library Cataloguing-in-Publication Data
A catalogue record for this book is available from the British Library.

ISBN   978-1-83772-009-5
eISBN   978-1-83772-010-1

The right of Paul Frame to be identified as author of this work has been asserted in accordance with sections 77, 78 and 79 of the Copyright, Designs and Patents Act 1988.

The University of Wales Press gratefully acknowledges the support of the Books Council of Wales in publishing this title.

THE LEARNED SOCIETY OF WALES
CYMDEITHAS DDYSGEDIG CYMRU

MIX
Paper | Supporting responsible forestry
FSC® C013604

Typeset by Marie Doherty
Printed and bound by CPI Group (UK) Ltd, Croydon, CR0 4YY

'We are all born like Moses with a Veil over the Face: This is it, which hinders the prospect of that Intellectual shining light, which God hath placed in us; And to tell a truth that concerns all Mankind, the greatest Mystery both in Divinity, and Philosophy is, How to remove it.'

Thomas Vaughan, *Anthroposophia Theomagica* (1650)

'A true story of science should not be a parade of famous names but should represent the ideas and achievements of the nameless majority of scientifically minded people.'

Seb Falk, 'Prologue' to *The Light Ages, A Medieval Journey of Discovery* (2020)

CONTENTS

SERIES EDITORS' PREFACE

Wales has a long and important history of contributions to scientific and technological discovery and innovation stretching from the Middle Ages to the present day. From medieval scholars to contemporary scientists and engineers, Welsh individuals have been at the forefront of efforts to understand and control the world around us. For much of Welsh history, science has played a key role in Welsh culture: bards drew on scientific ideas in their poetry; renaissance gentlemen devoted themselves to natural history; the leaders of early Welsh Methodism filled their hymns with scientific references. During the nineteenth century, scientific societies flourished and Wales was transformed by engineering and technology. In the twentieth century the work of Welsh scientists continued to influence developments in their fields.

Much of this exciting and vibrant Welsh scientific history has now disappeared from historical memory. The aim of the Scientists of Wales series is to resurrect the role of science and technology in Welsh history. Its volumes trace the careers and achievements of Welsh investigators, setting their work within their cultural contexts. They demonstrate how scientists and engineers have contributed to the making of modern Wales as well as showing the ways in which Wales has played a crucial role in the emergence of modern science and engineering.

RHAGYMADRODD GOLYGYDDION Y GYFRES

O'r Oesoedd Canol hyd heddiw, mae gan Gymru hanes hir a phwysig o gyfrannu at ddarganfyddiadau a menter gwyddonol a thechnolegol. O'r ysgolheigion cynharaf i wyddonwyr a pheirianwyr cyfoes, mae Cymry wedi bod yn flaenllaw yn yr ymdrech i ddeall a rheoli'r byd o'n cwmpas. Mae gwyddoniaeth wedi chwarae rôl allweddol o fewn diwylliant Cymreig am ran helaeth o hanes Cymru: arferai'r beirdd llys dynnu ar syniadau gwyddonol yn eu barddoniaeth; roedd gan wŷr y Dadeni ddiddordeb brwd yn y gwyddorau naturiol; ac roedd emynau arweinwyr cynnar Methodistiaeth Gymreig yn llawn cyfeiriadau gwyddonol. Blodeuodd cymdeithasau gwyddonol yn ystod y bedwaredd ganrif ar bymtheg, a thrawsffurfiwyd Cymru gan beirianneg a thechnoleg. Ac, yn ogystal, bu gwyddonwyr Cymreig yn ddylanwadol mewn sawl maes gwyddonol a thechnolegol yn yr ugeinfed ganrif.

Mae llawer o'r hanes gwyddonol Cymreig cyffrous yma wedi hen ddiflannu. Amcan cyfres Gwyddonwyr Cymru yw i danlinellu cyfraniad gwyddoniaeth a thechnoleg yn hanes Cymru, â'i chyfrolau'n olrhain gyrfaoedd a champau gwyddonwyr Cymreig gan osod eu gwaith yn ei gyd-destun diwylliannol. Trwy ddangos sut y cyfrannodd gwyddonwyr a pheirianwyr at greu'r Gymru fodern, dadlennir hefyd sut y mae Cymru wedi chwarae rhan hanfodol yn natblygiad gwyddoniaeth a pheirianneg fodern.

ACKNOWLEDGEMENTS

As the reader will learn when they delve further into the two volumes of *A World of New Ideas*, significant contributions were made by scientists of Wales in the period that it covers – the long eighteenth century from 1650 to 1820. A period bounded, for our purposes in this first volume, by the alchemical publications of Thomas Vaughan in the 1650s, and publication of the final volume of Abraham Rees's monumental forty-five-volume encyclopaedia in 1820. The aim, though, is not to puff-up scientists of Wales before slotting them into a seamless grand narrative. Rather, it is to recognise and illustrate their contribution – be it great or small, successful or unsuccessful – in the hope of eliciting a wider appreciation of their place in the history of science. In attempting this, *A World of New Ideas* is, at least in part, a synthesis of previous work. As such, it is heavily indebted to all those researchers whose individual efforts have contributed to its making. To all of them I offer my profound thanks, and I hope that they will find themselves properly referenced. I apologise in advance for any inadvertent omissions.

For help with my own original research, I send thanks to the librarians and archivists at Llyfrgell Genedlaethol Cymru/National Library of Wales, the British Library, the Royal Society Library, Beinecke Library at Yale University, the House of Commons Library, Lambeth Palace Library, the Oxford History of Science Museum, and to the authors of the various websites listed in the bibliography. My thanks also go to the University of Wales, Guild of Graduates for a 2017 Study Award that helped set my research in motion.

On a personal level, the book would never have been completed without the infectious enthusiasm of John Tucker. He has read each chapter of this volume a number of times over the course of several years, and his valuable comments and editing have substantially improved the text. Our regular meetings and numerous large cappuccinos were more important in keeping the project going than he will perhaps have realised. I am also happy to thank others who have provided illustrations, read all or parts of the text, and provided help with queries and snippets of valuable information. They are Nicola Bennetts, Kenneth Brassil, Mark Collins, Mary-Ann Constantine, Martin Griffiths, Andrew Hawke, Robert Wynn Jones, D. Densil Morgan, Stan Pridmore, Tracey Rihll, David Smith and, in particular, Gareth Ffowc Roberts. Thanks also to Bob and Heather Jones for accommodation during a research trip in London, to my neighbours for their encouragement when the will flagged, and to Geoff and Gareth at GPS Print and Design in Swansea. Finally, thanks go to Martin Fitzpatrick for gifting me a wonderful tranche of history of science-related books when this project was first mooted, and without which it might not have got off the ground.

In researching the book, and the second volume that will hopefully follow soon, something of a revelation has been the linguistic and translation skills possessed by many of the scientists of Wales mentioned. It also became clear early on that in order to obtain something approaching a balanced understanding of the development of science among the scientists of Wales in the long eighteenth century, records written in Welsh and Latin needed to be accessed. This has proven challenging, given that I do not speak or read my native tongue, let alone Latin. Luckily, help for this linguistically challenged Welshman came from my less linguistically challenged friends and colleagues Ken Brassil, Gareth Ffowc Roberts and, for Latin translations, the enthusiastic help of Dr Ineke Loots of Utrecht. Sadly, Ineke passed away near the end of the project and I dedicate the book, in part, to her memory.

My thanks also go to my anonymous reviewer, all members of the Scientists of Wales editorial board, and, in particular, to Llion Wigley at the University of Wales Press (UWP) for his extraordinary patience in the face of numerous delays and claims of 'almost done'. Also at

UWP, my thanks go to Steven Goundrey, Adam Burns, Dafydd Jones, Marie Doherty and Sarah Meaney for their editorial, design and production skills. Also to Georgia Winstone, Elin Williams and Maria Vassilopoulos.

Finally, any omissions or mistakes are mine and mine alone. I shall end with the words of the deist, philosopher, Newtonian and physician, Thomas Morgan (d.1743), one of those frustrating personages who may or may not have been a Welshman, but whose words, in his *Philosophical Principles of Medicine* (1725), are anyway universal:

> If I have here contributed any thing towards the setting of this most necessary part of natural Philosophy in a clearer Light, or towards the engaging some abler Pen to undertake a more complete Work of the same kind, my Design in it will be sufficiently answer'd: But if I should unhappily fail of any such Success, I doubt not but you will at least pardon the freedom of the Attempt, in consequence of the goodness of the Intention.

LIST OF ILLUSTRATIONS

LIST OF ABBREVIATIONS

AML-I, II	Hugh Owen (ed.), 'Additional Letters of the Morrises of Anglesey (1735–1786)', *Y Cymmrodor* XLIX, Parts I and II (London, 1947, 1949)
APS	American Philosophical Society
AV:NV	Donald R. Dickson (ed. and trans.), *Thomas and Rebecca Vaughan's Aqua Vitæ: Non Vitis (British Library MS Sloane 1741)* (Tempe AZ, 2001)
BBCS	*Bulletin of the Board of Celtic Studies*
BL	British Library
CHS-3	Katherine Park and Lorraine Daston (eds), *The Cambridge History of Science: Volume 3, Early Modern Science* (2006; Cambridge, 2008)
CHS-4	Roy Porter (ed.), *The Cambridge History of Science: Volume 4, Eighteenth-Century Science* (Cambridge, 2003)
CIM-I, II, III	Geraint H. Jenkins, Ffion Mair Jones and David Ceri Jones (eds), *The Correspondence of Iolo Morganwg*, 3 vols (Cardiff, 2007)
CR-I, II, III	*The Cambrian Register* (1796, 1799, 1818)
CWP	*Calendar of Wynn (of Gwydir) Papers 1515–1690* (Aberystwyth, 1926)

DWB	*Dictionary of Welsh Biography*
FSA	Fellow of the Society of Antiquaries
FGS	Fellow of the Geological Society
FLS	Fellow of the Linnean Society
FRS	Fellow of the Royal Society
FRSE	Fellow of the Royal Society of Edinburgh
Gunther	R. T. Gunther, *Early Science in Oxford, vol. XIV: Life and Letters of Edward Lhwyd* (Oxford, 1945)
Lhwyd	Brynley F. Roberts, *Edward Lhwyd: c.1600–1709: Naturalist, Antiquary, Philologist* (Cardiff, 2022)
ML-I, II	John H. Davies (ed.), *The Letters of Lewis, Richard, William and John Morris of Anglesey, (Morrisiaid Mon)* 1728–1765 (Aberystwyth, 1907, 1909)
MS	Manuscripts
(n.d.)	no date(s)
NLW	National Library of Wales
NLWJ	*National Library of Wales Journal*
(n.p.)	not paginated/no publication details
ODNB	*Oxford Dictionary of National Biography*
OED	*Oxford English Dictionary*
OHSM	University of Oxford, Museum of the History of Science
RSC	Royal Society Collections
SPCK	Society for the Promotion of Christian Knowledge
THSC	*Transactions of the Honourable Society of Cymmrodorion*
UWP	University of Wales Press
WHR	*Welsh History Review*

WTV	Alan Rudrum (ed.) with Jennifer Drake-Brockman, *The Works of Thomas Vaughan* (Oxford, 1984)

(Intro.)	Biographical Introduction
(*AMA*)	*Anima Magica Abscondita (1650)*
(*AT*)	*Anthroposophia Theomagica (1650)*
(*E*)	*Euphrates, or the Waters of the East (1655)*
(*F&C*)	*Preface to the Fama & Confessio (1652)*
(*LL*)	*Lumen de Lumine (1651)*
(*MA*)	*Magia Adamica (1650)*
(*MM*)	*The Man–Mouse (1650)*
(*SW*)	*The Second Wash (1651)*

PROLOGUE: WALES IN THE HISTORY OF SCIENCE

> An iron-founder from Carmarthen by the name of Philip
> Vaughan deserves a special reference in this history, since
> he was the first person to patent a ball bearing of great
> merit for carrying the loads on 'certain axle trees, axle
> arms, and boxes for light and heavy wheel carriages'.[1]
>
> *NASA Technical Memorandum 81689* (1981)

B all bearings are used in everything from roller skates to the Hubble telescope and the various 'rovers' that NASA has sent to Mars in recent years.[2] Yet, despite the importance of this piece of mechanical engineering science – first patented from Wales in 1794 – neither ball bearings nor their patentee figure prominently, if at all, in histories of Wales or of British science. Why?

Making the Invisible Visible

Since publication of Hugh Kearney's 1989 work *The British Isles: A History of Four Nations* some historians have presented the history of Britain from perspectives informed by the cultural differences between the four nations of the Isles – England, Ireland, Scotland and Wales.[3] Nevertheless, Britain's peripheral histories, in particular that of Wales, continue to suffer from the John of Gaunt syndrome: 'this scepter'd isle ... this England ... bound in with the triumphant sea'.[4]

When introducing *The Scientific Revolution in National Context* (1992) its editors, Roy Porter and Mikuláš Teich, state their (surely correct) conviction that:

> the twists and turns of global scientific change will not be understood without regard for questions of indigenous language, education, communication networks, institutions, economics, social relations, politics, religious confession, patronage, and other comparable elements that can be called its 'national context'.[5]

In a number of these areas Wales occupied a unique position relative to the wider British context in our period of interest – the long eighteenth century from 1650 to 1820. First and foremost, English was neither the indigenous language of the country nor the *lingua franca* of the majority (90 per cent) of its Cymraeg/Welsh speaking population. A significant proportion of this majority, unusually for the times, were literate in their mother tongue by the mid-eighteenth century, thanks to such institutions as the circulating schools established by Griffith Jones (see Chapter 6). Yet, beyond its Dissenter academies, the nation lacked (though not through want of trying) institutions of higher education comparable to those founded since the Middle Ages in England (Oxford and Cambridge in the twelfth century) and Scotland (St Andrews in 1411), and the early modern period in Ireland (Trinity College, Dublin in 1592). Wales also lacked, at least until the later eighteenth and early nineteenth centuries, the sort of cultural and scientific societies that helped maintain civic society in Scotland and Ireland. The Philosophical Society of Edinburgh, for example, was founded in 1737, with the Royal Society of Edinburgh following in 1783. The Dublin Philosophical Society was founded in 1683, and the Dublin Society of 1731 became the Royal Dublin Society in 1820. The Irish Academy appeared in 1785 and became the Royal Irish Academy in 1786, a title it retains in republican Ireland today.

Despite such differences, and his earlier noted comments with Mikuláš Teich, Roy Porter, though willing to devote a chapter to developments in the Scottish Enlightenment, chose not to 'focus on controversies taking place *within* and *about* Ireland and Wales' in his otherwise seminal volume *Enlightenment: Britain and the Creation of the Modern World* (2000) – a volume in which science plays a conspicuous role. He also felt justified in using the term 'English' somewhat interchangeably with 'British'. The use of national terms, he declared, 'will be clarified by context; if this laxity sometimes seems confusing or is galling to modern nationalist sensibilities, it reflects the realities of a time when "English" was commonly applied to people born anywhere in "our Isles"'.[6] It is almost *de rigueur* to begin a book of Welsh history with a *crie de cœur* over such sentiments, and despite a wish that it could be otherwise this book is no exception.

To criticise the approach taken by Roy Porter in *Enlightenment* is neither to deny the grains of truth within it, nor to pander to modern nationalist sensibilities. It is certainly reasonable to argue for the prominence of England and Scotland in the Enlightenment and scientific history of 'our Isles'. It is not reasonable for their prominence to largely exclude discussion of the remaining nations *in a book claiming to be about Britain*. Furthermore, the term 'English' may have been 'applied to people born anywhere in "our Isles"' in the eighteenth century, but doing so was not without controversy, as David Hume, a Scot, pointed out:

> Some [English] hate me because I am not a Tory, some because I am not a Whig, some because I am not a Christian, and all because I am a Scotsman. Can you seriously talk of my continuing an Englishman? Am I, or are you, an Englishman? Will they allow us to be so?[7]

In Wales, a small linguistically distinct country politically linked to its neighbour by conquest in 1282 and through being 'incorporated, united, annexed to and with [the] realm of England' in 1536, ideas of national identity in our period of interest were complex. And lest anyone should think that such distant historical events had little place in an eighteenth-century context, they need only recall the words of the attorney for English clergyman Thomas Bowles. In 1766, Bowles faced local attempts to remove him from his parishes in Anglesey, where few of his parishioners understood English. 'Wales is a conquered country', the attorney declared, 'it is the duty of the bishops to promote the English in order to introduce the language'.[8]

There certainly were Welsh-born scientists who referred to themselves as 'English', but some also had a more internationalist view of themselves. In a 1783 letter the Glamorgan-born/London-based Dissenting minister, mathematician and 'formative mind of the Enlightenment', Richard Price, referred to himself as '[a] native of England, but a citizen of the world'.[9] The latter sentiment reinforced in a letter to Thomas Jefferson in October 1788: 'I have learnt to consider myself more as a citizen of the world than of any particular country.'[10] Some with scientific interests, such as Glamorgan born Iolo Morganwg and the Morris brothers of Anglesey, whole-heartedly embraced their Welsh identity. For others, their identity does not seem to have simply depended on locale. The naturalist Edward Lhwyd was born at Loppington in Shropshire,[11] but he nevertheless changed his name from the anglicised Lloyd to the Welsh Lhwyd (via Lhuyd) and is said to have declared, 'I am not an Englishman but an Ancient Briton'. When the Abergele surgeon Roger Jones, who may have been born in Dodleston in Cheshire, submitted his medical thesis to the University of Leiden in 1735, he did so as a 'Cambro-Britannus'.[12] The claims of both men were presumably made on the basis of Welsh parentage, and they reflect the complexity of establishing nationality along the Wales-England border. Indeed, when Lhwyd sent out his 'Parochial Queries' in 1696 to all the parishes in Wales for his proposed 'Survey of *Wales*', he was careful to include both Shropshire and Herefordshire: 'where the Language and the Ancient Names of Places are still retain'd'.

Sentiment or emotion may also have entered into this equation. When faced with the overly pompous Lord Mornington, Governor General of India, the botanical and mineralogical collector Lady Henrietta Clive, Countess of Powys, blamed her feelings of anger at his manner on her 'Welsh spirit'. William Morgan, the Welsh-speaking, Bridgend-born/London-based father of modern actuarial science in Britain, blamed his 'Welsh temper' after punching and knocking-out the less than kind Limehouse apothecary to whom he had been apprenticed early in his career. Nor did those who left Wales forget their origins. Many returned during or at the end of their careers. William Morgan visited Wales regularly, as did his uncle Richard Price. Though her wish was not to be fulfilled, Lady Clive wanted to be buried in the family plot at Welshpool, and, while he was 'Born in America, in Europe bred, / In Africa travell'd, and in Asia wed',[13] Elihu Yale, university founder and Fellow of the Royal Society, was buried in the churchyard of St Giles in Wrexham, not far from his ancestral home at Plas Grono.

Implicit in Roy Porter's comments in *Enlightenment* is the advice once offered by the *Encyclopaedia Britannica* – 'For Wales, see England'. Whether intended or not, such comments reflect the nationalism of dominance, which usually passes unnoticed by the dominant and is too often acquiesced in by the dominated. The result is to render Wales, and any claim it might have to independent thought and attitudes, largely invisible in the wider British history of our period, and in the history of early modern science particularly. Hardly surprising therefore that when in 2004 the historian R. J. W. Evans asked the question, 'Was there a Welsh Enlightenment?', he answered by saying, 'My question seems rarely to have been asked. The latest reference books remain as silent on the subject as do – so far as I can see – writings within Wales'.[14]

Such absence can result in wrong or reduced perspectives. Thus, the ivy-league Yale University in the United States sponsors, at the time of writing (December 2023), a website citing as born in Scotland the already mentioned Wales-born 'formative mind of the Enlightenment' Richard Price; a man who not only received an honorary degree from the said university, in sole company on the appointed day with George Washington, but supplied it with many of its early scientific instruments.[15] The *Dictionary of Eighteenth-Century World History*, meanwhile, can conclude: 'Before its spectacular growth as a chief coal, iron and steel centre in the nineteenth century, Wales was chiefly a rural backwater lacking even a capital.'[16] That it lacked a capital and was chiefly rural is true. Caroline Franklin, in her essay 'Wales as Nowhere', has noted how in this period the country was 'geographically central within the British Isles' but 'nowhere from a metropolitan standpoint', and she concurs with Geraint H. Jenkins's summation of the period: '98% of the Welsh population were disenfranchised, 85% of the population of roughly half a million lived in villages of fewer than 1,000 inhabitants, and 90% of them spoke only Welsh.'[17] Nations as backwaters, though, are usually defined by those living in a distant 'civilising' centre and, however jokingly written, there is something of this attitude in a letter from London-based Joseph Banks, President of the Royal Society, to his north Wales friend John Lloyd, Fellow of the Royal Society. Is not Wales 'a *Damn'd* Country for a gentleman to live in', Banks wonders, 'for my part I am convinced that none but *Shentleman* should attempt it';[18] the use of *Shentleman* surely echoing the *c*.1747 illustration of 'Shon-ap-Morgan, Shentleman of Wales' riding to London on his goat.[19] Look closer at the idea of a national backwater, however, and the further away the civilising centre becomes. In a letter to King George III a Chinese imperial scribe is said to have declared his emperor to be 'not unmindful of the "remoteness of your tiny barbarian island, cut off as it is from the world by so many wastes of sea"'.[20] We might even imagine a local Welsh collector of Joseph Banks's botanical specimens grumbling to himself on a wet Snowdon cliff: 'that's the problem with London: it's so damned far away from everywhere.'[21] It is, though, invidious to single out Roy Porter's comments without also considering the attitude to science from within Wales itself.

Few historians have been as disparaging of the place of science in the history of Wales as J. Vyrnwy Morgan. Wales, he declared in *The Philosophy of Welsh History* (1914), has seen 'no advance towards creative effort in the sphere of the sciences, such as physics, chemistry and engineering' and it 'has no record worthy of attention in mechanical contrivances, or in inventions for the saving of time and of human labour'.[22] The reverend gentleman went further: this poverty of creative effort, he believed, resulted from the lack of a particular psychological disposition in the Welsh:

> Taking the Aristotelian view of the idea of development, a view which was revived and deepened by Hegel, *viz.*, that development means the unfolding of what has already a potential existence, we

cannot expect much from Wales in this direction, for the reason that the scientific element does not seem to be a constitutive element of the Welsh intellect. A highly credulous people are never an inquiring people, and the greatest historic intellects are the sceptical intellects.[23]

While it is comforting to think such opinions outmoded, we face reminders of them from a modern voice, A. N. Wilson, who reportedly opined not long ago that 'The Welsh have never made any significant contribution to any branch of knowledge, culture or entertainment'.[24] This book, and indeed the series of which it is part, debunks all such views as far as science is concerned. It also does no harm to remind ourselves of the Welsh biographical collections whose scientific and broader content are a rebuke to Morgan, Wilson and others, and to all ignorance revealed by arrogance.

In addition to the numerous biographies and achievements detailed in Wales's rich series of local society journals, John H. Parry published, in *The Cambrian Plutarch* (1834), memoirs of eminent Welshmen. They included three scientists – Thomas Pennant, Edward Lhwyd and Lewis Morris – two of whom had international reputations. Edward Lhwyd had been described as 'the best naturalist now in Europe'[25] by Hans Sloane, one of the founders of the British Museum and a one-time President of the Royal Society. Pennant would be described in 1948 as 'the leading British zoologist after [John] Ray and before [Charles] Darwin'.[26] In 1852, Robert Williams published his *Biographical Dictionary of Eminent Welshmen*, with Thomas Williams's monograph *The Science and Scientific Men of Wales* following in 1855. Two other extremely valuable biographical works appeared in 1908: T. R. Roberts's *Eminent Welshmen*; and T. Mardy Rees's *Notable Welshmen (1700–1900)*. The latter is particularly important since it detailed not only 'Welshmen who have distinguished themselves in England, America, and the colonies' but notable Welsh women as well.

In 1918, as the First World War ended, J. C. Morrice of Bangor published his *Wales in the Seventeenth Century: Its Literature and Men of Letters and Action*. In 1932–3 'The Contributions of Welshmen to Science' by T. Iorwerth Jones appeared in the *Transactions of the Honourable Society of Cymmrodorion*, for which, along with T. R. Roberts's earlier mentioned *Eminent Welshmen*, a debt of gratitude is owed to the National Eisteddfod, since both volumes were written in response to 'a subject set for competition'. Particularly valuable in Iorwerth Jones's work are the lists of published papers accompanying each scientist mentioned. Similar listings are found in John Cule's 1980 publication *Wales and Medicine: A source-list for printed books and papers showing the history of medicine in relation to Wales and Welshmen*, though here the biographical details are limited or absent. In 1956, O. E. Roberts published *Gwyddonwyr o Gymry* ('Scientists of Wales'), which included brief sketches of the contributions of Hugh Owen Thomas (1834–91) and his nephew Sir Robert Jones (1857–1933). As the founders of orthopaedic surgery in Britain they developed the 'Thomas splint' that proved enormously beneficial to the wounded of the First World War; it even merited a mention in an episode of the US Korean-War series *M.A.S.H.* (Series 7, episode 21 – C*A*V*E). *Gwyddonwyr o Gymry* also included Thomas Lewis (1881–1945) 'the father of clinical cardiac electrocardiography'. His electrocardiograph machine is today in the London Science Museum. A few years earlier, in 1953, *Y Bywgraffiadur Cymreig hyd 1940* had appeared, with an English language version – *The Dictionary of Welsh National Biography down to 1940* (hereafter *DWB*) – published in 1959. A supplemental volume covering the years 1941–70 appeared in 2001. Some of these biographical collections can now be accessed online; and though women – scientist or otherwise – remain almost invisible, this is an omission being corrected as the *DWB* continues to expand online.

According to another trenchant observation by Vyrnwy Morgan, 'No people have wasted so much of their energies on theology and theological subjects as the Welsh'.[27] It is certainly true that Dissenting ministers and clergymen dominate our biographical resources. As the *Guide to Welsh Literature c.1700–1800* notes: 'In a Calvinistic civilization like Wales, unfortunately at the same time pietistic, the finest intellects of the period as well as the most artistic abilities were channelled into the dominant literary genres, the hymn, the sermon and biographies of divines.'[28] Yet it is also the case that a considerable number of these fine theologically minded intellects also contributed, sometimes significantly, to science in the years between 1650 and 1820. It is true that, despite an ever-increasing number of scientists in our biographical resources, historians within and beyond Wales have, with a few exceptions, largely excluded science and its practitioners when telling the cultural, social, political and religious history of the country in this period. For example, in the index to one important volume the religious campaigner Howell Harris receives fourteen entries, compared to none for his elder brother Joseph who, as we shall see later, made important contributions to astronomy, navigation, economics and, as King's Assay Master at the Royal Mint, to a mid-eighteenth-century standardisation of weights and measures.

Whether such omissions reflect an aversion to science or simply the neglect of expatriate Welsh men and women who made their careers outside Wales is impossible to say. However, more recently this deficit in Welsh historiography has begun to be addressed. In 1990, R. Elwyn Hughes published an analysis of nineteenth-century Welsh-language prose writing on science in *Nid am un Harddwch Iaith: rhyddiaith gwyddoniaeth y bedwaredd ganrif ar bymtheg*. Following P. W. Carter's groundbreaking survey of the history of botanical exploration in each of the counties of Wales, the long botanic tradition in north Wales has been explored by Dewi Jones in *The Botanists and Guides of Snowdonia* (1996). An equivalent investigation into medicine appeared in Alun Withey's *Physick and the Family, Health, Medicine and Care in Wales, 1600–1750* (2011). In 1994, *Mathemategwyr Cymru* ('Mathematicians of Wales') was published by Llewellyn G. Chambers and recently Gareth Ffowc Roberts has enlivened the history of mathematics with his *Count Us In* (2016) and *For the Recorde: A History of Welsh Mathematical Greats* (2022), both having originally appeared in Cymraeg. We have also seen major research projects on individuals whose scientific interests have been more fully documented and revealed as a consequence. They include Robert Recorde (*c.*1510–58), via the published proceedings of a major conference held at Gregynog in 2008, together with other recent publications, including a volume in the *Scientists of Wales* series.[29] There have also been major studies of Edward Lhwyd (who now has a volume in the *Scientists of Wales* series) and Iolo Morganwg undertaken at the University of Wales Centre for Advanced Welsh and Celtic Studies in Aberystwyth, and another on Thomas Pennant in collaboration with the University of Glasgow. Overseas too there have been major publications.[30] In the journal tradition, which has long celebrated scientific achievement and biography, there have been recent papers on the scientific contributions of both Richard Price[31] and Joseph Harris.[32] To these we can add the Welsh Government publication *Cyflawniadau Cymru mewn Gwyddoniaeth, Technoleg a Pheirianneg/Welsh Achievements in Science, Technology and Engineering*. Newest of all is the *Scientists of Wales* series, which aims to 'resurrect the role of science and technology in Welsh history' and to which *A World of New Ideas* is a contribution.

Scope and Aims

A World of New Ideas is a two-volume synthesis of the work of scientists of Wales between 1650 and 1820. The period is referred to in both volumes as 'the long eighteenth century'. As such, it incorporates the last years of the seventeenth-century Scientific Revolution, the entire eighteenth-century Enlightenment, and the early years of the nineteenth-century Industrial Revolution. Specifically with regard to scientists of Wales, the start of the period, in 1650, is marked by the publication of Thomas Vaughan's first alchemical writings, and the end of the period, in 1820, by the appearance of the forty-fifth volume of Abraham Rees's monumental *Cyclopaedia; or, Universal Dictionary of Arts, Sciences, and Literature*.

In this first volume of *A World of New Ideas* we consider the contributions made in the long eighteenth century by those scientists of Wales living and working within the nations of 'The Isles': Wales, England, Scotland and Ireland. Volume 2 will consider those working within nations of 'The Wider World'. Wales produced a number of key scientists in this long eighteenth century. They include Edward Lhwyd, the founder of modern palaeontology; William Morgan, the father of actuarial science; and Richard Roberts, an inventor once likened to

Arkwright in his importance. Some readers may judge the discussion of celebrated names in the two volumes of *A World of New Ideas* as somewhat cursory. Aside from word count, and other publication pressures, there were two principal reasons for limiting discussion of scientific 'celebs'. First, they have usually left behind extensive archival material and so will deservedly be the subject of comprehensive academic study and, perhaps, a book of their own within this series. Second, a concentration on such figures inevitably curtails discussion of the contributions from less well-known scientists, on whom the history of science truly depends. As with the *Scientists of Wales* series generally, a primary aim of the two volumes of *A World of New Ideas* is to bring at least some of these lesser-known figures to wider notice. Nor should it be assumed that *A World of New Ideas* represents an exhaustive study. The aim is to illustrate the contributions made by as many scientists of Wales as possible; to draw preliminary conclusions as to the importance and value of their work; and to act as a stimulus to further research by pointing out areas where this might be particularly fruitful.

Practical Considerations

In attempting to reveal contributions made by scientists of Wales between 1650 and 1820 we need to briefly consider two practical issues – how the word 'science' is used in the two volumes, and the nature of the archival and other source material available.

Science in the way we know it today – as a group of discreet, highly specialised yet at the same time interrelated disciplines – took shape in the mid-nineteenth century. However, throughout most of the long eighteenth century – our subject here – we are actually dealing with the older and broader concept of 'natural philosophy'. A concept of science derived not only from the study of creation through traditional science subjects, such as mathematics, physics, botany and astronomy, but also through philosophy and theology; the last including a belief in a divinely ordered creation within which its creator might act through miracles and divine providence (the kindly care of God acting within the world). For simplicity and convenience, the word 'science' is used throughout both volumes to encompass this wider definition. The long eighteenth century was also a time of significant technological innovation and development and these too, along with the scientists of Wales who contributed to them, are included in our use of the word 'science'.

In relation to archival material, the difficulty of determining nationality on the basis of name or place of birth alone is often compounded in surviving archives by officialdom's use of 'English' for those of Welsh origin; the prevalence of a limited number of Welsh surnames (Jones, Davies, Williams, etc.) is a further complication.[33] Some individuals with Welsh-sounding names, who made notable scientific contributions, have no nationality recorded at all in surviving archives. In most cases they are simply subsumed into the broad history of British, and sometimes English or Scottish science. Some of these nominally 'stateless' scientists, and their contributions, will be summarised briefly in an appendix attached to the second volume of *A World of New Ideas* in the hope of eliciting further research into their origins.

The dominance of a famous name in ensuring the preservation of archival material can also create problems. For example, we have a large correspondence written to little-known John Lloyd of Wigfair from well-known Joseph Banks, but very little from Lloyd *to* Banks. Consequently, evidence for Lloyd's scientific interests, which were significant, comes from a generally one-sided correspondence and the 1816 sale catalogue of his library and scientific instruments: 'one of the best gentleman's collections in the kingdom'. Sadly, there have also been significant archival loses over the years. As elsewhere in Britain, the pursuit of science in Wales was often undertaken by the middling and gentry classes, and the loss of personal archives when an estate came to be disposed of remained a possibility. In other cases, archives of international importance have been lost to fire; most notably the substantial archives pertaining to Edward Lhwyd owned by Sir Watkin Williams-Wynn and by Thomas Johnes of Hafod. These were destroyed in a fire at a London book-binder and in the Hafod fire of 1807, respectively. Other archival letters appear to have been treated with little regard for their importance. Although not science-related, a letter sent by Benjamin Franklin to the Caerphilly-born political writer and educationalist

David Williams was simply given to an American visitor to Caerphilly Castle, as a memento of his visit there in 1878; fortunately, it ultimately found its way to the American Philosophical Society and so online.

Despite such problems, all is not lost. At home and abroad much archival material survives and unexpected finds can still be made; an example being the medical library of the seventeenth-century poet and physician Henry Vaughan (the Silurist). It was discovered in the 1950s 'among the treasures of the Library Company of Philadelphia'. Publication of online guides to the archives of collectors such as Hans Sloane, as well as publication or wholesale digitisation of the correspondence of science practitioners, such as Edward Lhwyd, Thomas Pennant, Iolo Morganwg, Richard Price and the Morris brothers of Anglesey, has also increased in recent years. These contributions, along with other print publication and digitisation projects, allow less well-documented scientists of Wales to be revealed more easily. Indeed, the quantity of material found has made it difficult to decide who and what contributions to include in this relatively brief synthesis.

Dates are assigned on the basis of our modern calendar and some quotations are modernised for easier reading.

Structure of Volume 1: The Isles

The text of *The Isles* comprises three parts and sixteen chapters. In 'Part I: Foundations', the relationship between scientific ideas and scientists of Wales is established, with the first chapter, 'A Welsh Renaissance: Science to 1650' (Chapter 1), providing a brief history of science up to the start of our period in 1650. It also notes the three main intellectual strands prevailing by that year. The first of these, Scholastic Aristotelianism, is considered mainly within Chapter 1. A second strand, hermeticism, and its alchemical and natural magic branches, are the subject of 'The Scholastic and the Alchemist' (Chapter 2). In 'Making New Worlds' (Chapter 3) we look briefly at a third intellectual strand – the ideas of René Descartes – and their relation to scientists of Wales. A fourth intellectual strand – the work of Isaac Newton – is the first to appear *within* the early years of our period of interest and it too is introduced in this third chapter. In 'New Worlds and the Challenge to Religion' (Chapter 4) we consider religious belief, as well as atheism and deism, in relation to science and, in the last chapter of Part I, we consider 'Wales and the Newtonian World' (Chapter 5).

In 'Part II: Gaining a Knowledge of Science', the first chapter, 'Learning about Science' (Chapter 6), is concerned with how and where science was taught to scientists of Wales in the long eighteenth century. In 'The Languages of Science' (Chapter 7) the three principal languages in which science operated in Wales – Welsh, Latin and English – are discussed. Finally, 'Spreading the Word' (Chapter 8) considers how ideas in science were being transmitted.

As suggested by its title, the chapters of 'Part III: Practical Science' consider the actual scientific contributions made by scientists of Wales. 'The Science of the Earth' (Chapter 9) is concerned with mapping, measuring and geology. 'The Science of Nature' (Chapter 10) concerns itself with botany, zoology and horticulture, while 'Looking Up' (Chapter 11) discusses meteorology and astronomy. The two parts of 'The Science of Us' includes 'Health and Social Well-being' (Chapter 12) and 'Nature and Economics' (Chapter 13). Two chapters on 'Science and Technology' then relate the contributions made by instrument makers, calculators and translators in 'Artisan Science' (Chapter 14), and by 'Innovators and Inventors' (Chapter 15). In the final chapter, 'Encyclopaedic Knowledge' (Chapter 16), we reflect on the contribution from scientists of Wales to the presentation of long-eighteenth-century scientific developments to the wider public. The chapter closes with a brief reflection on what is to follow in Volume 2.

Notes

1. Duncan Dowson and Bernard J. Hamrock, *History of Ball Bearings*, NASA Technical Memorandum 81689 (February 1981), pp. 28–9; *https://ntrs.nasa.gov/api/citations/19810009866/downloads/19810009866.pdf* (accessed 27 May 2024).
2. For technical details of one rover's ball bearings, see *www.acorn-ind.co.uk/insight/Bearings-in-Space-Curiosity-Rovers-Mission-to-Mars* (accessed 8 February 2018).
3. See Hugh Kearney, *The British Isles: A History of Four Nations* (Cambridge, 1989). For a discussion of the complexities of Britain's 'four-nation' history', see Norman Davies, *The Isles: A History* (1999; London, 2000), pp. xxiii–xli. Also, Mary-Ann Constantine and Dafydd Johnston, 'Introduction: Writing the Revolution in Wales',

in Mary-Ann Constantine and Dafydd Johnston (eds), *'Footsteps of Liberty and Revolt': Essays on Wales and the French Revolution* (Cardiff, 2013).

4. John of Gaunt in William Shakespeare's *Richard II* (act II, scene 1).

5. Roy Porter and Mikuláš Teich, 'Introduction,' in Roy Porter and Mikuláš Teich (eds), *The Scientific Revolution in National Context* (Cambridge, 1992), p. 6.

6. Roy Porter, *Enlightenment: Britain and the Creation of the Modern World* (2000; London, 2001), pp. xviii–xix.

7. Julie Flavell, *When London was Capital of America* (New Haven CT and London, 2010), p. 70, citing Janet Adam Smith, 'Some Eighteenth-Century Ideas of Scotland', in N. T. Phillipson and Rosalind Mitchison (eds), *Scotland in the Age of Improvement: Essays in Scottish History in the Eighteenth Century* (Edinburgh, 1970), pp. 108–10.

8. For a discussion of this, see Bethan M. Jenkins, *Between Wales and England: Anglophone Welsh Writing of the Eighteenth Century* (Cardiff, 2017), pp. 1–32. For the Bowles case, see Janet Davies, *The Welsh Language: A Pocket Guide* (1999; Cardiff, 2000), p. 33. Though Bowles retained his place, the judge declared that ignorance of Welsh should debar others from majority Welsh-speaking parishes.

9. W. Bernard Peach and D. O. Thomas (eds), *The Correspondence of Richard Price*, 3 vols (Durham NC and Cardiff, 1983–94), vol. 2 (1991), p. 191.

10. Peach and Thomas (eds), *Correspondence of Richard Price*, vol. 3 (1994), p. 182.

11. See *Lhwyd*, pp. 22, 251 n. 10.

12. For a discussion of this term, see Philip Schwyzer, 'The age of the Cambro-Britons: hyphenated British identities in the seventeenth century', *The Seventeenth Century*, 33 (2018), 427–39.

13. Quote from the inscription on Yale's tomb at Wrexham.

14. R. J. W. Evans, 'Was there a Welsh Enlightenment?', in R. R. Davies and Geraint H. Jenkins (eds), *From Medieval to Modern Wales* (Cardiff, 2004), pp. 142–59 (at p. 142).

15. Although a magnificent historical resource, a search 'By name' for 'Richard Price' at *franklinpapers.org* reveals his birthplace as Tynton, in Scotland, rather than his actual birthplace at Tynton, Llangeinor, Wales. Two efforts have been made to have this error corrected, to no avail.

16. Jeremy Black and Roy Porter (eds), *A Dictionary of Eighteenth-Century World History* (Oxford, 1994), p. 773.

17. Caroline Franklin, 'Wales as Nowhere: the *tabula rasa* of the "Jacobin" imagination', in Mary-Ann Constantine and Dafydd Johnston (eds), *'Footsteps of Liberty and Revolt': Essays on Wales and the French Revolution* (Cardiff, 2013), pp. 11–33 (at p. 11), citing Geraint H. Jenkins, 'Wales in the Eighteenth Century', in H. T. Dickinson (ed.), *A Companion to Eighteenth-Century Britain* (Oxford, 2002), pp. 392–402.

18. NLW MS 12415c, f. 7.

19. See Peter Lord, *Words with Pictures: Welsh Images and Images of Wales in the Popular Press, 1640–1860* (Aberystwyth, 1995), p. 49.

20. Quoted in Robert Hughes, *Nothing if Not Critical: Selected Essays on Art and Artists* (London, 1990), p. 177.

21. Quote from Adam Price, *Wales: The First and Final Colony* (Talybont, 2018), p. 141.

22. Rev. J. Vyrnwy Morgan, D.D., *The Philosophy of Welsh History* (London, 1914), p. 86.

23. Morgan, *Philosophy of Welsh History*, p. 83.

24. Vanessa Thorpe, 'Don't be so snobby about Wales, says shocked arts chief', *The Guardian*, 21 November 2004, *www.theguardian.com/uk/2004/nov/21/ britishidentity.artsnews* (accessed 27 May 2024).

25. See F. V. Emery, '"The Best Naturalist Now in Europe": Edward Lhuyd, F.R.S. (1660–1709)', *THSC*, 1 (1970), 54–69.

26. G. R. de Beer (ed.), *Tour on the Continent, 1765, by Thomas Pennant, Esq.* (London, 1948), p. vi.

27. Morgan, *Philosophy of Welsh History*, p. 89.

28. Branwen Jarvis (ed.), *A Guide to Welsh Literature c.1700–1800* (Cardiff, 2000), p. 325.

29. Jack Williams, *Robert Recorde: Tudor Polymath, Expositor and Practitioner of Computation* (London, 2011); Gareth Roberts and Fenny Smith (eds), *Robert Recorde: The Life and Times of a Tudor Mathematician* (Cardiff, 2012); Gordon Roberts, *Robert Recorde: Tudor Scholar and Mathematician* (Cardiff, 2016).

30. See, for example, Lawrence Henry Gipson, *Lewis Evans* (Philadelphia PA, 1939); and Ivor Hughes and David Ellis Evans, *Before we Went Wireless, David Edward Hughes his life, inventions and discoveries* (Vermont, 2011).

31. John V. Tucker, 'Richard Price and the History of Science', *THSC*, 23 (2017), 69–86.

32. Martin Griffiths, 'Joseph Harris of Trevecka, Scientist, Artisan, Servant of the Crown', *The Antiquarian Astronomer*, 6 (2012), 19–33.

33. For a fuller discussion of similar problems in Welsh historical research, see H. V. Bowen, 'Introduction', in H. V. Bowen (ed.), *Wales and the British Overseas Empire: Interactions and Influences 1650–1830* (Manchester 2011), pp. 1–14.

PART I

FOUNDATIONS

1

A WELSH RENAISSANCE: SCIENCE TO 1650

In truth, Eltut [Saint Illtud] was of all the Britons the most
accomplished in all the Scriptures, namely of the Old and
New Testaments, and in those of philosophy of every kind, of
geometry namely, and of rhetoric, grammar and arithmetic,
and of all the theories of philosophy. And by birth he was a
most wise magician, having knowledge of the future.[1]

From the *Vita Samsonis* (early seventh century)

Universi Nedd, llyna fawrson – Lloegr,
Llugorn Ffrainc, Iwerddon,
Ysgol hygyrch ysgolheigion
I bob seiens be bai seion

(University of Neath, famous in England, / A beacon
to France and Ireland, / A school sought by
scholars / For all science as if it were Zion)[2]

Lewys Morgannwg (1510–65) writing in praise of the last
Abbot of Neath Abbey, Lleisian ap Tomos (*fl.*1513–41)

To prepare for our investigation of the long eighteenth century, we
need to identify and *briefly* describe three intellectual strands or
currents of thought that helped shape the period's history. They are the
scholastic method of teaching and its association with the philosophy
of Aristotle; the advent of hermeticism, with its associated topics of
astrology, magic, numerology and alchemy; and the appearance of a
radical philosophy developed by French philosopher René Descartes.

So let us now situate the appearance of these three strands in the period(s) *before* our long eighteenth century opens.

Beliefs concerning the fundamental nature of Creation changed relatively little in the thousand years between Saint Illtud's (*c*.475–*c*.525) founding of a fifth-century monastic centre of learning at Llanilltud Fawr (Llantwit Major), and the 1539 demise of the Abbey and monastic 'University' at Neath during Henry VIII's dissolution of the monasteries. God remained central to everything. He had created the cosmos within six days and given man a dominion over the Earth at the centre of that cosmos. At the same time, western medieval Christian thinking saw perfect knowledge as having been given by God to the first man, the biblical Adam. He had then lost this perfect knowledge at the time of his expulsion, along with Eve, from the Garden of Eden. Since humanity (through Adam) had once possessed perfect knowledge, it was believed that only by looking back to earlier thinkers, those closer to the Adamic period, could such knowledge be regained.[3]

Among the earliest sources thought to contain this lost knowledge were the works of the Greek philosopher Aristotle (384–322 BCE). These had been rediscovered in Arabic sources in the twelfth century during the ongoing Christian re-conquest of Spain from the Moors, and they presented the medieval European world with the only complete philosophical system then known to survive from antiquity. Aristotle's philosophy covered the whole of creation, from the natural world to the arrangement of the heavens, and it reflected the extraordinary wide variety of his interests in the natural and human worlds.

Aristotle was a consummate observer and teacher and wrote extensively about science topics. However, the Church was initially suspicious of this pagan philosopher and only came to accept some of his ideas following their integration with Christian thinking; a synthesis principally undertaken in the thirteenth century by Thomas Aquinas (*c*.1225–74). As a result of this synthesis, any monk, nun, teacher or student looking skyward from the hills of Wales saw laid out before them a Christianised version of the cosmos. They stood on a globe placed at the centre of everything and composed of the first of the elements – earth. A layer of water – the second element – surrounded the globe and made up the world's oceans. Beyond the water came a layer of air – the third element – and beyond that a layer of fire – the fourth element – and the place of meteors and comets. Within this sub-lunar or inner realm these four elements – earth, water, air and fire – continually interacted with one another and all was in constant flux. Beyond it, in the outer realm, the Moon, together with the planetary bodies – Mercury, Venus, the Sun, Mars, Jupiter, and farthest of all Saturn – moved around the centrally placed globe of the Earth in a series of concentrically arranged spheres. Beyond the outermost planet, Saturn, lay the sphere of the fixed stars. In contrast to the elemental interactions and flux of the inner realm, everything in this outer realm was unchanging and formed of a so-called fifth element or essence, the aether or quintessence. Beyond this, Christian theologians added a further realm, the Empyrean Heaven, the abode of God.

The theological/philosophical synthesis produced by the likes of Thomas Aquinas would be taught at the growing number of universities being established in medieval and renaissance Europe. When being taught the synthesis, students took part in learned disputes on topics known to and often selected by their teachers – the scholars or school-men. Since the topics chosen for discussion often related back in some way to the works of Aristotle, albeit viewed now through the distorting lens of the Aquinian synthesis, this method of teaching became known as Scholastic Aristotelianism. It forms the first of the three intellectual strands that we will be concerned with in the next chapter as our period opens in 1650.

In medieval medicine, the ancient authority was the Greek physician Galen (129–199 CE). Here, the principal concern was maintaining a balance between the four fluids or humours found in the human body, and the temperaments that they sustained: blood being associated with a sanguine temperament; phlegm with the phlegmatic temperament; black bile with the melancholic temperament; and yellow bile with the choleric temperament. To help maintain this balance, bloodletting, enemas, induced vomiting and a mainly plant-based pharmacology comprised the principal arsenal of the medical practitioner. Elements of this arsenal are certainly present in Welsh manuscripts dating from the end of the fourteenth century to the beginning of the fifteenth century and said to relate to the Physicians of Myddfai, in Carmarthenshire, though their true provenance is more complicated.[4] Despite some advances, this regime would remain the basis for most medical care in Europe up to the early nineteenth century.

By the middle of the fifteenth century there had also appeared in Europe the works of another philosopher of ancient Greece – Plato (c.427–c.347 BCE). The rediscovery of his works came when Greek refugees fleeing the 1453 Ottoman conquest of Byzantium (Constantinople, now modern Istanbul) brought them into Italy, where renaissance scholars translated them into Latin, the universal language of medieval Europe. As with Aristotle, the Church found elements of Platonic philosophy inconsistent with Christian theology – for example, its advocacy of the transmigration of souls (a form of reincarnation), and the creation of the universe from pre-existing matter (in Christian teaching nothing existed before God created the universe; Aristotle held the universe to have always existed). Two Florentine scholars, Giovanni Pico della Mirandola (1463–94) and Marsilio Ficino (1433–99), were key figures in developing an acceptable 'theologia platonica'. They did so by integrating the Christian concept of an all-powerful, all-knowing Creator, who could act within the ordered world of his creation through miracles and divine providence, with works such as Plato's *Timaeus*, in which a creator (the Demiurge) produced a rational and purposeful universe out of an initial chaos onto which a mathematical and geometrical order had been imposed.[5] A background order to creation that was reflected in the geometrical structures that Plato assigned to the four sublunary elements: with fire seen as a tetrahedron; air as an octahedron; earth as a cube; and water as an icosahedron. The fifth element, the quintessence, which made up the outer realm of the cosmos beyond the moon, was assigned to a dodecahedron (see Figure 1).

A significant effect of this Christianised platonic philosophy was to imply that while the divine nature of miracles and providence effectively rendered them immune to investigation, the more ordered nature of God's creation, with its platonic geometrical and mathematical background, might be susceptible to investigation by humanity.

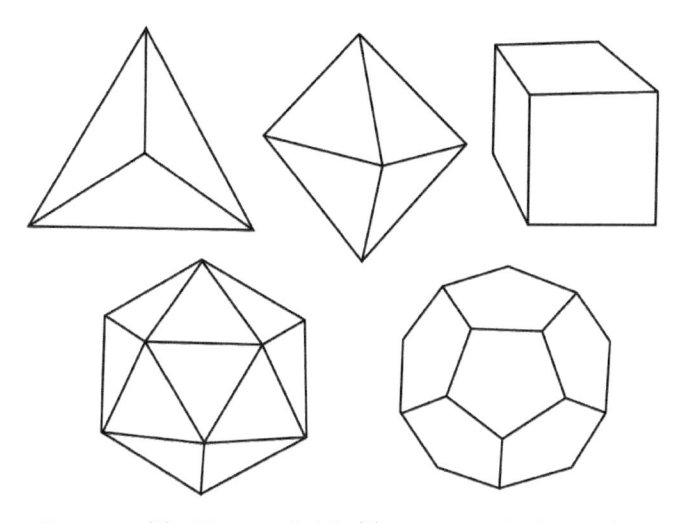

FIGURE 1 The Platonic Solids. Top row: tetrahedron = fire;
octahedron = air; cube = earth. Bottom row: icosahedron = water;
dodecahedron = the Cosmos.

Indeed, Mirandola and Ficino had incorporated into their synthesis a number of magical and mystical texts, which, in effect, provided the tools whereby such investigation could be undertaken. Believed to have been written closer to the Adamic period of perfect knowledge than even Plato's work, these texts included Christianised versions of Jewish Cabbala – a form of Jewish mysticism, which included such topics as numerology, astrology, magic and alchemy – and the seventeen treatises of the *Corpus Hermeticum*. These too described magic, astrology and alchemy and were believed to have been written by an Egyptian priest, Hermes Trismegistus who was also thought (wrongly, as it later turned out) to be a contemporary or precursor of Moses.[6] The resulting synthesis of Cabbala with the writings of the *Corpus Hermeticum* is termed Hermeticism and, like Scholastic Aristotelianism, it continued into our period of interest where it forms the second intellectual strand to be discussed in the next chapter, with a particular emphasis on its alchemical and natural magic branches.

By the sixteenth century, challenges to these various syntheses came about thanks to a widening of geographical and intellectual horizons.

First, through the voyages of Columbus to the Americas between 1492 and 1504, to India by Vasco da Gama between 1497 and 1502, and around the globe by Magellan between 1519 and 1522; voyages that opened up previously unknown parts of the world to questioning and often aggressively hungry European minds. Second, there were challenges being made to the established religious and political order: the rise of Martin Luther's Protestantism from 1517, for example, and Henry VIII's break with the Pope in Rome in 1534 and his declaring himself Supreme Head of the Church in England and Wales. Third, there were challenges from those who chose to interrogate creation in order to discover the order and structures behind it.

In 1543, just four years after Lleisian ap Thomas gave up the Abbey and 'University' at Neath, a scientific publication, *De Revolutionibus Orbium Coelestium* ('On the Revolutions of the Heavenly Spheres'), by Polish astronomer Nicolaus Copernicus (1473–1543), challenged the cosmological order and, thereby, the religious and political ones too. It did so by removing the Earth and humanity from their biblically assigned place at the centre of God's Creation. It replaced them with the Sun, around which the globe of the Earth and the other planets were now said to move. Some thirty years later, in 1572/3, came a further challenge when Danish astronomer Tycho Brahe (1546–1601) announced in his *De Nova Stella* that a new star had appeared within Aristotle's supposedly unchanging outer realm beyond the Moon (the star is now known to be a supernova – SN 1572). Challenges occurred in medicine too. In 1530, the Swiss surgeon and alchemist Paracelsus (1493–1541) questioned the theory of the four main elements in his *Opus paramirum*. He advocated a different theory, one in which the essential parts were sulphur, mercury and salt. He also disapproved of bloodletting and saw the human body as fundamentally a chemical system, rather than one defined by the four humours. As with the bodily humours though, the body still needed to be kept in balance, and so Paracelsus began treating sufferers with new chemical and mineral medicaments in addition to, or instead of, the already established plant and herb-based pharmacology. In 1543, a challenge also came to the anatomical conclusions previously reached by Galen, when Andreas

Vesalius (1514–64) published his *De Humani Corporis Fabrica* ('On the Fabric of the Human Body').[7]

Improvements in printing, since the introduction of movable type into Europe by Johannes Gutenberg (*c*.1400–68) in the mid-fifteenth century, allowed for an ever quicker spread of these new ideas and a wider dissemination of knowledge generally. For instance, in 1546 the first printed book in the Welsh language appeared. Written by Sir John Prise (or Price) of Brecon (*c*.1502–55), the sole surviving copy is missing its cover and is known only by its opening words: *Yny lhyvyr hwnn* ('In this book'). It contains a discussion of Welsh-language orthography and religious texts together with a calendar, astrological and almanac information, and a discussion and illustration of modern Hindu-Arabic numerals.[8]

Hindu-Arabic numerals had been introduced into Europe by the late tenth century but they appear to have gained ground in Britain only during the fifteenth century.[9] Published references to them remained comparatively rare, and that made by Tenby-born/London-based Robert Recorde (*c*.1510–58), in *The Ground of Artes* (1543), is now considered one of the very earliest.[10] Shortly afterwards, William Salesbury, of Llanrwst and London, published a *Dictionary in Englyshe and Welshe* (1547) that included astronomical terms, such as 'Kaer Gwydion' for Milky Way. Salesbury followed this in 1550 with an English translation of *De Sphaera* ('The Sphere') by Proclus (*c*.410–485); a translation that has been described as 'the first scientific book in the English language. (There had been translations of medieval science books, but this was the first from a classical source)'.[11] In 1556 Robert Recorde published a brief approving reference to Copernicus in *The Castle of Knowledge*. That comment, coming just thirteen years after Copernicus had published his heliocentric theory, has since 'led to an indecisive debate about Recorde's candidature for the designation of first Copernican in England'.[12] In 1557, with the London publication of *The Whetstone of Witte*, Recorde famously introduced the equals sign into mathematics. Indeed, by writing some of the very earliest English-language text books on the subjects of arithmetic (*The Grounde of Artes*, 1543), geometry (*The Pathway to Knowledge*, 1551), astronomy (*The Castle of Knowledge*, 1556) and algebra (*The Whetstone of Witte*, 1557), works of a practical nature that are often ignored in pure mathematics, Recorde helped facilitate a wider knowledge of mathematics and its role in the world's work. As Tucker (2012) suggests, Recorde's works not only founded mathematics for practitioners in the British Isles, they also represent an intellectual milestone in the history of modern Britain.[13]

As we move into the seventeenth century, William Vaughan (1575–1641) of Carmarthenshire produced a broad and lengthy definition of 'science' in *The Golden-grove* (1600), his pedagogic guide to morals, politics and literature:

> [F]irst, it is used for every certain knowledge of a thing. So we say that the snow is white, the crow black, the fire hot. [S]econd, the name of Science is taken for every true habit of the mind separated from the knowledge of the senses; in which signification Hippocrates proved Phisick to be a science. [T]hird, it is used more properly for every habit gotten by demonstration, separated from the habit of action: in this sense supernatural philosophy is named the chief science. [F]ourth, the name of science is taken more strictly for a habit gotten by demonstration separated from wisdom; in which last signification Natural philosophy, & the Mathematics are called Sciences, and supernatural Philosophy is termed human Divinity.

Today, the fourth clause of this definition is used in the *OED* for the earliest example of the word 'science' as:

> a branch of study that deals with a connected body of demonstrated truths or with observed facts systematically classified and more or less comprehended by general laws, and incorporating trustworthy methods (now esp. those involving the scientific method and which incorporate falsifiable hypotheses) for the discovery of new truth in its own domain.[14]

Seen in this way, Vaughan's definition prefigures the hugely influential *Advancement of Learning* (1605) and *Novum Organum* (1620) by Francis

Bacon (1561–1626), in which he laid out the 'scientific method' for the scrutiny of nature through empirical observation and experiment.

The observation of the heavens was one area that benefited at this time from a major technological development – the invention of the telescope by Hans Lippershey in Holland in 1608. As news of the invention reached Galileo, in Padua, in 1609, an actual 'Dutch Trunk' arrived at Trefenty in Carmarthenshire in 1610. Cornishman William Lower (*c*.1569–1615) and Carmarthenshire-born John Prydderch (*c*.1582–*c*.1624) then used it to make some of the very earliest telescopic observations of the Moon.[15] Then, in 1609 and 1619 another challenge to the supposedly unchanging nature of the outer realm of planets came when Johannes Kepler (1571–1630) introduced his laws of planetary motion; among them the revelation that the planets not only moved about the sun, but that they did so in an elliptical path and not the circular one long taught by Scholastic Aristotelianism.

In 1638, Lewis Roberts (1596–1640) of Beaumaris made an early contribution to the science of economics in *The Merchants Mappe of Commerce* (1638) and *The Treasure of Traffic or a Discourse on Foreign Trade* (1640). He also worked in England for the family of William Harvey (1578–1657) whose discovery of the circulation of blood presented yet another challenge to the ancient medical wisdom of Galen, and the more recent ideas of Vesalius. But then, with the publication of *Discourse on Method* (1637), the *Meditations* (1642) and *Principia Philosophiae* (1644), the French-born philosopher René Descartes (1596–1650) profoundly challenged all previous thinking with a 'mechanical' philosophy that forms the third intellectual strand pertaining as our period opens in 1650. It will be described more fully in Chapter 3.

By 1647, in the world at large, Australia and New Zealand remained unexplored but had been visited by western travellers. In North America, Petrus Stuyvesant, the Dutch West India Company's one-legged recently married representative, arrived in New Amsterdam to find its European residents 'grown very wild and loose in their morals'.[16] New Amsterdam will become 'British' New York in 1664. In the Bay of Bengal, the East India Company is trading with a base at Hugli some twenty miles north of what will become, by 1682, Calcutta.[17] In continental Europe there is another year to run before the Peace of Westphalia brings to an end the horrors of the Thirty Years War.

In Britain, meanwhile, near the end of 1647, the Breconshire born political writer and historian James Howell reflected on the situation in England from his confinement in London's Fleet prison: ''Tis tru we have had many such black days in England in former ages', he writes, 'but those parallel'd to the present are to the shadow of a mountain compar'd to the eclipse of the moon.'[18] Imprisoned by order of Parliament for having royalist sympathies (or so he claimed, others suggest debt), Howell's black days stemmed from the Civil War raging in England and Wales since 1642. This five-year battle for dominance between the forces of Parliament and those of King Charles I remained unresolved by the surrender of the king to Scottish troops at Newark in May 1646. Despite the king's captivity, tensions were again rising by the time Howell wrote. In spring 1648 a second Civil War erupted in largely royalist Wales,[19] and in the south-east and north of England. By November, Parliament had regained control, but then came calls for 'capital punishment upon the principal author' of two wars that had created turmoil and left thousands dead or maimed. Proceedings against the king began on 20 January 1649. Ten days later, God's anointed representative on earth – an intransigent one to the end – lost his head and set his country on the path to republicanism.

Such is the background leading up to the opening of our period in 1650 and, within that background, we have identified the three principal intellectual strands that will continue into the long eighteenth century – Scholastic Aristotelianism; Hermeticism and its associated subjects of astrology, magic, numerology and alchemy; and a new 'mechanical' philosophy propounded by Descartes. These did not advance as a linear progression. All were the subject of dispute and in the next two chapters we will discover their story principally by reference to contributions made by one particular scientist of Wales – Thomas Vaughan.

Notes

1. Quoted from the early seventh-century *Vita Samsonis* in Philip Morris, *Llanilltud: The Story of a Celtic Christian Community* (Talybont, 2020), p. 31 citing Thomas Taylor (trans.), *Vita Samsonis* (SPCK, 1925), I.VII.

2. Lewys Morgannwg (*fl.*1510–65) writing in praise of the last Abbot of Neath Abbey, Lleisian ap Tomos (*fl.*1513–41); as quoted in Prys Morgan (ed.), *Writers of the West/ Llenorion y Gorllewin: 4, West Glamorgan/Gorllewin Morgannwg* (Carmarthen, 1974), pp. 12–13.

3. See Peter Harrison, *The Fall of Man and the Foundations of Science* (2007; Cambridge, 2009); John Henry, *A Short History of Scientific Thought* (Basingstoke, 2012); Philip Ball, *Curiosity: How Science Became Interested in Everything* (London, 2012). For general medieval science, see Irven M. Resnick and Kenneth F. Kitchell Jr. *Albertus Magnus and the World of Nature* (London, 2022); and Seb Falk, *The Light of Ages, A Medieval Journey of Discovery* (2020).

4. See Diana Luft, *Medieval Welsh Medical Texts: Volume 1, The Recipes* (Cardiff, 2020).

5. See details at *https://plato.stanford.edu/archives/sum2019/entries/plato-timaeus/* (accessed 28 February 2022).

6. Even though the writings in the *Corpus Hermeticum* were shown by Isaac Casaubon in 1614 to actually date from early in the Christian era, this does not appear to have seriously hampered their later mid-seventeenth century popularity.

7. See Toni Mount, *Medieval Medicine, Its Mysteries and Science* (2015; Stroud, 2016), pp. 204–11.

8. R. Geraint Gruffydd, 'Yny Lhyvyr Hwnn (1546): The Earliest Welsh Printed Book', *BBCS*, 22/2 (1969), 105–16. The book is available online at NLW Digital Gallery. See also James Pierce, *The Life and Work of William Salesbury: A Rare Scholar* (Talybont, 2016), pp. 83–8.

9. Nia M. W. Powell, 'The Welsh Context of Robert Recorde', in Gareth Roberts and Fenny Smith (eds), *Robert Recorde: The Life and Times of a Tudor Mathematician* (Cardiff, 2012), pp. 123–44 (at p. 130).

10. See John V. Tucker, 'Foreword: Robert Recorde and the History of Computing', in Jack Williams, *Robert Recorde: Tudor Polymath, Expositor and Practitioner of Computation* (London, 2011), pp. v–xi (at p. vi and n. 2), also main text pp. 77–101.

11. Pierce, *Life and Work of William Salesbury*, p. 150, citing Francis R. Johnson, *Astronomical Thought in Renaissance England* (Baltimore MD, 1937), pp. 121, 133.

12. Jack Williams, *Robert Recorde*, pp. 147–51 (at 147).

13. See Tucker, 'Foreword', in Williams, *Robert Recorde*, pp. v–xi (at pp. vi–vii).

14. 'science, n.' *OED Online*, September 2022, *www.oed.com/view/Entry/172672* (accessed 5 December 2022) at definition 4b.

15. Bryn Jones, 'Despite the Clouds: A History of Wales and Astronomy', *Antiquarian Astronomer*, 8 (2014), 66–96. On 21 June 1610, replying excitedly to a letter from the scientist Thomas Harriot detailing the recent findings of Galileo (including the satellites of Jupiter), William Lower notes that the 'Traventane philosophers' were considering a work by Kepler. Harriot, who had sent Lower a telescope in

1610, was himself a correspondent of Kepler (see Early English Letters Online). According to Jean Jacquot, 'Thomas Harriot's Reputation for Impiety', *Notes and Records: The Royal Society Journal of the History of Science*, 9/2 (1952), 164–87, the Carmarthenshire 'Traventane philosophers' were discussing Chapter 11 of Kepler's *De Stella Nova Serpentarii* (1606) concerning the size and structure of the universe, and the implications for Kepler's work of Galileo's recent discoveries. After Lower's death at Trefenty in 1615, John Prydderch kept up his friendship with Thomas Harriot, and later acted as an executor of Harriot's will (see F. Jones, 'The Squires of Hawksbrook', *THSC* (1938), 339–55, (at 343–4)). Allan Chapman in *Stargazers: Copernicus, Galileo, the Telescope and the Church* (London, 2014), p. 270 suggests that the 'Traventine' philosophers, who met to discuss Kepler and study the moon through a telescope, may have been the first astronomical society in Britain.

16. G. Burrows and Mike Wallace (eds), *Gotham: A History of New York City to 1898* (New York and Oxford, 1999), pp. 42–3.

17. See John Keay, *The Honourable East India Company* (1991; London, 1993), p. 149.

18. Quoted in Geoffrey Parker, *Global Crisis: War, Climate Change and Catastrophe in the Seventeenth Century* (New Haven CT and London, 2013), p. xxiii; citing James Howell, *Epistolae Ho-elianae or Familiar Letters* (London, 1650), ed. by J. Jacobs (London, 1890).

19. See, for example, Lloyd Bowen, *John Poyer, the Civil Wars in Pembrokeshire and the British Revolutions* (2020; Cardiff, 2022).

2

THE SCHOLASTIC AND THE ALCHEMIST: THOMAS VAUGHAN

Certainly for us to think, that we can find the *Truth* by meer *Contemplation* without *Experience*, is as great a *madness*, as if a *Man* should *shutt* his *Eyes* from the *Sun*, and then believe hee can *travaile* directly from *London* to *Grand Cairo*, by *fansying himself* in the *right way*, without the *Assistance* of the *Light*. It is true, that no man enters the *Magicall Schoole*, but hee *wanders first* in this *Region* of *Chimæras*: for the *Inquiries* which we make before wee attain to *Experimentall Truths*, are *most* of them Erroneous. Howsoever wee should bee so *rational*, and *patient* in our *Disquisitions*, as not *imperiously* to *obtrude* and *force* them upon the *world*, before wee are *able* to *Verifie* them.[1]

Thomas Vaughan, *Lumen De Lumine: or A new Magical Light discovered and Communicated to the World* (1651)

Despite the widening of geographic and intellectual horizons related in the previous chapter, our first intellectual strand – teaching by the method known as Scholastic Aristotelianism – still predominated in the universities of Britain and continental Europe as our period opens in 1650. For many, however, this method remained one that simply reiterated and recycled the same ideas, rather than being a search for new knowledge by questioning old certainties.[2] For one scientist of Wales writing in 1650 – Thomas Vaughan (1621–66) – the method had simply produced:

an age of intellectual slaveries; if they [the scholastics] meet any thing extraordinary, they prune it commonly with distinctions, or daub it with false glosses, till it look like the traditions of Aristotle. His followers are so confident of his principles they seek not to understand what others speak, but to make others speak what they understand. It is in nature, as it is in religion; we are still hammering of old elements, but seek not the America that lies beyond them.[3]

In his determination to seek the 'America' of new ideas that lay beyond this scholastic stagnation, Vaughan, like others in Europe at the time, wanted to stop 'vain babbling' and 'building castles in the air'. Instead, he, and they, wanted to adopt the empirical system of observation and the gaining of experience and knowledge through experimentation advocated by 'the noble Verulam'. This was Francis Bacon, first Baron Verulam, and it is perhaps no coincidence that the title page of Vaughan's first published work, *Anthroposophia Theomagica* (1650), bears a biblical quotation used by Bacon in his *Novum Organum* of 1620: 'Many shall run to and fro, and knowledge shall be increased' (Daniel, Chapter 12, verse 4). Bacon's own title page illustrates a ship entering the Pillars of Hercules between Gibraltar and North Africa 'after exploring an unknown world' – the Americas.[4]

Thomas Vaughan and his better-known twin brother, the poet and physician Henry Vaughan, the Silurist (1621–95), were born to Thomas and Denise Vaughan in 1621. Little is known of the father, other than his appearance in the records as a Justice of the Peace in 1624, and as an under-sheriff in 1646. The mother, Denise Vaughan (née Morgan), had inherited a small estate at Newton by Usk in Breconshire and it was there, in the parish of Llansantffraid, that the brothers Thomas and Henry were born.[5] Educated locally, before adolescent intellectual yearnings (and the absence of a university in Wales) took him beyond Offa's Dyke, Thomas Vaughan continued his education in Jesus College, Oxford. He attained a BA degree in early 1642 before being ordained, probably around 1645, and made rector of his parent's home parish of Llansantffraid.[6]

Between 1650 and 1652 Vaughan published eight works under the pseudonym 'Eugenius Philalethes'; a ninth followed in 1655. All were written in English – 'a language the author was not born to'[7] – and the first, *Anthroposophia Theomagica* (1650), has a preface dated 'Oxonii [16]48'. In it, having already 'rambl'd over all those inventions which the folly of man call'd sciences',[8] Vaughan declares his intention to quit 'this book-business' by which, given his own prolific book production, he appears to mean formal university-based study. His purpose in doing so: 'to study nature [rather] than opinion'.[9]

Why he turned so completely to an active study of nature is a matter of conjecture. Perhaps it provided him with stability and purpose when facing the sort of despondency and insecurity that can arise from familial tragedy – the death of his younger brother William, possibly in the second civil war; the nature of his times – as a royalist, he found himself on the losing side in a civil war that had not ended political and religious turmoil; and, as we have already seen, there was his profound disenchantment with the intellectual condition of his time. Whatever the reason for his turn to nature, it is evident that the natural world and its mysteries had been an interest since his youth in Wales, as he recalled in 1655:

> I have sojourn'd now for some years, in this great Fabrick, which the fortunate call their World: and certainly I have spent my time like a Travailer, not to purchase it, but to observe it. There is scarce anything in it, but hath given me an occasion of some thoughts; but that which took me up much, and soon, was the continual action of fire upon water. This speculation (I know not how) surpris'd my first youth, long before I saw the University, and certainly Nature, whose pupil I was, had even then awaken'd many notions in me, which I met with afterwards, in the Platonick Philosophie.[10]

Vaughan believed God had planted in us 'a most fervent desire to know' and, using Baconian ideas of empirical observation, experience and experimentation, he wanted to put into practice the advice he offered the scholastics in *Anthroposophia Theomagica*:

to use their hands, not their fancies, and to change their abstractions into extractions; for verily as long as they lick the shell in this fashion, and pierce not experimentally into the center of things, they can do no otherwise then they have done; they cannot know things substantially, but only describe them by their outward effects and motions, which are subject, and obvious to every common eye.[11]

Penetrating to the centre of things by experimentation he felt to be crucial for obtaining practical results. 'We have been abused with Greek Fables and a pretended knowledge of causes, but without their much desired effects,' he argues:

We plainly see, that if the least disease invades us, the school-men have not one notion, that is so much a charm, as to cure us: and why then should we imbrace a philosophy of mere words, when it is evident enough, that we cannot live but by works.[12]

In his opposition to such narrow scholasticism Vaughan entered on an examination of nature via Hermeticism, the second intellectual strand pertaining as our period opens in 1650. He did so through two of its principal branches – alchemy and natural magic.

The Torture of Metals

An image and a quest usually accompany modern perceptions of the alchemist. They are of the adept in his laboratory of bubbling retorts searching for the elusive 'Philosopher's Stone', the universal agent with a reputed power to transmute base metal into gold. Vaughan, though, did not see himself 'as some sort of conjuror of cheap tricks'. As he wrote in 1655: 'Alchymie in the common acceptation, and as it is a torture of metals, I did never believe; much less did I study it.'[13] Realising this to be somewhat disingenuous he admits, later in the same passage, that for three years he had waved his own principles and followed those 'who will hear of nothing but metals'.

Although Vaughan accepts the use of gold in alchemical experiments, and the possibility of the philosopher's stone as an agent of transmutation, he rails from his very first publication against those seeking the stone solely in order to produce gold. The 'study' and the 'noodle' of such enquirers, he suggests, were 'stuff'd with old receipts … if you bring him to the field, and force him to his polemics, if you demand his reason, and reject his recipe, you have laid him as flat as a flounder'.[14] The thought is prophetic. As late as 1782, one James Price (actually London-born James Higginbotham who adopted the surname Price as a condition of receiving an inheritance from a deceased relative) claimed an ability to transmute mercury into silver and gold using various powders. With Price having apparently done this before witnesses, the highly sceptical Joseph Banks (1743–1820), as president of the Royal Society to which Price had been elected in 1781, demanded a demonstration before the scientific community. Price intended to oblige at his home in July 1783 but drank poison and died on the appointed day.[15]

True alchemical experimentation, of a kind later leading to the development of chemistry, and that in this 'transition' period between medieval and modern science is often referred to as 'chymistry', actually had as its principal goals – 'achieving metallic transmutation, producing better medicines, improving and utilizing natural substances, [and] understanding material change'.[16] Furthermore, even though the origins of alchemy lay in the third to the ninth century, it underwent a major resurgence in the first half of the seventeenth century, reaching the peak of its new popularity just as Vaughan published his first works in 1650. Indeed, it has been said that between 1650 and 1680 'more alchemical books appeared in English than in all the time before and after those dates'.[17] Vaughan involved himself in this alchemical resurgence in two ways: through practical and philosophical alchemy.

Practical Alchemy

In his approach to practical alchemy Vaughan followed the example of Paracelsus by centring his work on medical alchemy and the development of new medicines and medicaments; otherwise known as iatrochemistry. His experiments in this field he detailed (along with his dreams) in Latin and English entries in a notebook titled *Aqua Vitae: Non Vitis*.[18] Unpublished in his lifetime the manuscript volume is inscribed as being 'From the Books of Thomas and Rebecca Vaughan' with a date of 28 September 1651, the day of their marriage. Although her maiden name is not mentioned in surviving archives, there is evidence to suggest Rebecca was the daughter of Dr Timothy Archer (1597–1672) of Meppershall in Bedfordshire.[19] Even though dated entries in the notebook relate to the years 1658–62, and so after Rebecca's early death on 17 April 1658 (she would be buried at Meppershall), Donald R. Dickson, in his 2001 biographical introduction to the text of *Aqua Vitae*, shows it to contain evidence of Rebecca playing 'a significant role in their alchemical experiments at London in the 1650s'.[20]

Vaughan expresses his attitude to women as alchemists in *Magia Adamica* (1650): 'For my part I think women are fitter for it than men, for in such things they are more neat and patient, being used to a small chimistrie of sack-possets, and other finicall sugar-sops.'[21] His hermaphroditic attitude to female/male relationships being considered more generally in *Anthroposophia Theomagica*: 'For life is nothing else but an union of male and female principles, and he that perfectly knows this secret, knows the mysteries of marriage, both spiritual and natural, and how he ought to use a wife.'[22] Yet, despite the obvious love for Rebecca expressed in *Aqua Vitae*, Vaughan still uses the term 'obedient wife',[23] suggesting intellectual and alchemical equality did not completely overturn 'the commonplace assertion and widespread acceptance of a hierarchical gender order'.[24]

Detailed discussion of the recipes and experiments in the *Aqua Vitae* notebook is beyond our scope. Suffice to say, many are resolutely 'chymical', such as this recipe for 'Aqua fortis' (strong water) or 'mainly nitric acid (HNO_3), used as a corrosive agent':[25] 'Take talc, saltpeter, rock salt, a half pound of each; [take] the same amount of crude alum, or one pound. Distill with a bare flame, and you will have it.'[26] Other entries reveal evidence of his wife's alchemical participation in what Dickson in his introduction calls 'the most important development in medicine in a millennium, the advent of chemical medicaments'.[27] One such entry comes in Thomas's list of his wife's belongings, in which he notes 'a great glass full of eye-water, made … by my dear wife, and my sister Vaughan, who are both now with God'.[28] The deceased sister is probably his sister-in-law, the wife of twin brother Henry. As a practising physician Henry also had an interest in medical alchemy. So much so, he translated two alchemical texts of the German physician Heinrich Nolle (1583–1630?) into English as the *Hermetical Physick* (1655) and *The Chymist's Key* (1657). Thomas wrote an introduction for the latter work and acted as its publisher.[29] That Thomas Vaughan also practised physic, as well as producing medicines, is evidenced in a letter from his brother Henry to their 'honoured cousin' John Aubrey, whose great-grandfather, William Aubrey (*c*.1529–95), came from an old Breconshire family and had been a friend of mathematician and alchemist John Dee (1527–1608). 'My brothers imploym^t was in physic and Chymistrie', Henry wrote, 'My profession also is physic'.

Donald R. Dickson notes that '[m]ore than fifty of the experiments (about half)' in *Aqua Vitae* 'clearly involved the preparation of medicines – either involving herbal preparations … or mineral compounds used as medicaments'.[30] He also notes that in making these preparations Vaughan records the details of his experiments, the apparatus needed to conduct them, and the relevant weights and measures of material used. Though he is less exact in recording the duration of his experiments, he does sometimes revisit them in order to amend them with new information. For Dickson, the evidence of the notebook reveals that the experiments of Thomas and Rebecca Vaughan in the 1650s 'were empirically rigorous, even in comparison to many of those performed before the Royal Society the following decade',[31] by the likes of Robert Boyle and, after 1662, Robert Hooke.

The mid-1600s also saw a significant increase in experimental philosophers working within proto-scientific societies and laboratories that lay outside the dominion of the established universities. Vaughan soon involved himself in two such societies and in an alchemical laboratory. The first society was 'the most illustrious and truly reborn Brothers of the Rosy Cross', to whom Vaughan dedicated his first published work

– *Anthroposophia Theomagica* (1650). Better known as the Rosicrucians, the Brothers of the Rosy Cross claimed to be a pan-European brotherhood aiming for the reform of scholastic teaching. They wanted instead the introduction of Baconian-style observation and experiment and towards this end they published two manifestos – the *Fama* and the *Confessio*.[32] These appeared in Europe in 1614 and 1615. Originally appearing in German and (in the case of the *Confessio*) Latin, Thomas Vaughan published the first ever English translations of them in 1652, adding, in the process, a lengthy preface of his own composing (see Figure 2).[33]

He did not actually translate the works, but quite possibly obtained manuscript translations already owned by his patron, Sir Robert Moray, a Scot, Freemason and soon to be an important founder member of the Royal Society (founded 1660).[34] Beyond the Rosicrucian appeal for a new type of philosophical enquiry, Vaughan is likely to have been attracted

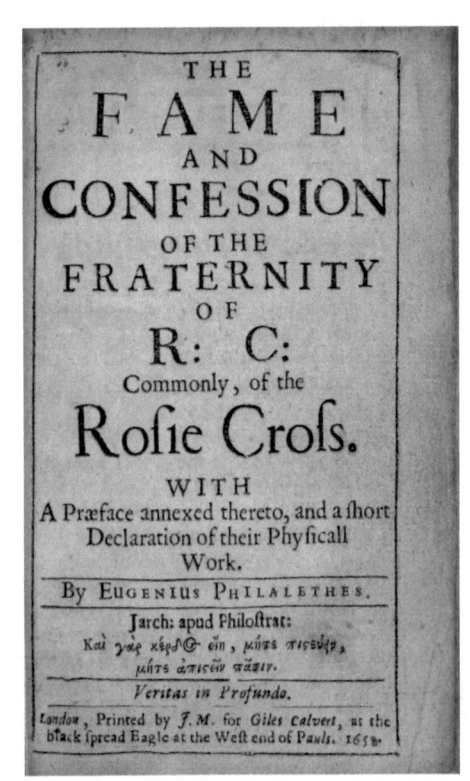

FIGURE 2 Title page to *The Fame and Confession of the Fraternity of R: C: commonly, of the Rosie Cross*, prefaced and published by Thomas Vaughan (London, 1652); the first English translation. Source: the Open Library, Internet Archive (Original: Getty Research Institute).

by their adherence to the sort of alchemical and natural magic branches of hermeticism that he too espoused. Nevertheless, and for reasons that are unclear, Vaughan, like others at the time, denied actual membership of the Rosicrucians while simultaneously lauding their aims.[35]

Vaughan's involvement with the second society, and his entry into a laboratory, came about through changing personal circumstances. In about 1650, as he published his first works, and with Wales and England under the control of Cromwell and Parliament, Vaughan's royalist sympathies caught up with him. Commissioners acting under Parliament's *Act for the Better Propagation and Preaching of the Gospel in Wales, Ejecting Ministers and Schoolmasters, and Redresse of Grievances* were authorised, in Geraint H. Jenkins's memorable description, 'to rid Wales of the "dumb dogs", drunkards, fornicators, pluralists, and malignants who had brought the ministry into disrepute'. They 'set about their tasks with brisk, unfussy zeal'.[36] Vaughan found himself charged as 'a common drunkard, a common swearer, no preacher, a whoremaster, and "in armes personally against the Parliament"'.[37] As to his preaching skills and sexual peccadilloes, there is no evidence. On being 'a common drunkard' and 'a common swearer' we are on firmer ground given the sometimes scatological language of his later pamphlet war with the Cambridge philosopher Henry More (see Chapter 3), and his admission of having had a 'distemper in the Liver' and a dream in which he saw a 'certaine person, with whom I had in former times revell'd away many years in drinking'.[38] On the final charge, of taking up arms against Parliament, it is likely he was the Captain Thomas Vaughan listed as being taken prisoner following the defeat of royalist forces at the Battle of Rowton Heath, near Chester, on 24 September 1645. As a consequence of these charges Thomas lost his living at Llansantffraid. By 1650 he had moved to Kensington, in London, and into the house of a man he describes in a dedicatory epistle to his third work, *Magia Adamica* (1650), as 'my best of friends'.

Vaughan's friend was royalist Englishman, and founding member of the Royal Society, Thomas Henshaw (1618–1700). Following studies at Oxford and in private tuition with the eminent clergyman, mathematician and alchemist William Oughtred (1574–1660),[39] Henshaw had abandoned the practice of law and established himself at Pond (or

Moat) House in Kensington; a place equipped with a laboratory and a large and important alchemical library. Vaughan joined him there as, in all probability, did Rebecca Vaughan following her marriage to Thomas in September 1651. This small group, which may also have included Henshaw's mother who, like her late husband, appears to have been alchemically inclined,[40] aimed to 'lay their heads and hands together to see what they can produce'. How long Vaughan remained at Kensington is unclear; but until 1653 and a visit back to Wales in that year is possible.[41] Equally uncertain is whether the alchemical experimentation recounted in Vaughan's *Aqua Vitae* notebook was undertaken at the Pond House laboratory, though it is clearly likely. What *is* certain is that the London-based Polish émigré and collector of scientific intelligence, Samuel Hartlib (*c.*1600–62), noted sometime between May and October 1650 that the agriculturalist, alchemist and physician Robert Child (1613–54) was 'endeavouring to form a chymical club with Hinshaw [*sic*], Webbe, Vaughan etc'.

Hartlib refers to the resulting 'chymical club' as a 'Model of [a] Christian Learned Society'. He also suggests that it remained something of a secretive organisation so that the members of the Christian Learned Society at Kensington tended 'to devotion and studies' while 'separating them[selves] from the world by leading a severe life'.[42] Yet he also notes that the Society aimed:

1 to collect all English Philosophical books or other chymists
2 all ms [manuscripts]
3 to translate and publish them in one volume
4 to make all philosophers acquainted one with another and to oblige them to mutual communications.[43]

Such aims, and in particular the making of all philosophers acquainted with one another, would clearly entail some degree of openness beyond the secrecy Hartlib assigns to the society's endeavours at Kensington. This mix of secrecy and openness is also found in Vaughan's alchemical writing, whether in relation to practical alchemy or, as we shall see in the next section, philosophical alchemy. In adopting this amalgam of openness and secrecy Vaughan resembles many other alchemical practitioners of the time, and both trends are reflected in his alchemical notebook which, according to Dickson, 'stands Janus-like in this transitional age: with its codes and exotic language, it hearkens backwards to traditions of esoteric secrecy; with its attempts at precision, it looks forward to modern laboratory practice'.[44] The recipe quoted earlier for 'Aqua Fortis' exemplifies the open alchemical description in the notebook, but Vaughan's more usual presentation embraces the kind of allegorical language found in this formulation for 'Blood of the Dragon and Sophic Mercury':

> Dissolve the dragon joined with Pluto with oil of sal ammoniac, and cohobate, until the whole turns into a thick oil. Coagulate this oil with unnatural fire, and fix. Dissolve what has been fixed in the oil of sal ammoniac, distill the solution, and you will have it. It was proved in the days of my sweetest wife.[45]

The term 'dragon' is a complex one in alchemy, but 'Blood of the Dragon' could suggest that Vaughan is referring to the red-coloured resin obtained from the *Dracaena draco* tree; a native of the Canary Islands and Morocco. 'Sophic Mercury' is an equally complex term but his use of it suggests common mercury that has been refined.[46] His use of allegory is even more prevalent in his published works. There, his sentences sometimes seem designed to save the author trouble and/or to deflect the reader from what might otherwise prove significant or revelatory explanations, as in these examples from *Magia Adamica*:

> I could cite many more magical and mystical places, but in so doing I should be too open, wherefore I must forebear.
> But these are things of a higher speculation than the scope of our present discourse will admit of.[47]

While literary secrecy and deflection helped maintain the image of the alchemical adept as a specialist with occult knowledge, they impact negatively on the production of (verifiable) scientific knowledge; perhaps even obscuring at times the extent of the alchemist's own contribution. Vaughan, for example, follows Paracelsus in seeing nitre or saltpetre as a fundamental component of matter and, when Vaughan suggests in *Euphrates* (1655) that 'the earth is full of nitre', he also uses the phrase 'the fire and the soul of nitre'. This suggests some knowledge of the 'vital spirit' or 'food of life' given off when saltpetre is heated and that, following the work of Priestley and Lavoisier in the late eighteenth century, would eventually be called oxygen. At about the time Vaughan published *Euphrates* the alchemist and proto-chemist Robert Boyle was moving to Oxford to begin his own 'work on air and saltpetre',[48] and Vaughan's experiments on saltpetre were certainly known about. In June 1651, the alchemist William Backhouse had informed Elias Ashmole that Vaughan was working 'upon the spirit of saltpetre / and of late he added May-dew to it'.[49] Whether Vaughan made any *new* observations on the nature of saltpetre, and its by-product, we will probably never know, for although he declares in *Anthroposophia Theomagica* 'I should amaze the reader if I did relate the several offices of this body [air]', he also says 'but it is the Magicians Backdoor, and none but Friends come in at it'.[50]

This tendency towards secrecy in his writings and, to a degree, in his social milieu is one that he shared with many other alchemists/scientists of the time (indeed, secrecy is prevalent even today in commercial and industrial science). While such secrecy does, to a degree, restrict the wider value of his alchemical/scientific endeavours, they do not invalidate the clear evidence that his notebook provides of his involvement in practical alchemical/scientific experimentation.

Philosophical Alchemy

In his 1655 work *Euphrates: or, the Waters of the East* Vaughan provides us with a source for his philosophical outlook: 'Nature whose pupil I was, had even awoken'd notions in me, which I met with, afterwards, in the Platonick Philosophie.' As we saw in the previous chapter, the 'divine Plato', as Vaughan calls him, had proposed that an order had been applied to Creation by its Creator. We have also seen that Vaughan's mode of enquiry into this order came principally through practical alchemy and its particular application to the production of medicaments. His attitude to the more philosophical aspects of the subject, however, and to the very nature of creation itself, involved him in another branch of the hermetic intellectual strand – that of natural magic.

Magic in these terms is far removed from conjuring and card tricks, or the dark magic that was also believed to exist. In keeping with the platonic idea of a background order to creation, natural magicians believed in the existence of hidden or occult powers – those lying behind the observable effects of magnetism, for example, or the manner in which fire acts upon water that had so entranced the young Thomas Vaughan. Believed to have been implanted into the things of creation by God, these occult powers represented natural as opposed to supernatural powers. As such, they were capable of being investigated and understood by Natural Magicians, who aimed to get closer to the Creator by such enquiries and to use any powers they might discover 'to perform wonders' through practical application.

As a follower of the alchemical and natural magic branches of the hermetic tradition, Vaughan saw this magic as 'nothing else but the Wisdom of the Creator revealed and planted in the creature', and he made no apology for basing his philosophy on it: 'That I should profess Magic in this Discourse, and justify the professors of it withal, is impiety with many, but religion with me. It is a conscience I have learnt from Authors greater than my self, and Scriptures greater than both.'[51] The authors greater than himself included not only the already mentioned Hermes Trismegistus and Paracelsus, but also the Polish alchemist and philosopher Michael Sendivogius (1566–1636), and the major proponent of natural magic, Heinrich Cornelius Agrippa von Nettesheim (1486–1535), who Vaughan describes as 'that Grand *Archimagus*, as the antichristian *Jesuits* call him? He is indeed my author, and next to God I owe all that I have unto him'.[52]

From the writings of such authors, Vaughan believed that the matter from which creation had been made was imbued with a universal spirit. A divine activating force, imbibed directly from the Creator at the creation, so that 'the world, which is God's building, is full of spirit, quick, and living'. In this so-called 'vitalist' interpretation of creation the universal spirit was 'the cause of multiplication, of several perpetual productions of minerals, vegetables, and creatures ingendred by putre-faction [the idea of spontaneous generation]: all which are manifest, infallible arguments of life'.[53] No surprise, therefore, that when attempting to describe the structures and processes of creation in *Anthroposophia Theomagica*, and his later works, Vaughan integrates natural magic and alchemical ideas with the greater scriptures of the bible and its story of Genesis. Thus:

> *God* before his *work* of *Creation* was wrapp'd up, and contracted in himself ... But when the decreed *Instant* of *Creation* came, then appeared *Aleph Lucidum* [the manifested state of God], and the first *Emanation* was that of the *holy Ghost* into the *bosom* of the *matter* ... No sooner had the Divine *Light* pierced the *Bosom* of the Matter, but the *Idea*, or Pattern of the whole Material World appeared in those *primitive waters*, *like* an *Image* in a *Glasse*: by this Pattern it was that the Holy Ghost fram'd and modelled the Universal Structure. This Mystery or appearance of the *Idea* is excellently manifested in the *Magicall Analysis* of Bodies (For he that knows how to imitate the *Proto-Chymistrie* of the Spirit, by Separation of the Principles wherein the Life is imprisoned, may see the Impresse of it Experimentally in the outward natural vestiments.)[54]

Similar ideas can be seen at work in the writings of another Welsh alchemist active at the same time as Vaughan – Bassett Jones (*fl.*1634–59). Hailing from Michaelston-super-Ely, in Glamorgan, Jones attended Jesus College Oxford. After having studied physics and chemistry at the Dutch university of Franecker, in Friesland,[55] he published, in Latin, in 1648, a prose alchemical treatise entitled *Lapis Chymicus Philosophorum Subjectus*. A little later (between *c.*1648 and 1656), he produced, in English, a substantial poetic work *Lithochymicus, or a Discourse of a Cymic Stone*. Based partly on the earlier *Lapis Chymicus*, the 'paraphrastically Englished' manuscript of the poem *Lithochymicus* survives in the British Library.[56] It comprises some 2,917 lines in English, together with a few in Welsh, and it has been described as 'the most ambitious alchemical poem of the seventeenth century, and arguably the most ambitious English alchemical poem of them all'.[57]

Jones, like Vaughan, admired the Rosicrucians and as Robert M. Schuler writes in his introduction to the first published edition of the poem (1995), it 'would have been welcome in [the] milieu of Paracelsian medicine, hermetic poetry and mysticism' of the time – a milieu characteristic of Thomas Vaughan. Schuler also notes Jones's belief in 'a universal system of nature linking human beings with God through the macrocosm-microcosm relationship' (i.e., the universe seen as the macrocosm and oneself as the microcosm). Some of these relationships appear in one of the drawings in Jones's *Lithochymicus* manuscript (see Figure 3).

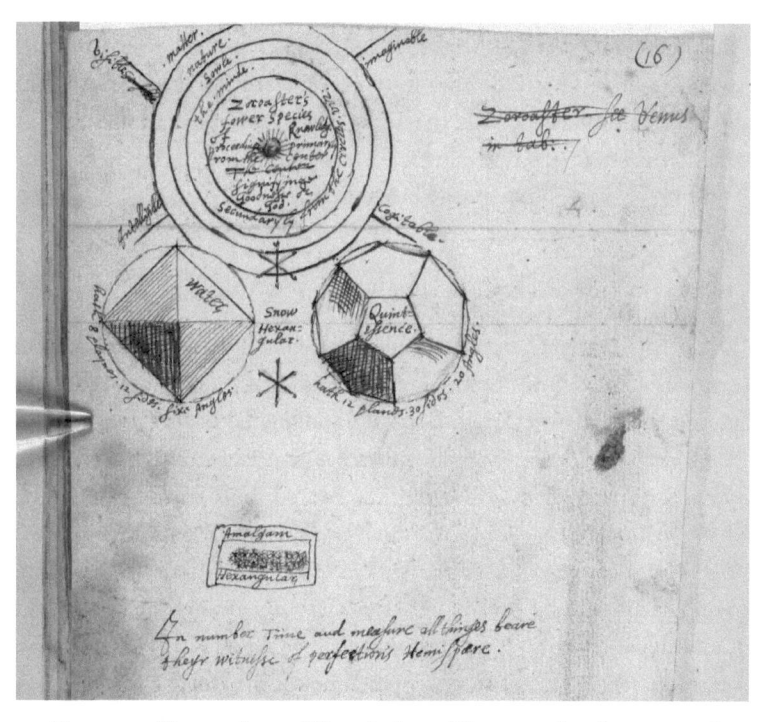

FIGURE 3 Bassett Jones 'Hirogliphic of forms under the name of Zoroaster' from his manuscript 'Lithochymicus or A Discourse of a Chymic Stone'; BL MS Sloane 315, fol. 13v, © The British Library.

At its heart lies the 'goodness of God', reflected in this case by the sixth-century BCE Persian prophet Zoroaster, a member of the so-called *prisci theologi*. Supposedly representing a line of ancient theologians to whom God, in earlier times (those closer to the Adamic period of perfect knowledge) had revealed the true knowledge of the cosmos, the genealogy of the *prisci theologi* extended from Zoroaster and Hermes Trismegistus to Moses and Plato, and from Plato to Christianity. Some writers also included other ancient philosophers, such as Pythagoras, within the lineage.[58]

In Jones's illustration we see Zoroaster surrounded by 'four species of knowledge' that also signify the goodness of God – mind, sense, nature and matter – together with 'beams sparkling' from these circles of knowledge. These four beams are labelled the intelligible, the cogitable, the imaginable, and the visible. Two of the elements are also illustrated, using the geometrical forms assigned to them by Plato: water as an icosahedron, and the quintessence, representing the entirety of the heavens, as a dodecahedron; the hexagonal snowflakes descending from the heavens perhaps reflecting the close link between the macrocosm and microcosm – between the heavens and us.[59]

'I know my reward is Calumnie' – Thomas Vaughan

Just as Thomas Vaughan[60] had challenged Scholastic Aristotelianism, not everyone agreed with the hermetic, natural magic and vitalist ideas he presented in his own publications. No sooner had his first two works – *Anthroposophia Theomagica* and *Anima Magica Abscondita: Or, a Discourse of the universall Spirit of Nature* – appeared in 1650 than they were being critiqued, that same year, in *Observations Upon Anthroposophia Theomagica and Anima Magica Abscondita*. Written under the pseudonym Alazonomastix Philalethes, the *Observations* were the work of the Cambridge Platonist and philosopher Henry More (1614–87), another person destined to be a founding member of the Royal Society.[61]

Vaughan immediately replied to More's critique in *The Man-Mouse* [More] *Taken in a Trap, and Tortur'd to death for gnawing the Margins of Eugenius Philalethes* [Vaughan's pseudonym] (1650). He then proceeded to answer More's 'observations' point by point (forty-nine for *Anthroposophia* and twenty-seven for *Anima Magica*).

More returned to the fray in *The second lash of Alazonomastix, laid on in mercie upon that stubborn youth Eugenius Philalethes* (1651). To which Vaughan replied in *The Second Wash: or the Moore Scour'd once more, being A Charitable Cure for the Distractions of Alazonomastix* (1651). The 'debate' ended after this, but the tone of the titles gives evidence of the argument's often virulent nature; a virulence bordering on the scatological at times, as two brief examples illustrate:

> *Vaughan on More*: He is indeed a scurvy, slabbie, snotty-snowted thing. He is troubled with a certain splenetic looseness, and hath such squirts of the mouth, his readers cannot distinguish his breath from his breech ... Fie Upon thee thou Quack! Never was there in the world such an impudent, ignorant scribbler![62]

> *More on Vaughan*: But I find I have so nettled him unawares, as if his senses lay all in his backsides, and had left his brains destitute ... Verily if I had thought his retentive faculty had been so weak, I would not have fouled my fingers with meddling with him.[63]

Whatever Vaughan's opinion of him, Henry More did make some pointed scientific observations, most notably concerning Vaughan's suggestion in *Anthroposophia Theomagica* that the rise and fall of the tides resulted from a form of respiration of the Earth. More saw this as evidence of Vaughan's adoption of the concept of a 'World Soul', or the 'world animal' as he chose to describe it. Here, More had a point. When describing the world as being of 'God's building' and 'full of spirit, quick and living', Vaughan had also declared that the 'texture of the universe clearly discovers its animation' and so:

> The earth which is the visible natural basis of it, represents the gross, carnal parts. The element of water answers to the blood, for in it the pulse of the great world beats; this most men call the flux and reflux, but they know not the true cause of it [presumably a

reference to the tides].[64] The air is the outward refreshing spirit, where this vast creature breathes, though invisibly, yet not altogether insensibly. The interstellar skies are his vital, aethereal waters, and the stars his animal sensual fire.[65]

Despite the fact that Vaughan is correct in saying that the true cause of the flux and reflux of the tides was not known at the time,[66] it is hard, even allowing for a possible use of allegory, not to see his depiction of the Earth as leading to the concept of the 'world soul'. That concept, though, was neither new, nor unique to Vaughan. It can be traced back to ancient times, and it had also been postulated by Vaughan's 'archimagus', Cornelius Agrippa: 'All philosophy affirms … that the world has a soul, which soul is intelligent.' As we shall see in the next chapter, even Isaac Newton, in some of his early writings, held to the same idea.

Henry More, though, reserved his principal criticisms for Vaughan's hermetic, natural magic and vitalist ideas. These, he suggests, were opposed to reason, examples of philosophical enthusiasm, and represented a:

desire to be filled with high swoln words of vanity, rather than to feed on sober truth, and to heat and warm ourselves rather by preposterous and fortuitous imaginations, than to move cautiously in the light of a purified minde and improved reason.[67]

While the comment might reflect More's fear of a return to the philosophical enthusiasm and 'high swoln words of vanity' that had characterised many publications during the recent upheavals of the Civil War, he had also, by the time of his debate with Vaughan, become an early advocate of our third intellectual strand – the philosophy of René Descartes.[68] This was a philosophy whose principal tenets were certainly in conflict with aspects of Vaughan's hermeticism.

Notes

1. *WTV(LL)*, pp. 325–6.
2. Today, modern scholarship is reassessing the significance of Aristotle and sees the scholastic method of teaching as not entirely impervious to new ideas, so long as fundamental tenets were not challenged. See, for example, T. E. Rihll, *Greek Science*

(Oxford, 1999; Cambridge, 2011); Armand Marie Leroi, *The Lagoon: How Aristotle Invented Science* (2014; New York, 2015); and Anthony Gottlieb, *The Dream of Reason* (2000; London, 2016), at chapter 12, 'The Master of Those Who Know: Aristotle'.

3. *WTV(AT)*, p. 51.
4. See David Wootton, *The Invention of Science* (2015; London, 2016), p. ii.
5. *AV:NV*, p. ix.
6. See Brigid Allen, 'The Vaughans at Jesus College, Oxford, 1638–48', *Scintilla*, 4 (2000), 68–78.
7. *WTV(AT)*, p. 94. Vaughan's first language being Cymraeg (Welsh); he also had English, Latin, Greek and 'some knowledge of Hebrew'. He is described in Wood's *Athenae Oxoniensis* (1721) as 'an understander of some of the Oriental languages', *WTV*(Intro.), p. 2 and n. 2.
8. *WTV(AT)*, p. 54.
9. *WTV(AT)*, p. 54. See also Robert Wilcher, '"Thalia" and the "Father of Lights": Nature and God in the Works of Henry Vaughan and Thomas Vaughan', *Scintilla*, 16 (2012), 9–36.
10. *WTV(E)*, p. 521.
11. *WTV(AMA)*, p. 108.
12. *WTV(F&C)*, p. 483.
13. *WTV(E)*, p. 513.
14. *WTV(LL)*, p. 346.
15. *ODNB*.
16. Lawrence M. Principe, *The Secrets of Alchemy* (Chicago IL and London, 2013), p. 107.
17. *AV:NV*, p. xiv n. 27 citing John Ferguson, 'Some English Alchemical Books', *Journal of the Alchemical Society*, 2 (1913), 5.
18. As Dickson notes (*AV:NV*, p. 3 n. 2): 'Vaughan uses aqua vitæ for common alcohol, but in the title he plays on the biblical metaphor of the waters of life', hence the 'non vitis/not of the vine' part of the title.
19. *AV:NV*, pp. xviii–xxiv.
20. *AV:NV*, p. xxv.
21. *WTV(MA)*, p. 220.
22. *WTV(AT)*, p. 72.
23. *AV:NV*, p. 240.
24. Dorinda Outram, 'Gender' in *CHS*-3, p. 806.
25. *AV:NV*, p. 250.
26. *AV:NV*, p. 137.
27. For details of the Vaughan's as experimental philosophers and their place in iatrochemistry, see *AV:NV*, pp. xxxi–xlix.
28. *AV:NV*, p. 244.
29. For the text see *WTV*, pp. 557–9.
30. *AV:NV*, p. xli and xliii–xliv.
31. *AV:NV*, p. xlvi.
32. See Donald R. Dickson, *The Tessera of Antillia* (Leiden, 1998), pp. 62–88 for details of the Rosicrucians, and pp. 207–18 for the Rosicrucians and Vaughan.

Also, Michael Srigley, 'Thomas Vaughan, the Hartlib Circle and the Rosicrucians', *Scintilla*, 6, (2002), 31–54; Thomas Willard, *Thomas Vaughan and the Rosicrucian Revival in Britain, 1648–1666* (Leiden, 2022).

33. *WTV(F&C)*, pp. 476–510.
34. See *WTV*(Intro.), pp. 22–4.
35. See Francis Yates, *The Rosicrucian Enlightenment* (1972; London and New York, 2002), p. 192 for another example, and Chapters 16 and 17 for the Rosicrucian's generally.
36. Geraint H. Jenkins, *The Foundations of Modern Wales 1642–1780* (Oxford, 1987), p. 50.
37. *WTV*(Intro.), pp. 7–8.
38. *AV:NV*, pp. 45, 234.
39. Oughtred invented the slide rule and introduced into mathematical notation 'x' for multiplication and the use of sine and cosine.
40. Samuel Hartlib described Henshaw's father as a 'great chymist' and 'so is his mother who is yet alive'; see *WTV*(Intro), p. 12.
41. See Dickson, *Tessera of Antillia*, p. 189 and n. 30.
42. See Dickson, *Tessera of Antillia*, p. 191.
43. Sheffield University, *Hartlib Papers*, 'Ephemerides', 1650, part 3, May–October, f. 28/1/61B. Robert Child introduced the American alchemist George Starkey to Robert Boyle.
44. *AV:NV*, p. xlix.
45. *AV:NV*, p. 185.
46. For 'Dragon's Blood' see, Toni Mount, *Dragon's Blood & Willow Bark, the Mysteries of Medieval Medicine* (Stroud, 2015), p. 119. For the complexity of the term 'dragon' in alchemy, see *AV:NV* p. 253; for 'sophic mercury', see pp. 258–9.
47. *WTV(MA)*, p. 176 and 177.
48. See Steven Shapin and Simon Schaffer, *Leviathan and the Air-Pump: Hobbes, Boyle, and the Experimental Life* (1985; Princeton, 2018), p. 71, and *WTV*, p. 730.
49. *AV:NV*, p. xliii n. 132. Newman *CHS-3*, p. 516 notes that the 'earliest sustained series of experiments carried out by the Royal Society consists of an attempt made in 1664–5 to analyze May-dew and to extract from it the Sendivogian *sal nitrum*'.
50. *WTV(AT)*, p. 65. For details of saltpetre see William R. Newman, 'From Alchemy to "Chymistry"' in *CHS-3*, pp. 513–15.
51. *WTV(MA)*, p. 150.
52. *WTV(AT)*, p. 84.
53. *WTV(AT)*, p. 52.
54. *WTV(AT)*, pp. 57–9.
55. Founded in 1585 the university had been attended by Descartes between 1629 and 1630.
56. BL Sloane MS 315, f. 1r–91v.
57. See Robert M. Schuler, *Alchemical Poetry 1575–1700: From Previously Unpublished Manuscripts* (1995; Abingdon, 2013), p. 211.

58. See D. P. Walker, *Spiritual and Demonic Magic: From Ficino to Campanella* (1958; Pennsylvania, 2003), p. 23; Frances A. Yates, *Giordano Bruno and the Hermetic Tradition* (1964; Chicago IL and London, 1991), p. 15 n. 1.

59. The snowflake had been described in Europe by Kepler in his *De Niva Sexangular* ('The Six-Cornered Snowflake') of 1611. Descartes also illustrated them in his *Les Météores* of 1637 on weather and other atmospheric phenomena.

60. *WTV(AT)*, p. 70.

61. Henry More [as Alazonomastix Philalethes], *Observations Upon Anthroposophia Theomagica and Anima Magica Abscondita* (London, 1650).

62. *WTV(MM)*, p. 237 and *WTV(SW)*, p. 447.

63. Henry More [as Alazonomastix Philalethes], *The Second Lash of Alazonomastix* (London, 1651), pp. 8–9.

64. See also *WTV(MM)*, p. 253.

65. *WTV(AT)*, pp. 52–3.

66. Galileo had argued for movement of the earth as a cause and although More correctly argued for the moon having a part to play, this would not be confirmed until Newton's work on gravitational attraction in 1687.

67. Quoted in Frederic B. Burnham, 'The More-Vaughan Controversy: The Revolt Against Philosophical Enthusiasm', *Journal of the History of Ideas*, 35/1 (1974), 33–49. Further criticism of Vaughan came from George Starkey (1628–65), an American alchemist who came to England in 1650 just as Vaughan published his first works. Starkey wrote under the pseudonym Eirenaeus Philalethes, which, with its similarity to Vaughan's own pseudonym (Eugenius Philalethes), resulted in considerable confusion among historians until the twentieth century. Starkey became friendly with Robert Boyle and it was Boyle who, in 1651, reported to the Polish émigré and collector of scientific intelligence Samuel Hartlib (*c.*1600–1662) that Starkey 'is about to refute Vaughan as likewise to translate a Chymical Booke into Eng. out of Latine'. Satirical rather than scientific criticism came from later writers, such as Samuel Butler in his poem *Hudibras*, published between 1664 and 1678; Jonathan Swift in *A Tale of a Tub* (1709); and from 'Torrescissa', the author of the undated alchemical poem *Hermetical Raptures*: 'Ah! poor Eugenius! how cams't thou to fall, / From Anthroposophia magicall, / Into the burning lake? Nay then I see / (Though men said one) thou hadst ne're an eye. / Yet thou disputest *More* & *More* ... / ... But I'me afraid / That thou hast preacht thy congregation madd', see Schuler, *Alchemical Poetry*, p. 596.

68. Not only is More credited with inventing the word Cartesianism to describe the philosophy of Descartes, he also, despite his own reservations concerning some of the philosophy's tenets, corresponded with Descartes and taught the philosophy at Cambridge from 1648. See John Henry, *Henry More* at *https://plato.stanford.edu/entries/henry-more*.

3

MAKING NEW WORLDS

Since Des Cartes led the way Every New Philosopher
thought himself wise enough to make a world.[1]

John Vaughan, FRS of Golden Grove, in
Carmarthenshire, Third Earl of Carbery

René Descartes (1596–1650) was born in France, later moved to
Holland (1629–49), and eventually ended his days in Sweden.[2] As
a mathematician and scientist he contributed to a variety of subjects,
including optics and geometry, but it is his contribution to philosophy
that is our main concern here. The modern literature on that contribution, and the nature of the challenge it presented, is vast.[3] Consequently,
our remit extends no further than a brief outline of the main elements
of the philosophy, their relation to the already existing hermetic ideas of
Thomas Vaughan and, later in the chapter, to our soon to appear fourth
intellectual strand – the world of Isaac Newton.

Making a New World

In works such as *A Discourse on the Method of Rightly Directing One's
Reason and of Seeking Truth in the Sciences* (1637), the *Meditations* (1641)
and *Principia Philosophiae* (1644), Descartes determined, to his own
satisfaction, that God existed and had created and set in motion the
universe. Matter, as the very stuff of that Creation, existed simply as
'extension' – its occupying of all available space through length, breadth
and depth – and it was composed of minute particles, often referred
to as atoms or corpuscles. Then, in the words of Geoffrey Sutton: 'As

things settled down (and God gracefully withdrew), the world began to take its present shape, following the laws of motion.' Eventually the particles of matter organised themselves into independent spheres of motion, or vortices and 'these regular, stable whirlpools set the pattern for heavenly motion'.[4] Thereafter, everything in Creation, including its human and animal components, ran as a sort of mechanical mechanism.

Descartes's mechanistic philosophy presented problems then and now, but one critical problem was how do mechanical processes give rise to non-mechanical ones such as thought and consciousness. What was it that connected matter with thought and the body with the soul? To solve this, Descartes suggested the corporeal and non-corporeal realms were linked by what he called 'animal spirits', and that these entered the brain through its pineal gland.[5]

This new mechanical world presented a significant challenge to the hermetic view of creation espoused by the likes of Thomas Vaughan. If, as Descartes's philosophy implied, God had largely withdrawn from the world and all phenomena, even biological ones, could be explained by mechanical processes and the effects of matter in motion, then the hermetic belief in matter imbued with vital spirits and occult powers derived from God is rendered null and void. In the alchemical and natural magic worlds of both Thomas Vaughan and Bassett Jones, God not only created the cosmos, he also formed a vital element within it and was continually linked to humanity through the macrocosm/microcosm relationship. It was this essential unity between God, Creation and Humanity that many in the 1650s saw as being seriously challenged, even broken, by Descartes's remote creator and the new mechanical world.

A Changing World

Although aware of Descartes's ideas, Thomas Vaughan, in his published works, rejects what he calls 'the whymzies of Des Chartes', and he chastises Henry More for being a 'Hackney of Des Chartes' ('hackney' being used here in its figurative sense of a hired drudge).[6] At which point it is all too easy for us, looking back, to see Vaughan as the voice of outdated ideas, and More as the promoter of a Cartesian philosophy that came to form part of the route to modern science, but this is too sweeping and simple a conclusion.

As Vaughan's critique of well-established Scholastic Aristotelianism indicates, he was not averse to abandoning old ideas, even when those ideas may have come to him from a man he clearly respected, and who possibly taught him at home in Breconshire. This was the Revd Thomas Powell, a near neighbour during Vaughan's youth in Wales, and whose *Elementa Opticæ: Nova, Facili, & Compendiosâ Methodo Explicata* ('Elements of Optics: A New, Easy, and Compendious Method Explained') contains Latin panegyrics addressed to him from the poetic quills of both Thomas Vaughan and his twin brother Henry. Powell's *Elementa Opticæ* was published in 1651 (see Figure 4), precisely the time that Thomas Vaughan was publishing his criticisms of scholasticism and that 'scabby sheep' Aristotle; a view of the Greek philosopher not shared by Powell, as the 'Epilogue' to his *Elementa Opticæ* makes clear:

FIGURE 4 Title page to *Elementa Opticæ* by Thomas Powell (London, 1651). Courtesy of Wellcome Collection.

We acknowledge (and it should not be denied) that we have crafted this small work according to the principles and dogmas of Aristotle. We ought not to follow any other guide than him to whom the saner world allots the sovereignty of philosophy.[7]

Nor should we see Vaughan's rejection of the 'whymzies of Des Chartes' as implying an unwillingness to consider and adopt challenging new ideas. For example, in *Euphrates* (1655), Vaughan willingly expresses an idea that he knows will be unpopular while, at the same time, giving us a fascinating insight into his religious/scientific reasoning and the importance to him of God as an active presence in the world:

if by Divinity, we understand the Doctrine of Salvation, as it is laid down in Scripture, then verily it is a *Mixt Doctrine*, involving both *God* and *Nature*. And here I doubt not to affirm, That the *Mysterie* of *Salvation* can never be fully understood without *Philosophie*, not in its just latitude, as it is an *Application* of *God* to *Nature*, and a *Conversion* of *Nature* to *God*, in which *two Motions* and their *Meanes*, all spiritual and natural knowledge is comprehended.

To speak then of God *without Nature*, is more than we can do, for we have not known him so: and to speak of Nature *without God*, is more than we may do, for we should rob God of his glory, and attribute those Effects to Nature, which properly belong to God and to the spirit of God, which works in nature. We shall therefore use a mean form of speech, between these extremes, and this form the Scriptures have taught us, for the Prophets and Apostles have used no other. Let not any man therefore be offended, if in this Discourse we shall use *Scripture* to prove *Philosophie*, and *Philosophie* to prove *Divinity*, for of a truth our *knowledge* is such, that our *Divinity* is not without *Nature*, nor our *Philosophie* without *God*. Notwithstanding, I dare not think but most men will repine at this course, though I cannot think, wherefore they should, for when I joyne *Scripture* and *Philosophy*, I do but join *God* and *Nature*, an union certainly approved of by God, though it be condemned of men.[8]

This union of God and nature was easily interpreted as blasphemous. In the Bible, God, at the time of the Creation, had granted man dominion over certain parts of nature: 'And God said, Let us make man in our image, after our likeness: and let them have dominion over the fish of the sea, and over the fowl of the air, and over the cattle, and over all the earth, and over every creeping thing that creepeth upon the earth.'[9] Consequently, any idea of God and nature being joined as one would effectively render God *subordinate* to man. Some years later the propagation of a similar argument for the unity of God and Nature, by the philosopher and lens grinder Baruch Spinoza (1632–77), would lead to his work – *Tractatus Theologico-Politicus* ('Theological-Political Treatise', 1670) – being described as a book forged in hell. Spinoza was roundly condemned and ostracised by Christians and his own Jewish community in Amsterdam.[10]

Despite Vaughan's dismissal of Descartes's ideas, there are some areas where he is in agreement with them. For example, Vaughan accepts, as does Descartes, the impossibility of a vacuum.[11] Air, Vaughan declares in *Anthroposphia Theomagica*, is necessary for respiration and nourishment and, being 'spread through all things, hinders vacuity, and keeps all the parts of nature in a firm, invincible union'.[12] Vaughan writes this in 1650, when the possibility of a vacuum's existence had been a subject of debate and experimentation in science since Evangelista Torricelli's work on the subject in 1643.

While Vaughan appears unaware of the work of Torricelli, he is clearly aware of other scientific developments of his own and earlier times. At one point in their 1650 pamphlet debate Henry More had challenged Vaughan over his attitude to Galileo, and the advent of the telescope. Vaughan replied by saying: '[Y]ou ask me, if I did ever look at a Galileo's Tube?' before adding, 'Doest thou think … I never saw these spectacles, these trunks and tricks which are grown the very Hackneys and Prostitutions of Art?'[13] Although the reply appears dismissive of the telescope, its actual intent is irony. Vaughan makes clear his knowledge of 'Galileo's Tube', and its value to science, in *Magia Adamica* (1650). There he writes of 'spots and darkness' on the Sun, 'as it hath been discovered by telescope';[14] sunspots had been discovered by Galileo and Christoph Scheiner in 1611, though with priority for the discovery still unclear.

Vaughan's attitude to the heliocentric ideas of Copernicus is more difficult to determine. The rejection of Copernican heliocentrism is said to be relatively common among hermetic philosophers.[15] Vaughan, however, makes no reference in his published works to the arrangement of the planets beyond noting that the early earth-centred cosmos, of Ptolemy and Aristotle, had made it necessary for the scholastics to imagine, within the heavens, 'so many wheels there with their small diminutive epicycles that they have turn'd that regular fabric to a rumbling confused labyrinth'.[16] Epicycles were necessary to explain retrograde motion, whereby distant planets, when viewed from the Earth, appeared to move backwards with reference to the background stars. This was a result of the distant planets moving more slowly than the Earth as they orbit the Sun. The issue had been largely explained by Johannes Kepler's third law of planetary motion (1619), whereby the greater the distance of a planet from the Sun, the slower will be the speed of its orbit. That Vaughan knew of Kepler's work is apparent from a short Latin poem Vaughan composed at some point in his career. Titled *In Ephemerides J Kepleri* it was printed posthumously in *Thalia Rediviva* (1678), a collection of works by Vaughan's brother, Henry.[17] Kepler published his *Ephemerides*, or what became known as the *Rudolphine Tables*, in 1627. They showed the daily position of the sun, moon and planets and were prepared using observations made by Tycho Brahe (1546–1601) and those of Kepler himself.

In this age of scientific transition and change we can see from the above that Vaughan adopts or maintains ideas that he considers to be right, based on his own understanding of them, while rejecting others. In doing so, he is as much a man of his times as Henry More, who, like Vaughan, was a Platonist in philosophy and someone who also postulated his own particular version of a vitalist creation.[18] He also fervently believed in spirits and witchcraft.[19] Other founder members of the Royal Society who had heard of or knew Vaughan, present a similar complexity with regard to the scientific ideas of their time. Robert Boyle, a founder of modern chemistry, and who knew of Vaughan through Samuel Hartlib, had a lifelong interest in alchemy. However, as a devout Christian, he also interested himself in the relationship between alchemy and the world of spirits and angelology. As Principe (2000) notes: 'Boyle's belief in the ability of the Philosophers' Stone to attract intelligent spirits meant that the Elixir was a potential ocular demonstration against atheism.'[20] Elias Ashmole, who founded the Ashmolean Museum in Oxford and noted some of Vaughan's experiments at Kensington, had a passion for astrology, a subject whose 'false magic' Vaughan studiously avoided.[21]

Nor was Isaac Newton immune from such complexity. Although he does not seem to have been particularly interested in Thomas Vaughan's alchemical work, he did, as a lifelong alchemical experimenter, possess a copy of Vaughan's translation of the Rosicrucian *Fama* and *Confessio*,[22] and he wrote some passages that resonate strongly with the alchemical writings of Vaughan. For example, in *Of Nature's obvious laws & processes in vegetation*, written sometime between 1670 and 1675[23] (so before the advent of the gravitational theory of *Principia Mathematica* in 1687, see below), Newton 'presents his view that the earth is itself a living creature and uses its respiration to account for gravity'.[24] In fact, as Newman (2019) also records in his magisterial *Newton the Alchemist*, Newton revealed in *Of Nature's obvious laws* a 'commitment to the Sendivogian model of the organic earth, which he unforgettably describes as a living being whose respiration accounts both for the renewal of the atmosphere and for the gravitation of falling bodies'.[25] As we have seen, Thomas Vaughan wrote of a breathing, living earth and counted Sendivogius among his important teachers (see Chapter 2).

Like these scientists, and others at this time, Vaughan interrogated and commented (however obscurely) on ideas he agreed or disagreed with, just as a modern scientist might do with the claims made for string theory, the expanding universe, or the multiverse. We should also remember that at the time Vaughan was writing (1650s), the publications of Descartes were little more than ten years old. That Vaughan chose to continue to follow a hermetic, natural magic and vitalist tradition that was destined to be left behind in the years ahead is a consequence of misjudgement on his part, not a lack of scientific interest, knowledge or thought. He is also doing this according to his own guiding lights and, beyond his faith in a Creator, knowing precisely what those guiding lights were is difficult. As a man who declared, 'I am neither papist, nor Sectary, but a true, resolute Protestant in the best sense of the Church of England',[26] did the gulf between Protestant and Catholic at this time influence his rejection of the work of the Catholic Descartes? Does his urging of us to be 'patient in our disquisitions' argue against the philosophical enthusiasm imputed to him by Henry More?

We need to remember too that, as our period opens in 1650, the likes of Vaughan, More and Powell were publishing their works and arguments at a time of transition not only in science but also in politics and religion. They lived through the rise of Puritanism, the political factionalism and religious sectarianism of the Civil War, the execution of a divinely appointed king, the subsequent republican and Cromwellian interregna, and the Restoration in 1660, under Charles II, of a monarchy still not universally welcomed. At the same time, exploration of the New World and discoveries relating to astronomy and medicine had challenged established views of the working of creation and of humanity's place within it. Indeed, it has been argued that it was precisely the uncertainty induced by this changing world that gave rise to a desire for a new equilibrium in life and society. An equilibrium some sought in a renewed interest in the Christian-based hermetic, alchemical and natural magic writings of the past, and of Vaughan's present. The latter evidenced by the rise in alchemical publications seen between 1650 and 1680, as already noted in Chapter 2.[27]

Yet despite this renewed interest in hermetic and alchemical ideas, attitudes to them were perceptibly changing. To take just one example, in comparison with the prolific alchemical publications Thomas Vaughan produced in the 1650s, Isaac Newton felt it necessary in the 1660s and 1670s to keep much of his alchemical work to himself. Partly, as he informed the secretary of the Royal Society, Henry Oldenburg, in 1676, because hermetic secrets might cause 'immense damage to ye world'.[28] But it might also have been because attitudes to hermetic undertakings were changing in the light of the new mechanical world of Descartes (and, later, of Newton himself). As the antiquary, lawyer, and chymist, Meredith Lloyd (*fl*.1655–77), wrote on 14 May 1664 to

Sir Thomas Myddelton (1586–1666) of Denbighshire, who was establishing a laboratory near Chirk:

> Our pretence must be for curing the sick & indeed it is the chief work. I need not tell you what reports will be raised in the Country when we go upon things that are unknown to seeming great masters in Sciences as well as the vulgar; they will talk of the Philosopher's stone & so bring us all to a disreputation with credulous & ignorant persons, and it may be some near friends may disgust it.[29]

René Descartes died in Stockholm in March 1650, just as Vaughan's first works appeared in print. Vaughan died in 1666, six years after the founding of the Royal Society. In the previous year, 1665, he had moved to Oxford, possibly in the employ of the king who had gone there in September as part of his travels to avoid the plague then rampant in London. Vaughan's patron, friend and founding member of the Royal Society, Sir Robert Moray, was nearby. In a letter written from 'Alberry' on 19 December 1665, Moray records how he had 'retired from Court for a time to amuse myself in a private place, where I am about some chemical experiments'.[30] Albury, as the village is known today, lies nine miles east of Oxford and Moray's 'private place' is likely to have been the house of the clergyman and army officer Samuel Kem (1604–70).[31] At some point Vaughan must have joined Moray at Kem's house, for it was there that Vaughan died in February 1666 after 'operating strong mercury, some of which by chance getting up into his nose marched him off'. Vaughan was buried in Albury with his funeral costs covered by Moray, to whom he bequeathed his books and manuscripts.[32] He would not be the last Welsh scientist to expire from inhaling noxious fumes.

Vaughan had died on 27 February, and it is evident from a subsequent event that his opinion on matters alchemical had been valued by Moray. On 14 March, two weeks after Vaughan's demise, Moray relayed one of those opinions to the first meeting of the Royal Society to be held following the suspension of its activities during the plague of 1665–early 1666. Meeting at Gresham College in London, the Fellows who were present reported on the scientific topics in which they had been engaged during the Society's enforced recess. Among them, 'Mr Hook' gave an account of his experiments 'weighing of bodies, in a very deep well, and above ground'; he found no difference in their weight. Robert Moray then 'gave an Account of his Imployment in trying of Ores, brought him out of a mine in Wales', and this was followed by a discussion of the state of work by Dr Clark and Mr Boyle regarding 'Injection into Veins' and 'the transfusing of the blood of one animal into another'. Moray then reported 'that Baptista Porta, affirmed, to have changed Sublimate into Water, and made use of it to harden iron-tools with that liquor; and that one Mr Thomas Vaughan had affirmed to him it would do'.[33] Giambattista della Porta (1537–1615) was the author of *Magia naturalis* in which Book XIII, Chapter V, is titled 'On the Tempering of Steel'. It includes the use of 'Liquors that will temper Iron to be exceeding hard', a method whose efficacy it seems Vaughan had tested.

In questioning the scholastic and Aristotelian orthodoxies of their day, both Thomas Vaughan and Descartes had wanted, in their different ways, to 'make a world'. Vaughan through a hermetic vitalist cosmos investigated through alchemy and Descartes through a purely mechanical philosophy of colliding particles moving in closed vortices. Yet, while Descartes's ideas fundamentally challenged established science, they remained purely theoretical. They were derived from a process of rational thinking about the subject, rather than the sort of Baconian observation and experiment so vigorously championed by Vaughan (and others) as the foundation for future progress. Indeed, the lack of empirical observation and experiment in Descartes's philosophy is almost certainly the reason why Vaughan chose to dismiss 'the whymzies of Des Chartes'. As Vaughan had written: 'why ... should we imbrace a philosophy of mere words, when it is evident enough, that we cannot live but by works'.[34] He was also wise enough to see that in science, as elsewhere, both rational thought *and* empirical experimentation and observation are important – 'we should be so rational, and patient in our disquisitions, as not imperiously to obtrude and force them upon the world, before we are able to verify them'.[35] Although Vaughan did not live to see it, a profound work of observation and experiment, allied to rational thought, would soon arrive to challenge all the worlds that had gone before.

Making Another World

Isaac Newton's *Philosophiae Naturalis Principia Mathematica* ('Mathematical Principles of Natural Philosophy') was published in Latin in 1687, with the first English translation appearing some forty-two years later, in 1729. It comprises our fourth intellectual strand and the first to appear from within our period of interest (1650–1820). In the 'Preface' to the 1687 first edition of *Principia*, Newton makes clear that it presents a challenge to Aristotelian and hermetic ideas through its aim of reducing the mechanical 'phenomena of nature to mathematical laws'; what many modern historians of science have seen as the beginnings of the mathematisation of science.[36] As Newton also records, the 'rational mechanics' he intended to present in the *Principia* would be based on empirical observation and mathematical demonstration and would be applied to the heavenly bodies in what later writers dubbed 'celestial mechanics'. Thus, we today have Newton's three standard Laws of Motion and a universal theory of gravity. The latter, in particular, representing a significant departure from the thinking of Descartes: for when matter fills every available space through extension – length, breadth and depth – and its constituent particles move by individual collision within closed circular vortices, as they do in Descartes's formulation, the sort of action-at-a-distance that Newtonian gravity represents is rendered impossible.

While the new mechanical, mathematical and empirical world that Newton presented undoubtedly challenged the purely rational and mechanical one envisaged by Descartes, the world of the *Principia* did not immediately overthrow that of Descartes. In a 1711 letter to the Cambridge Plumian Professor of Astronomy and Experimental Philosophy Roger Cotes, the Anglesey-born/London-based mathematician William Jones Sr, to whose works we will come later, noted how 'the Germans and French have in a violent manner attacked the philosophy of Sir Isaac Newton, and seem resolved to stand by Des Cartes'.

In replying, Cotes, who was busy editing the second Latin edition of the *Principia* (1713), recorded an instance of home-grown opposition:

> The controversy concerning Sir Isaac's philosophy is a piece of news that I had not heard of. I think that philosophy needs no defence, especially when attacked by Cartesians. One Mr. Green, a fellow of Clare Hall, seems to have nearly the same design with those German and French objectants, whom you mention.[37]

Whatever the extent of the opposition to Newton in France his ideas were certainly present there, just as Cartesian ideas were in Britain. Like Newton too, Descartes faced criticism, some of it invited by Descartes himself. 'It being his custom to communicate all things to be strictly examined by his friends and correspondents, before he committed them to the press', as John Davies of Kidwelly asserts in a preface to his 1654 publication *Reflections upon Monsieur Des Cartes's Discourse*; his translation of an anonymous French manuscript that he had obtained, which was critical of Descartes's ideas.[38] As Geoffrey Sutton says in his analysis of public science in France in this period: 'It was not until 1737 ... that there was the first significant evidence that Newton's mathematical mechanics, with its theory of universal gravitation offered advantages over the Cartesian system.'[39] Newton's new world would eventually prevail, though perhaps not as quickly as Alexander Pope implied in his famous but rejected lines for Newton's tomb in Westminster Abbey:

> Nature and Nature's laws lay hid in night:
> God said, Let Newton be! And all was light.[40]

In this chapter we have seen Scholastic Aristotelianism and the world of Aristotle lauded by the likes of Thomas Powell and challenged by the hermetic vitalist world of Thomas Vaughan. In turn, the worlds of Powell and Vaughan were challenged by the new mechanical world rationally envisaged by Descartes. All these worlds were then challenged by the rationally envisaged, mechanically operating, and empirically and mathematically proven, new world of Isaac Newton. All these ideas prevailed and were in flux together in the last half of the seventeenth century, and they continued to be so until the Newtonian world finally predominated in the eighteenth century. At the same time, the advent of mechanical worlds from which God appeared to have gracefully withdrawn presented a considerable challenge to religion and its place within these new worlds. That challenge, one that resonated among scientists of Wales throughout the long eighteenth century, is the subject of our next chapter.

Notes

1. Scott Mandelbrote, 'Isaac Newton and Thomas Burnet: Biblical Criticism and the Crisis of Late Seventeenth-Century England', in James E. Force and Richard H. Popkin (eds), *The Books of Nature and Scripture: Recent Essays on Natural Philosophy, Theology, and Biblical Criticism in the Netherlands of Spinoza's Time and the British Isles of Newton's Time*, International Archives of the History of Ideas, 139 (Dordrecht, 1994), pp. 149–78 (at p. 155).

2. For the life of Descartes, see Desmond Clarke, *Descartes: A Biography* (Cambridge, 2009); also various papers by E. J. Aiton.

3. For general discussion of Cartesian philosophy, see, for example, John Henry, *A Short History of Scientific Thought* (Basingstoke, 2012); and Ritchie Robertson, *The Enlightenment* (London, 2020). For detailed discussion, see Gary Hatfield, *René Descartes, https://plato.stanford.edu/entries/descartes* (accessed 21 May 2024); also papers by E. J. Aiton on the vortex theory.

4. See Geoffrey V. Sutton, *Science for a Polite Society: Gender, Culture and the Demonstration of Enlightenment* (Oxford, 1995), pp. 83–100 (at p. 83).

5. See Ritchie Robertson, *The Enlightenment* (2020), pp. 46–9.

6. *WTV(AMA)*, p. 137, *WTV(SW)*, p. 443.

7. From Thomas Powell, *Elementa Opticæ* (London, 1651), p. 109. My thanks to Dr Ineke Loots for the translation from the Latin.

8. *WTV(E)*, pp. 520–1.

9. Genesis, 1. 26.

10. See Steven Nadler, *A Book Forged in Hell: Spinoza's Scandalous Treatise and the Birth of the Secular Age* (Oxford, 2011). Although Spinoza is believed to have had an interest in alchemy, there is no certain proof that his 1670 comment was influenced by Vaughan's 1655 writing on God and Nature. However, it is notable that in 1661 Spinoza was visited by Henry Oldenburg of the Royal Society. Oldenburg was, of course, a close associate of Vaughan's mentor and sponsor Robert Moray. Having visited Spinoza in Rijnsburg, in 1661, Oldenburg wrote noting their discussion of God, extension, of infinite thought, the connection between soul and body, and the principles of the Cartesian and Baconian philosophies.

11. In Descartes's view, since matter, via its constituent particles, occupied all available space through extension, a vacuum is rendered impossible.

12. *WTV(AT)*, p. 62.
13. *WTV(MM)*, p. 271.
14. *WTV(MA)*, p. 156.
15. Ann Blair, 'Natural Philosophy', in *CHS*-3, p. 375.
16. *WTV(AT)*, pp. 61–2.
17. *In Ephemerides J Kepleri* by Thomas Vaughan: 'Ecce! Mori properat dum prodigus annus, et horas / Urget sydereis in sua fata rotis, / Das, *Keplere*, novam temeris *Echineida* cœlis; / Et stupet ad remoram machina tota tuam. / Nunc duraturo radias, *Aurora*, rubore; / Et præsens hic est, præteritusque dies.' ('Lo, while the spendthrift year hastens to its end and drives on the hours to their fate with the circling of the stars, you, Kepler, impose a new echineis on the heedless heavens; and the whole machine stands amazed at your remora. Now, Dawn, you glow with a blush that will last, and both today and yesterday are here.') Both the Latin text and the translation are from *WTV*, pp. 569 and 747, respectively.

   Of particular interest here is Vaughan's use of the words 'Echineida' and 'remoram'. These are a play on the species *Echeneis remora*, the sucker-headed fish often seen attached to the bodies of sharks and rays. In ancient times the fish was believed capable of slowing or even stopping ships, and so the word 'remora' subsequently came to mean hindrance or delay. It seems likely, therefore, that Vaughan's use of the word in the poem relates to Kepler's third law of planetary motion – that the greater the distance of a particular planet from the Sun, the slower the speed of its orbit. Yet, while this helps in revealing Vaughan's awareness of developments in early modern science, the poem is unclear as to Vaughan's actual acceptance, or otherwise, of Kepler's law.
18. See *Stanford Encyclopedia of Philosophy*, *https://plato.stanford.edu/entries/henry-more/* (accessed 21 May 2024).
19. See, for example, Noel L. Brann, 'The Conflict between Reason and Magic in Seventeenth-Century England: A Case Study of the Vaughan-More Debate', *Huntington Library Quarterly*, 43/2 (1980), 103–26.
20. Lawrence Principe, *The Aspiring Adept: Robert Boyle and his Alchemical Quest* (1998, Princeton NJ, 2000), p. 213, for more details of Boyle's attitude to 'Spiritual Alchemy', see pp. 188–213.
21. *WTV(MA)*, p. 174.
22. Newton's copy of Vaughan's Rosicrucian work is inscribed as being a gift from Oliver Doyley or Doiley of Oxford University – 'a minor member of the "neo-Platonic circle of Cambridge"' – suggesting the possibility that he 'represented an undercurrent of interest in chymistry and perhaps Rosicrucianism among the Cambridge Platonists', another member of which was Henry More; see William R. Newman, *Newton the Alchemist* (Princeton NJ and Oxford, 2019), p. 110. As for Vaughan's other works: in later years Georg Stiernhielm, 'the father of Swedish poetry', is said to have owned and annotated copies of *Anthroposophia Theomagica* and *Anima Magica Abscondita* and, following translation of Vaughan's works into German, they were read by the likes of Goethe. Later still, his works became entwined in the mystical religious world of Theosophy, as established by

the writings of Helena Blavatsky (1831–91), so that the first complete edition of them, edited by Arthur Edward Waite, would be published under the auspices of the *Theosophical Society of England and Wales* in 1919, a date not without significance coming directly after a Great War when there was, perhaps, a desire among some for a return to calmer times through spiritual renewal and the certainties of religion. A similar proposal is made by some historians to explain the resurgence of interest in hermeticism, alchemy and natural magic post the religious and political enthusiasms of the Civil Wars of the mid-seventeenth century.

23. William R. Newman, *Newton the Alchemist: Science, Enigma, and the Quest for Nature's 'Secret Fire'* (Princeton NJ and Oxford, 2019), p. 139 n. 6.

24. Newman, *Newton the Alchemist*, p. 150.

25. Newman, *Newton the Alchemist*, p. 167.

26. *WTV(AT)*, p. 92.

27. For general book publication, see Lauren Kassell, 'Secrets Revealed: Alchemical Books in Early-Modern England', *History of Science*, 49/1 (2011), 61–87 and Appendix A1–A38, available online. For publication of John Dee's diary (1659), see Edward Fenton (ed.), *The Diaries of John Dee* (1998; Charlbury, 2000), p. ix.

28. Newman, *Newton the Alchemist*, p. 19. See also pp. 434–6.

29. Nesta Lloyd, 'Meredith Lloyd', *Journal of the Welsh Bibliographical Society*, 11 (1975–6), 133–92 (at 172).

30. *WTV*(Intro.), p. 24, citing R. C. Winthrop (ed.), *Correspondence of Hartlib, Haak, Oldenburg and Others of the Founders of the Royal Society, with Governor Winthrop of Connecticut, 1661–1672* (Boston, 1878), p. 20.

31. For Moray, see David Stevenson (ed.), *Letters of Sir Robert Moray to the Earl of Kincardine* (Aldershot, 2007); and the *ODNB* entry for Samuel Kem. Between 1665 and 1667, when Moray and Vaughan would have been at his house, Kem was probably away in charge of a privateer during the second war with the Dutch.

32. For details see *WTV*(Intro.), pp. 23–6.

33. RSC JBO/2/87 (alternate ref. JBO/2/235).

34. *WTV(F&C)*, p. 483.

35. *WTV(LL)*, pp. 325–6.

36. I. Bernard Cohen and Anne Whitman (trans.), *Isaac Newton, The Principia, Mathematical Principles of Natural Philosophy: A New Translation* (Los Angeles CA and London, 1999), p. 381.

37. Lord Teignmouth, *Memoirs of the Life: Writings and Correspondence of Sir William Jones* (London, 1804); see p. 9 for Jones to Cotes, 25 October 1711; see p. 10 for Cotes to Jones, (n.d.). The letters are also available online at Early English Letters Online.

38. J. D. [John Davies, trans.], *Reflections upon Monsieur Des Cartes's Discourse of a Method for the Well-guiding of Reason, and Discovery of Truth in the Sciences* (London, 1654), 'To the Reader' (n.p.). Of the French author of the *Reflections* Davies notes, 'I cannot do him so much honor as to tell the world his name, for I never knew it, having met with them accidentally in a *Manuscript*'.

39. Sutton, *Science for a Polite Society*, p. 258.

40. Quoted in A. Rupert Hall, *Isaac Newton: Adventurer in Thought* (Oxford, 1992), p. v.

4

NEW WORLDS AND THE CHALLENGE TO RELIGION

Dû a Digon: Hêb Dhû, Hêb Dhim
('God and Plenty: Without God, Without Anything')

Welsh proverb used on the title page of Thomas
Vaughan's *Anima Magica Abscondita*, 1650

Nullius in Verba ('Take nobody's word for it')

Motto of the Royal Society, founded 1660–3

How if all our faith, and Christ, and Scriptures,
should be but a think-so?

John Bunyan, *Grace Abounding to the Chief of Sinners*, 1666

In his 1651 work *Lumen de Lumine* Thomas Vaughan had posed a question when criticising those practising alchemy to find gold: 'How many are there in the world, that study nature to know God?'[1] As we shall see in this chapter, the relationship between religion and the new mechanical worlds of Descartes and Newton would preoccupy many scientists of Wales in the long eighteenth century from 1650 to 1820.

Science and Religion

The language of science certainly made early inroads into long-eighteenth-century prose having a religious context. The *OED*, for instance, credits Thomas Vaughan's writing with the first use in English not only of the word *theosophy* (1650) – 'any system of speculation which bases the knowledge of nature upon that of the divine nature' – but also *mineralize* (1655) and *telescope* (1650).[2] Another early use of *telescope* comes in the preface to a 1652 civil war tract, *Mercurius Cambro-Britannicus, or, News from Wales*, by Caernarfonshire-born cleric and controversialist Alexander Griffith (*c*.1601–76):

> This sober honest *Mercury* [a news-sheet][3] coming to my hands, I thought it no great *Error* if I gave it that entertainment which I sometimes give even the *Phrantick* Bedlam *Pamphlets*: I must confess it was to me a kind of *Eye-salve*, for I looked formerly at the wrong end of the *Perspective* [a telescope], and the transgressions of our Welsh *itinerants*, palliated with the name of *Saints*, seemed but small *Atoms* in a large Sun-shine. This *Book* is a new *Telescope*; it discovers what we could not see before; and the *Spots* in this Spiritual *Moon*, are *Mountains*.[4]

Griffith aimed his pamphlet at those promoting Parliament's 1650 *Act for the Better Propagation and Preaching of the Gospel in Wales, Ejecting Ministers and Schoolmasters, and Redresse of Grievances*. Not only had the act removed him from his living at Glasbury in Breconshire, just as it had Thomas Vaughan at nearby Llansantffraid (see Chapter 2), but he felt the act was transforming Wales so that: 'A man on a Lord's day may ride twenty miles through a county and not see a church door open, supplied with a constant, able, godly minister.'[5] Religious belief and worship were an integral part of people's lives, yet, as we saw in the Prologue, its form and content could be divisive. Unrest, violence, even civil war might result from the sort of unbridled religious 'enthusiasm' to which Griffith's phrase '*Phrantick* Bedlam *Pamphlets*' surely alludes.

To escape these divisions, and any state-imposed sanctions designed to curtail or control them, many saw emigration as a way-out. For example, the thirteen American colonies, including the 30,000 acre 'Welsh Barony' in Quaker William Penn's 'holy experiment' at Pennsylvania, garnered a number of Welshmen who, with their descendants, would make significant contributions to science in what later became the United States of America (see Volume 2).[6] For those remaining at home, and who had an interest in the subject, science itself might offer a refuge since, as the twentieth-century Wales-born mathematician, philosopher and Nobel prize winner Bertrand Russell notes, the authority of science 'is intellectual, not governmental. No penalties fall on those who reject it; no prudential arguments influence those who accept it. It prevails solely by its intrinsic appeal to reason'.[7] Yet the orthodox in the Church and political establishments of the 1650s still viewed with suspicion and concern any appeal to reason that could undermine their authority – one Oxford cleric describing the newly founded Royal Society (1660) as 'a set of fellows ... formed into a kind of diabolical society, for the finding out new experiments in vice'.[8] Aside from a self-interested fear that enquiry into the natural world would lead to a similar enquiry into the natural order of what was a hierarchical society, the political and religious establishments worried too over an associated return of the sectarian enthusiasms that had characterised the Civil War period in Wales and England. This was a particular concern between 1650 and 1700, given the political tensions accompanying the restoration of the monarchy under Charles II in 1660, and the subsequent Glorious Revolution of 1688, when Catholic James II fled into exile to be replaced by his Protestant daughter Mary, and her husband, William of Orange.

One scientific work, published in Latin (1681) and English (1684), appeared to justify this state and ecclesiastical vigilance. This was *The Sacred Theory of the Earth* by Yorkshire-born headmaster and philosopher Thomas Burnet (*c*.1635–1715).[9] In it, Burnet challenged the biblical account of Noah's Flood. He did so by asserting that the biblically revealed forty days and nights of universal rainfall could never account for the scale of the deluge in Noah's time. Burnet's 'mechanical' solution to the problem, as Edward Lhwyd outlined it in a letter to the Revd Mr Lloyd, a schoolmaster of Ruthin in Denbighshire, was to argue that 'ye antediluvian earth was only a shell over the ante-diluvian sea; which shell cracking and breaking, it [the shell] sunk into ye abyss or sea that was under it; and so happened the deluge'.[10] Lhwyd had clearly been considering Burnet's idea of the fracturing earth-shell since at least 1692, when he noted it in relation to the *'prodigious heaps'* of jumbled stones, 'many of them of the largeness of those of Stonehenge', seen on the summit of 'Glyder' in Snowdonia (Glyder Fawr and/or Glyder Fach). Even though Lhwyd thought these heaps of rock might support Burnet's hypothesis of the fracturing of the earth 'for my part', he declared, 'though I admire his learning and ingenuity, yet I must confess I cannot (as yet) reconcile his opinions either to scripture or reason'.[11] By 1695, Lhwyd had become even more dismissive of Burnet's theory, as well as those of two other English writers who had vouchsafed their own ideas, John Woodward (1665–1728) and William Whiston (1667–1752).[12]

John Woodward's *Essay Toward a Natural History of the Earth* appeared in 1695, with William Whiston's *A New Theory of the Earth* appearing a year later. Woodward, as a skilled geologist, argued that while Burnet's theory was correct overall, the initial earth-shell break up would still have required the 'Assistance of Super-natural Power'. Whiston, meanwhile, introduced a comet to produce the necessary earth-shell break up, the force of whose tail would have produced the cracking of the earth, and whose associated vapours would help produce the rains mentioned in the Bible. Despite Woodward's appeal to a supernatural power to explain the earth-shell fracturing, all three theories effectively challenged religious orthodoxy by questioning the ability of the biblically revealed forty days of rain to instigate a flood on the scale of that associated with Noah.

Lhwyd's approach to the work of Burnet, Woodward and Whiston is particularly interesting in that he does not criticise their ideas solely on the challenge they make to biblical and theological orthodoxy. His critique also relies on reason and common sense. In a letter to the physician and naturalist Martin Lister (1639–1713, FRS 1671), of late 1695, Lhwyd concluded that Whiston's 'hypothesis will probably appear more ridiculous than that of Dr Burnet or Woodward: so I hope that ere long we shall have this wrangling philosophy laugh'd out of countenance, and ye plain Natural History better esteem'd of.'[13] In an earlier letter to Lister, in March, Lhwyd analysed in some detail Woodward's work with regard to its scriptural accuracy. He concluded that despite Woodward's undertaking 'to confirm by his observations the History of Moses', the 'hypothesis agrees not with ye Scripture'. Lhwyd then goes further: 'when we consider how far it may agree with reason and common sense; we find so many absurdities in it, that to me it seems scarce worth our consideration'. Sadly, he does not enumerate what those absurdities might be, or the nature of his objection to them. Instead, he concludes by relaying what he hopes might come from Woodward's work, despite its evident drawbacks. First, that enquiry into 'Formed Stones [fossils], &c. will not hence forward be thought so trivial'; second:

> the invalidity of this Hypothesis, as well as of that of Dr Burnet; will make men prefer Natural History, to these romantic theories, which serve to no other use, but to give us some show of ingenuity in ye inventors; who are yet in my opinion to be less valued than the authors of ingenious romances, for whereas those deliver their writings as fables; these do not only fully believe what they write themselves, but endeavour to possess others with their same persuasions.[14]

Lhwyd appears to want less theory and more of the practical natural history in which he was becoming a leading figure, and for the continuation of which, in 1695, he was seeking financial support via his *Design of a British Dictionary, Historical and Geographical; with an Essay, Entitled, Archæologia Britannica: and a Natural History of Wales.* This was an appeal for funds from the gentry and nobility for a work planned to include everything from language comparisons, persons memorable in *British* History, as well as natural history and the enumeration and description of earths, stones, mineral bodies, formed stones (fossils), plants and animals.[15] Such pecuniary concerns, though, should not override the fact that his assessment of the theories of Woodward, Burnet and Whiston approaches to the scientific in its appeal to practical natural history, reason and common sense.

Orthodox theologians, meanwhile, saw within these challenges to biblical orthodoxy, and the scientific method of their appraisal, the seeds of scepticism towards biblical tenets generally. Almost inevitably, they believed such scepticism would undermine not only scriptural orthodoxy and belief in biblical revelation, but the religious, moral and political order based on them. It was, after all, King James I who had declared 'No Bishop, no King' and commoner Quaker William Penn who suggested 'No cross, no crown'.

For historians, as Geraint H. Jenkins has perceptively noted, there is also the problem of judging how far any supposedly blasphemous statements 'were the fruit of scepticism or open infidelity'. How, he asks, do we interpret the comments of William Morgan, a Gentleman of Neath, who declared before the court of Great Sessions in 1698: 'This world was not made by God, but was made before there was a God; nor do I believe the Scripture which is an old book; for we are not to believe old books. And Moses was either a fool or a liar, and he made the Scripture, which is but a fable.' Are these the fulminations of an unbalanced mind, Jenkins asks, or do they represent 'a rational denial of orthodox Christianity'? If the latter, one wonders what might have constituted the new books for Morgan. As Jenkins concludes: 'There must have been disbelievers, particularly among the educated who, when called before their betters to answer for their oral indiscretions, may well have voiced the apathy and atheism that was shared by a small, but often vocal, minority.'[16] There were certainly those in Wales who saw the new science and its questioning of biblical orthodoxy as sowing seeds of deism or, even worse, atheism.

Of Deists

In contrast to a simple atheistic disbelief in God, deists accepted the existence of a creator, albeit a remote one, and of a creation run on uniform and rational laws, but they rejected biblical revelation and the miracles and providence whereby God acted in the world. Instead, deists favoured the revelation of God's existence being found in the book of nature and through human reason. Wales had contributed to early deistic ideas through the writings of the founding father of deism in Britain, the Welsh-speaking, England-born Edward Herbert (1583–1648, First Baron Herbert of Cherbury), whose family hailed from Montgomery.

His *De Veritate, prout distinguitur a revelatione, a verisimili, a possibili, et a falso* ('On Truth, as it is distinguished from revelation, the probable, the possible, and the false'; 1624) is the foundational text of deism in Wales and England.

One man who certainly saw deists among the later nobility and educated gentry of Wales was the Old Prophet ('yr Hen Broffwyd'), Edmund Jones (1702–93). Born in the parish of Aberystruth, in Monmouthshire, he wrote *A Relation of Apparitions of Spirits in the County of Monmouth and the Principality of Wales* as late as 1780 specifically 'to prevent a kind of Infidelity which seems to spread much in the Kingdom, especially among the Gentry and Nobility'. Deists among these groups, Jones declares, were making reason their 'God … in the place of God's Spirit the guide to all truth'. They even put reason in place of scripture, making them 'worse than the ancient Sadducees' (a social elite within biblical Judaism who, among other things, denied the resurrection of the dead). In so doing, the deists were not only denying the evidence of Jones's own work – a diligent listing of spirits and apparitions based on the testimony of Ministers of the Gospel – but the very 'being of God who is a Spirit, and the Father of Spirits'.[17] As Thomas Vaughan's bête noire Henry More had put it in *An Antidote against Atheism* (1653), 'No Spirit, no God'. As the 1780 date of Edmund Jones's publication reveals, a defence of religion by asserting the undeniable existence of spirits continued throughout much of the eighteenth century in Wales.

Of course, deist beliefs were not restricted to the gentry and nobility. Although based in England for most of his working life, the mathematician William Jones Sr, whose links to Newton were discussed in Chapter 3, and to whose own works we will return later, was born to a relatively poor family in a stone cottage on Anglesey. As the largely self-educated son of a small-time farmer, he later made notable contributions to mathematics, became a Fellow and Vice President of the Royal Society and, for a time, acted as a Baptist minister at Wantage in Oxfordshire. Yet this Baptist preacher, from a poor background, was not only a member of the Cabbala Club founded by Freemason, FRS, and possible student of the occult, John Byrom (1692–1763, FRS 1724), but also one of the 'heathen stamp' who joined the Infidel Club founded by the eminent antiquarian, mathematician, Freemason and outright atheist Martin Folkes (1690–1754, FRS 1714). Folkes was a man who, according to the antiquarian Revd William Stukeley (1687–1765), not only 'thinks there is no difference between us & animals' but propagated 'the Infidel system with great assiduity, & made it even fashionable at the Royal Society, so that when any mention is made of Moses, of the deluge, of religion, Scriptures, &c., it is generally received with a loud laugh'.[18]

As one of the most noted deists of the eighteenth century, David Williams (1738–1816) also came from a fairly impoverished background. Born in the area of Waunwaelod, near Caerphilly, his father acted there as a relatively unsuccessful supplier of equipment to the developing mining industry. Williams went on to establish, in 1776, at Margaret Street, in London's Cavendish Square, the very first place of public deistic worship in Europe. It used a deist liturgy he had written in conjunction with fellow deist and polymath scientist, the American Founding Father Benjamin Franklin (1706–90). Its declared aim was 'to admit every honest man in the world to join us', and to enlarge 'the limits of intellectual liberty'. Williams made clear the all-encompassing nature of this endeavour in his published lectures:

> That good men of all nations and all religions; that believers in Moses, Christ, and Mahomet, Free-thinkers, Deists, and even Atheists who acknowledge beneficent principles in nature, may unite in a form of public worship, on all the great and important truths of piety and morality.

The services at Margaret Street attracted visitors including followers of the French philosopher and Freemason Claude Adrien Helvétius, and the 'militant atheist' Baron d'Holbach. The eminent English and Swedish botanists and natural philosophers Joseph Banks and Daniel Solander merely 'peeped into the chapel, and got away as fast as they decently could'.[19]

Meanwhile, back in Wales, the Vale of Glamorgan diarist William Thomas (1727–95) describes one John Bradford, a 'fuller and dyer' buried at Bettws near Bridgend in 1785, as 'a great disputer and a Nominated

Deist, or a free thinker'.[20] In 1745, the great evangelical Methodist preacher Howell Harris (1714–73) of Trefeca, in Breconshire, having visited his scientific brother, Joseph Harris (1703–64), who was an Assay Master at the Royal Mint in the Tower of London, frankly admitted: 'my brother is a deist'. Perhaps a not entirely unexpected development given that, by the time of Howell's visit to London, Joseph had associated with the eminent polymath scientist Edmond Halley (1656–1742, FRS 1678), a man many suspected of atheism. Even more surprising than Joseph's deism, though, is *preacher* Howell's own revelation of a past 'temptation to atheism'.[21]

Of Atheists

In 1677 an 'Englished' version of Latin epigrams written by 'Cambro-Britannicus' John Owen (1564?–1628?) appeared in print. Placed on the Catholic index of forbidden books in 1654, *Epigrammatum* was originally published in ten parts between 1606 and 1613. The epigrams it contains are not scientific in content, but they do touch on the vagaries of alchemy – 'The Chymist Gold decocts, till (leaving none) / he loseth all his Gold to find a Stone'; echo the nature of the earth as a living entity – 'The Earth's Bones are Stones, skin, [its] surface, Metalls, [its] Nerves, / the Grass for Hair, for Blood the Water serves'; and are particularly harsh on atheists – 'Thou Sayest, What's above concerns us not: / True: for the lowest hole in Hell's thy lot'.[22]

Welsh atheists, let alone those whose atheistic beliefs resulted from the new worlds related by the likes of Descartes and Newton, appear to be thin on the ground in the long eighteenth century. We have already met William Morgan of Neath and his denial that God made the world (see above). Another example is that of John Richard, a blacksmith buried in Cardiff in 1788, who William Thomas records as being 'once a very atheistical man in his discourse and expression'.[23] Equally difficult to source are works written against atheism in which science is mentioned. In his *Historia Atheismi, breviter delineata à Jenkino Thomasio Cambro-Britanno, V.D.M. & M.D.* ('*History of Atheism, briefly delineated by Jenkin Thomas, Welshman*'; Basel, 1709), its author, Jenkin Thomas (d.1755, a.k.a. Jenkin Thomas Philipps), traced the history of atheism from antiquity to his own times and severely criticised the atheistic works, as he saw them, of Thomas Hobbes and Baruch Spinoza. We will meet Jenkin Thomas more fully in Volume 2. Caernarfonshire-born Gryffith Williams (1589?–1672), meanwhile, wrote his *Truth Vindicated, Against Sacrilege, Atheism and Prophaneness* (1666), while acting as Bishop of Ossory, in Ireland. It is a long turgid piece less concerned with atheism than politics and the position of the Anglican Church, the king, the parliament and the 'brain sick sectaries [the multiple religious sects that arose during the Civil War]'. Science is referred to only briefly, in a late section on 'the vanity of every man', and then via a classical allusion to the Greek philosopher Socrates appearing before the Temple of Apollo at Delphi. Inscribed onto the temple were three Delphic maxims, including 'know thyself'. In Williams's telling, Socrates, standing before the temple, concluded that the readiest way to come to God was indeed to know himself; as a consequence, he 'left the course and practice of other philosophers' who, in order to 'make them[selves] happy', searched 'into the motions of the heavens, and the influence of the planets and applied their studies … to understand the causes of all things'. There was, Socrates conceived, in the opinion of Williams, 'no folly comparable to this, to be painful and diligent to know all other things, and to be ignorant and know nothing of himself'.[24]

One Welsh writer who clearly saw the new worlds envisaged by science as a parent to atheism was the Carmarthenshire-born soldier, spy(?), and historian David Jones (*fl.*1675–1720). In commenting in his *A Compleat History of Europe* (1698) on the 1692 death of alchemist/chemist Robert Boyle, Jones is keen to show that while Boyle was addicted to the study of Natural Philosophy, he was also far from being tainted 'with that Atheism, that is too too [*sic*] usual for such speculative Heads'.[25] Jones's remark is certainly a truism for Boyle who, as a committed Christian, funded translations of the Bible into a number of different languages, including Malay, Turkish, Irish and Welsh. As 'Gilbert, Lord Bishop of Sarum' recalled in his sermon at Boyle's London funeral, the scientist 'was at 700*l* charge in the edition of the Irish Bible, which he ordered to be distributed in Ireland; and he contributed liberally both to the impressions of the Welsh Bible, and of the Irish Bible for Scotland'.[26] A series of letters in 1681 between Boyle and Henry Jones (1605–82), Bishop of Meath in Ireland, who came from a Welsh family,[27] makes the connection to the Irish and Welsh Bibles clear. In an undated letter, probably coming between those written on the same subject by Boyle on 3 September and Jones on 5 November, Boyle writes:

> I perceive, that, God permitting, there will be time to retrieve the papers that I sent your lordship and that miscarried about the method used in getting subscriptions for the Welsh Bible; but by reason of the absence of the manager of that impression, I cannot yet get copies of his papers.[28]

The Welsh subscription method was to be used in Ireland and the absent manager is almost certainly Thomas Gouge (1605–81), who had died on 29 October. An Englishman and philanthropist, he founded the Welsh Trust in 1674 to supply bibles and a level of religious education in the country. The Trust's Bible of 1678, edited by Stephen Hughes (1622–88), was the first publication in Wales to make use of subscription with the cost of the 8,000 copies produced being £2,000. Seven thousand were sold at 4s. 2d, and the remainder given free to the poor.[29] Boyle also notes in his letter to Jones that with any Irish material 'I send not the Bookes over in quires to save the charge of having them well bound in sheep's leather, as, if I mistake not, the Welch Bibles were'.

But, as David Jones also says, Boyle went further in promoting his Christian cause by leaving 'a legacy to have a monthly Sermon preached against Atheism'.[30] These were the *Boyle Lectures* inaugurated in 1692 by their first lecturer, the English scholar and theologian Richard Bentley (1662–1742). Bentley's lectures were published the following year (1693) under a title that hints at an attempt to challenge both deism and atheism by integrating religious belief with the study of creation – *The Folly and Unreasonableness of Atheism Demonstrated from the Advantage and Pleasure of a Religious Life, the Faculties of Human Souls, the Structure of Animate Bodies, and the Origin and Frame of the World.*

It was presumably while preparing his lectures for the press that Bentley addressed a number of queries to Isaac Newton who, as we saw in Chapter 3, had not long before outlined a new mechanical world described with mathematical precision in his *Philosophiae Naturalis Principia Mathematica/Mathematical Principles of Natural Philosophy* (1687). Newton replied to Bentley's queries on 10 December 1692, saying:

> when I wrote my treatise about our system, I had an eye upon such principles as might work with considering men, for the belief of a deity; nothing can rejoice me more than to find it useful for that purpose … [T]o your second query I answer that the motions which the Planets now have could not spring from any natural cause alone but were imprest by an intelligent agent.[31]

As the seventeenth century gave way to the eighteenth, many latitudinarian clergy (those willing to accept minimal doctrinal requirements so as to encourage unity in the Church) began to use their published work, and even the early Boyle Lectures, as platforms to reinforce this unity between Newtonian science and theology. They did so through the avenue of natural theology, where an understanding of God is achieved by employing human reason to study creation without any reliance on biblical revelation. It is often associated, though not exclusively, with the intelligent design argument, whereby the evidence of design we see within the natural and physical worlds is taken to indicate the existence of an intelligent designer – namely God. Newton, who 'was firmly committed to natural theology',[32] contributed further to this accommodation between science and religion when he added his *General Scholium* to the second edition of the *Principia*, in 1713.

The *Scholium* contains a clear outline of intelligent design – 'This most elegant system of the sun, planets, and comets could not have arisen without the design and dominion of an intelligent and powerful being'. Furthermore, divine providence (the kindly care of God acting within the world) had to operate within this system – 'For a God without dominion, providence, and final causes is nothing other than fate and nature'.[33] Such ideas coming from Newton's quill must have seemed a providential godsend to those hoping to change the mind of an atheist or counter the Cartesian and deistic image of a remote creator.[34] As we shall see in Chapter 5, this sort of natural theology, together with the mechanical aspects of Newton's ideas, would play their part in the work of scientists of Wales throughout the long eighteenth century.

Notes

1. *WTV(LL)*, p. 325.
2. See *OED* entries and *WTV(AT)*, pp. 53 and 600, n.53.149 for 'Theosophie'; *WTV(E)* pp. 524 and 728–9, n.524.439 for 'Mineralize'; *WTV(MA)* pp. 156 and 637, n.156.568–9 for 'Telescope'. The earliest entry in the *OED* for 'telescope' is J. Bainbridge in 1619, but he uses the terms 'Telescopium' and 'Trunk-spectacle' not telescope.
3. One official government news sheet had been called *Mercurius Politicus*.
4. Alexander Griffith, *Mercurius Cambro-Britannicus, or, News from Wales* (London, 1652), 'Preface'.
5. Quoted in D. Densil Morgan, *Theologia Cambrensis, Protestant Religion and Theology in Wales: 1 From Reformation to Revival 1588–1760* (Cardiff, 2018), pp. 153 and 166, n. 188.
6. Geraint H. Jenkins, *Protestant Dissenters in Wales 1639–1689* (Cardiff, 1992), p. 66.
7. Bertrand Russell, *A History of Western Philosophy* (1946; London, 1984), p. 480.
8. Larry Stewart, *The Rise of Public Science: Rhetoric, Technology, and Natural Philosophy in Newtonian Britain, 1660–1750* (Cambridge, 1992), p. 6.
9. Thomas Burnet, *The Theory of the Earth; Containing an Account of the Original of the Earth, and all Changes which it hath already undergone, or is to undergo till the consummation of all Things* (London, 1691).
10. Gunther, pp. 198–203 (at p. 200).
11. Gunther, pp. 156–60 (at p. 159).
12. For full discussion of these works and their significance in this debate, see Chapter 3 in Stewart, *The Rise of Public Science*.
13. Gunther, pp. 282–3 (at p. 283).
14. Gunther, pp. 268–9.
15. See Dewi W. Evans and Brynley F. Roberts, *Edward Lhwyd Archæologia Britannica, Texts and Translations* (Aberystwyth, 2009), pp. 35–9.
16. Geraint H. Jenkins, *Literature, Religion and Society in Wales 1660–1730* (Cardiff, 1978), p. 99.
17. Edmund Jones, *A Relation of Apparitions of Spirits in the County of Monmouth and the Principality of Wales* (Trevecca, 1780), 'Preface', pp. ii–iii.
18. For main details, see Michael J. Franklin, *Orientalist Jones: Sir William Jones, Poet, Lawyer, and Linguist, 1746–1794* (Oxford, 2011), pp. 41–53 (quote at p. 47); also, Brian Bowers, 'A Baptist Man of Science', *The Baptist Quarterly* (2000), 305–7;

and *ODNB* entries for Folkes and Byrom. Bowers indicates that Baptist minister William Jones of Wantage, who wrote on longitude, is actually our William Jones Sr from Wales. This has been accepted in this work, although others have argued that they are two separate people. There is certainly an entry in 'Early Records of Baptist Churches', *The Baptist Magazine*, 10 (1818), 135, which states that Jones Wantage resigned his ministry there in 1737 and died the following year. However, in the Royal Society there are a series of letters from William Jones of Wantage discussing longitude. These are mostly written in the early 1730s, but one 'copy' letter is dated to 1740; so, after his supposed death in 1738. There is also a mention in the *Baptist Magazine* article of a son, Richard Jones, in relation to Jones Wantage. However, this could equally relate to a previously unknown son of William Jones Sr from the first of his two marriages. In this book it is currently suggested that the two Jones are the same person, and that Jones did not die at Wantage but simply moved elsewhere. A cursory comparison has been made between the confirmed signature and writing of William Jones Sr (from his archive in Cambridge University and a signature in one of his library books) with the letters of Jones Wantage at the Royal Society, and they show a number of correspondences. So, at the present time, there seems no reason to see them as two separate people. However, the issue is not entirely clear, and it needs further detailed research, particularly with regard to their handwriting, for any conclusive attribution of the Wantage letters to be made.

19. J. Dybikowski, *On Burning Ground: An Examination of the Ideas, Projects and Life of David Williams* (Oxford, 1993), particularly chapter 3 (quote at p. 75) and p. 288.
20. R. T. W. Denning (ed.), with J. B. Davies and G. H. Rhys (transcribers), *The Diary of William Thomas 1762–1795* (Cardiff, 1995), p. 339.
21. Tom Beynon (ed.), *Howell Harris's Visits to London* (Aberystwyth, 1960), p. 81 for Joseph's deism, and p. 184 for Howell's temptation to atheism.
22. Thomas Harvey, *John Owen's Latine Epigrams Englished by Tho. Harvey, Gent.* (London, 1677). See epigrams 9 (alchemy), 38 (atheism), 44 (earth's body).
23. Denning, *Diary of William Thomas*, p. 362.
24. Gryffith Williams, *Truth Vindicated Against Sacrilege, Atheism, and Prophaneness* (London, 1666), p. 2 of *The Vanities of Everyman* section.
25. (David Jones), *A Compleat History of Europe* (1698; London, 1699), p. 475; this is a precursor volume later appearing as part VI of his eighteen-volume *A Compleat History of Europe*, published as an annual series between 1705 and 1720.
26. Gilbert, Lord Bishop of Sarum, *A Sermon Preached at the Funeral of the Honourable Robert Boyle; at St Martin's in the Fields Jan 7 1691/2* (London, 1692), p. 26.
27. Henry Jones's father, Lewis Jones (d. 1646), went from Merioneth to Ireland where he married a sister of James Ussher (who famously dated the age of the earth to 4,006 years); he later became Bishop of Killaloe. Michael Jones, the brother of Henry, fought for Parliament at the battle of Rowton Heath, near Chester, during which royalist and alchemist Thomas Vaughan was taken prisoner. Henry Jones, who some sources say was born in Wales, others say Ireland, became Bishop of Clogher in 1645 and Vice Chancellor of the University of Dublin in 1646, to which

he gifted the Books of Kells and Durrow. He also designed the oak staircase in the new library. In 1657 he became a principal trustee of an educational trust, which probably explains his letters with Boyle relating to Wales. Henry became Bishop of Meath in 1661.

28. For the sequence of letters between Boyle and Jones, see Michael Hunter, Antonio Clericuzio and Lawrence M Principe (eds), *The Correspondence of Robert Boyle* (6 vols, London, 2001), V, pp. 268–71 (quote at p. 269).

29. Eryn M. White, *The Welsh Bible* (Stroud, 2007), pp. 60–1 and 69–70. The sale represents a significant loss of some £500. Even if they had sold all 8,000 copies at 4s. 2d each they would still have made a loss, one that was presumably covered by the Welsh Trust (thanks to Gareth Ffowc Roberts for this comment).

30. (Jones), *History of Europe*, p. 475.

31. See Newton to Bentley 10 December 1692, *www.newtonproject.ox.ac.uk* (accessed 1 July 2020).

32. Stephen D. Snobelen, 'To Discourse of God: Isaac Newton's Heterodox Theology and His Natural Philosophy', in Paul Wood (ed.), *Science and Dissent in England 1688–1945* (Aldershot, 2004), p. 43.

33. I. Bernard Cohen and Anne Whitman, *Isaac Newton: The Principia, Mathematical Principles of Natural Philosophy – A New Translation*, (Los Angeles CA and London, 1999), p. 940, 942 and n. ll.

34. For more detailed discussion of Newton's religious attitudes and the 'General Scholium', see Stephen D. Snobelen, '"God of Gods, and Lord of Lords": The Theology of Isaac Newton's General Scholium to the *Principia*', *Osiris*, 16 (2001), 169–208; Snobelen, 'To Discourse of God', in Wood (ed.), *Science and Dissent in England*, pp. 39–65.

5

WALES AND THE NEWTONIAN WORLD

> Of this I can assure the Reader, that in perusing the following
> Pages, he will find no Theories asserted, that are not grounded
> upon Experience, supported with the most evident Matters
> of Fact; and I would not receive the Law of Attraction; or the
> Gravitation of Matter, as a natural Principle resulting from the
> inherent Laws of Nature, first discover'd by that Ornament of
> the *English* Nation, Sir ISAAC NEWTON, till, with the utmost
> Care, Diligence and Application, I had enquir'd into the Properties
> and Tendencies of Matter, and perceiv'd it impossible rightly to
> explain any natural Phænomenon concerning the Descent of
> Bodies, without admitting that Principle as a *Dat.*[Dative][1]
>
> Nicholas Robinson, *A New Theory of Physick and Diseases,*
> *Founded on the Principles of the Newtonian Philosophy* (1725)

Not much is known about the appearance of Newtonian ideas among the scientists of Wales, and to map when such ideas first appear and subsequently spread within Wales is a project requiring considerable further research. What follows here is no more than an introduction to these issues. Quickly apparent, however, even from the most cursory research, is the rapid adoption by scientists of Wales in the early eighteenth century of two Newtonian ideas. The first – the mechanical nature of creation – is discussed below in 'Early Welsh Newtonians' and 'Newtonian Science and Medicine'. The second idea –the continued importance of natural theology to ministers of religion with scientific interests – is explored below in 'Newtonian Science and Religion' and 'On the Probability of Miracles'. The chapter then concludes with the section 'New Worlds, Shifting Foci', in which we reflect on the place of Newtonian science among the scientists of Wales by the mid-eighteenth century and later.

Early Welsh Newtonians

Probably the earliest interaction by a scientist of Wales with Newtonian science (and with Newton himself) is that of Anglesey-born, London-based mathematician William Jones Sr. By 1711/12 a simmering dispute between Newton (1642–1727, FRS 1672) and the German philosopher and mathematician Gottfried Leibnitz (1646–1716, FRS 1673), over priority for the invention of calculus, needed to be resolved. To that end, on 6 March 1712 the Royal Society appointed a committee to enquire into the matter. What Newton called 'a numerous Committee of gentlemen of different nations' originally had six members: William Jones Sr from Wales; Dr John Arbuthnot, a Scot, and inventor of the character John Bull; and the English members Dr Abraham Hill, a merchant and scientist; the astronomer Edmond Halley; John Machin of Gresham College; and a Mr Burnet, probably the theologian and cosmologist Thomas Burnet, who knew Newton.[2] All were Fellows of the Royal Society. To these: 'Robarts, a contributor to the [Philosophical] Transactions, was added on the 20th [March], Bonet, the Prussian minister, on the 27th, and de Moivre, Aston, and Brook Taylor on the 17th April.'[3] Of the additional members, only Bonet and de Moivre were from overseas.

William Jones Sr had been admitted FRS in November 1711 following publication of his *Analysis per quantitatum series, fluxiones ac differentias* (1711). In that work, Jones utilised a transcript of one of Newton's early mathematical manuscripts 'De Analysi per Æquationes numero terminorum infinitas' ('On analysis by equations unlimited in the number of their terms'; *c*.1669), together with various letters relating to the Newton/Leibnitz quarrel. These had come into Jones's possession in 1708, when he acquired the manuscripts and papers of the mathematics teacher and book editor John Collins (1625–83, FRS 1667). Collins was a correspondent not only of Newton, who he first met in

1669, but many other mathematicians and scientists, including Leibnitz, who he met in London in 1676. Jones, having obtained Collins's papers, made them available for scrutiny by the Royal Society committee on the calculus. In the committee's final report – *Commercium epistolicum* (1712) – priority for the invention of calculus was given to Newton. A conclusion appearing somewhat inevitable given the assertion by some that Newton actively selected, annotated and edited the papers involved. He is also said to have written much of the final report himself; despite another suggestion that William Jones Sr helped 'to take care of the impression'.[4] Today, the issue of priority is settled by concluding that Newton and Leibnitz arrived at the calculus independently.

An early Welsh Newtonian to benefit from access to the Newton manuscripts in the possession of William Jones Sr was Joseph Harris (1702–64), the brother of preacher Howell Harris, both of whom we met briefly in Chapter 4. As an Assay Master at the Royal Mint, which was then in the Tower of London, Joseph Harris employed his (presumably) spare time carrying out various optical and other experiments. Harris died before he could publish his results but, thanks to the editorial efforts of interested friends and colleagues, some of the results he had already written up on optics appeared eleven years after his death in 1764 under the title *A Treatise of Optics: Containing elements of the science; in two books* (1775, see Figure 5).

As Wade (2005) notes, Harris's optical work was among those that 'extended further in the visual domain' Newton's ideas on the subject.[5] For example, in his important *Opticks: Or, a Treatise of the Reflections, Refractions, Inflections and Colours of Light* (1704), Newton made a passing reference to an experiment that he had undertaken in relation to the optic nerve and the optic chiasm. Located in the front part of the brain, the optic chiasm is where the optic nerves intersect and cross. Today, in humans, it is known that about half the optic nerve fibres coming from each eye cross to the opposite side of the brain, while the rest remain on the same side. This decussating (the action of intersecting or crossing) occurs within the optic chiasm, but it was little understood in Newton's time. However, Wade shows that Newton actually made 'the first representation of partial decussation at the optic chiasm, and proposed a

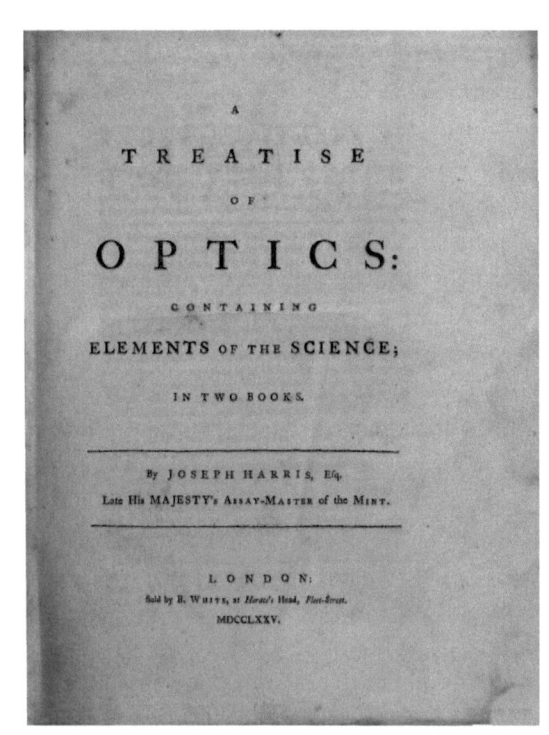

FIGURE 5 Title page to *A Treatise of Optics: containing elements of the science; in two books* by Joseph Harris (London, 1775) with Harris's illustration of the decussating at the optic chiasm originally described, but not illustrated, by Isaac Newton. Courtesy of Google Books/Eighteenth Century Collections Online (Original: Bavarian State Library).

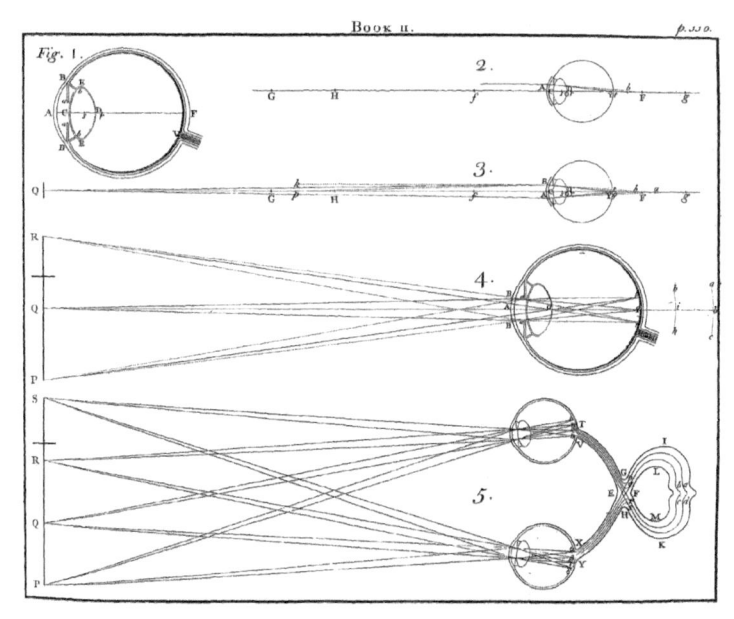

theory of binocular single vision based upon it'. Newton did this in an unpublished manuscript written *c*.1682. However, in his *Opticks* of 1704 he made only a passing reference to the experiment and provided no explanatory diagram. As a consequence, Wade continues: 'In presenting this as a query [in *Opticks*], rather than the report of an experiment (as in the unpublished manuscript), its speculative nature would have been reinforced.'[6] Fortunately, the 1682 manuscript eventually passed into the possession of William Jones Sr, who at some point made it available to Joseph Harris. It would be Harris who then provided, in his posthumous *Treatise of Optics* (1775), a transcript of Newton's actual experiment together with an illustration based on Newton's description.[7]

Wade also notes that Harris, in his *Treatise of Optics*, provided another quote from Newton's unpublished manuscript of *c*.1682.[8] It indicates that Newton undertook experiments with cut sections of optic nerve in order to try and observe something of the 'animal spirits' postulated by Descartes as the basis for nerve and muscle function (see Chapter 3). According to Newton:

> tho' I tied a piece of the optic nerve at one end, and warmed it in the middle, to see if any airy substance by that means would disclose itself in bubbles at the other end, I could not spy the least bubble; a little moisture only, and the marrow itself squeezed out.[9]

Having been unable to isolate any so-called 'animal spirits', Newton's opinion, as Joseph Harris records it, was 'that the sensations caused by the impulses of light upon the retina, are communicated to the brain or sensorium, by the mediation of a fine æther'.[10] Newton himself describes a purely mechanistic interpretation of vision: 'there are pipes filled with a pure transparent liquor passing from the eye to the sensorium, and the vibrating motion of the æther will of necessity run along thither.' Writing at a time when theories of vision and the mode of functioning of the eye were little understood, Newton also concludes that 'vision thus made, is very conformable to the sense of hearing, which is made by like vibrations'.[11] Developing Newton's ideas was not Harris's only contribution. As Griffiths (2012) has noted, Harris also:

made determinations of the limits of the human eye, delving into biology and experiment to ascertain the optical structures of the eye and measuring its resolving power. He determined that the diameter of the eye was 0.95 of an inch and the distance from the posterior surface of the lens and the retina averaged 6.44 tenths of an inch (16.36mm). From this he worked out that 1 second of arc subtended to 0.000157 of an inch (0.0048mm) on the retina, giving one of the most accurate estimates of visual resolution then available.[12]

Newtonian Science and Medicine

According to the psychiatrist George Makari in his *Soul Machine: The Invention of the Modern Mind* (2015), those seeking to abandon the increasingly queried but long held four-humours theory in the field of medicine:

> often turned to a mechanical system of physiology and pathology. Medical mechanists, who first emerged in the 17th century, insisted the body was a contraption. The jaws were pincers, the stomach a container, veins were water mains, the heart a piston, the lungs a bellows, the liver a filter, and the corner of the eye a pulley. Reason had no place among these chains and gears.[13]

Consequently, Makari suggests, Newtonian physicians emerged who argued that medical advances had to be linked:

> to new knowledge in math and physical science. Fashionable models of flesh pulleys and bony rods took hold in London. Many doctors embraced Newton and pursued the hydraulics of the flesh. Nicholas Robinson even sought to apply Newton's theory of ether to the soul's actions on the body.[14]

The adoption of Newtonian science by Wales-born, London-based physician Nicholas Robinson (1697–1755)[15] has already been displayed in the opening quotation to this chapter, and by the title of the work from which the quote is taken – *A New Theory of Physick and Diseases, Founded on the Principles of the Newtonian Philosophy* (1725). In another of his medical publications – *A New System of the Spleen, Vapours, and Hypochondriack Melancholy: Wherein all the Decays of the Nerves, and Lowness of the Spirits, are mechanically Accounted for* (1729) – Robinson not only reinforced his mechanical credentials, he also involved himself in another issue consequent on the advent of the new mechanical Cartesian and Newtonian worlds.

By the early eighteenth century, the dualistic separation of body and soul, introduced by Descartes in the mid-seventeenth century (see Chapter 3), had clearly been accepted by Nicholas Robinson. As he writes in *A New System of the Spleen*:

> By the clearest Reflection, we perceive, that this noble Creature Man consists of two parts, Matter and Thought, or Soul and Body, evidently distinct in all their Operations, though both subsist under the same Form, and jointly constitute that Essence, that denominates him a rational Creature.[16]

However, he also accepts, as did many others, that there is a problem with this view since:

> what kind of Mechanism or Arrangement of Corpuscles is necessary to make up a System of Matter to be directed by Thought, or the Influence of the Will, is a Question that, I must confess, puzzles my Philosophy, and can only be resolved by the Supreme Author of Nature, who has thus fearfully and wonderfully made us.[17]

The very act of thinking about these sorts of physical and mental problems introduced into the eighteenth-century world a further issue – that of the mind itself. As Makari (2015) notes:

> the old dichotomies of body and soul now became a three-way contest between body, soul, and mind, with the last term existing somewhere between scientific discourse with its prerequisites of

materialism, mechanization, and quantification, and the metaphysical credos of an immaterial human essence.[18]

Makari also suggests that like some other theologians and doctors, Robinson, in defending the immaterial soul in his *A New System of the Spleen*, 'felt obliged to dismiss the mind'.[19] Robinson certainly gave to the soul some abilities we might perhaps see as more attributable to the mind: 'We perceive further, by Experience, that the Soul sees through the Eyes, hears through the Ears, and is capable of being affected with Matter and Motion from those Objects, that are apply'd to the Surface of the Body.'[20] In fact, Robinson appears to dismiss the mind by seemingly reuniting it with the soul:

> It is true, we cannot conceive the *Modus Operandi*, that is, how Thought can work upon Matter, or Matter upon Thought; but we very well know, that a disorder'd body will make a disorder'd Mind; and *vice versa*, God Almighty, our supreme Author, having ordain'd that the Affections of the Body should accompany the Mind through all the various Scenes of Pleasure and Pain, Joy and Grief, while the Soul is confin'd to this Globe.[21]

Meanwhile, within Wales itself, the Caernarfon physician Thomas Knight (*c.*1697–1775) revealed his adoption of a mechanical approach to both medicine and medication in *A Dissertation of the Operations of Chalybeat Medicines in Human Bodies* (1731):

> The Principles and Method of Reasoning, introduced by Sir Isaac NEWTON, and happily applied by others to the animal Structure and Medicine, teaches us to account how Medicines operate by their mechanical Properties. Hence it is that our Rationale's are seldom less than demonstrable, and when it happens otherwise, it is because we draw our Conclusions from false Principles.[22]

In a later work – *A Critical Dissertation upon the Manner of the Preparation of Mercurial Medicines, and their Operation on Human Bodies* (1734), which was dedicated to members of the Royal College of Physicians – Knight includes a section on 'A Certain Method of introducing the same metal [mercury] in its essential or native Dress internally into the blood; and the Manner of it Mechanically explain'd; not publish'd before'; all to be accounted for with 'Exactness and Accuracy, by the Rules of Mechanism and the Laws of Motion'.[23]

Newtonian Science and Religion

On St David's Day 1731, the Reverend Walter Williams, rector of Llanddetty, a small Anglican church on the banks of the Usk in Breconshire, delivered a 'Discourse' to the *Society of Ancient Britons* in St Paul's Church, Covent Garden. His subject was Psalm 33 verse 12 – 'Blessed is the nation whose God is the Lord' – and he began with a classic outline of natural theology in accord with the content of Newton's *General Scholium*:

> That there is a God, a superior Being, from whom we derive our Existence, and on whom alone we have our sole Dependance [*sic*], appears from undeniable Arguments; but more especially from considering the Works of the Creation: for by observing the surprizing [*sic*] and beautiful Frame of Heaven and Earth, and attending to the Dictates of our own natural Reason, we cannot but be convinced that there is a supreme Cause and Source of all Things; One that directs and superintends the Affairs of Mankind.
>
> For what, but an intelligent Cause *could call forth Light and Beauty out of Darkness and Deformity*? could direct the Motion of Matter? could *spread out the Heavens like a Curtain; and lay the Beams of his Chambers in the Waters*? What, but a wise and knowing Agent cou'd plant within us a thinking self-moving Principle, by which we can reason and recollect past and scatter'd Ideas, and direct ourselves as we judge proper; and thereby become the Glory and Ornament of the Creation?

God, the reverend gentleman went on, had imposed an uncontrollable dominion across all his works as an 'intelligent first cause'. Williams then became more specific as to examples of the power and providence we might everywhere discover should we descend to a more particular consideration of God's works. 'What, but the perfection of wisdom and goodness could paint the fields with such variety of colours, or could make the grass to grow upon the mountains, and herbs for the service of men?' And, as with Newton, an all-wise Providence was at work for Williams; though in terms that appear to include something of the vitalist view of creation expressed in the works of Thomas Vaughan: 'All things *live, and move, and have their Beings in Him*; They could neither Submit nor Act, should He cease to supply them with Strength and Vigour; but would soon relapse into their primitive Nothing, should He withdraw his supporting influence.'[24] Just a few years after Williams gave his 'Discourse', the International Evangelical Revival of 1735 to 1750 made its appearance in Wales. It was led by charismatic Methodist proselytisers such as Daniel Rowlands (1713–90), Howell Harris (1714–73) and William Williams Pantycelyn (1717–91). Although more widely remembered today as a hymnist who gave Chapel services and Welsh rugby a memorable anthem ('Guide me O thou great redeemer', also known as 'Bread of Heaven'), Pantycelyn made extensive use of scientific imagery in his work; most notably in the 1,367 verses of *Golwg ar Deyrnas Crist/A View of Christ's Kingdom* of 1756. Morus (2018) has noted how Pantycelyn 'rhapsodized about the speed at which the planets moved around the sun and about the sheer scale of the planets Saturn and Jupiter'. He even speculated on extraterrestrial life.[25] Eifion James, in his excellent biography of Pantycelyn,[26] notes the extensive reading required for the work's composition, much of it done on horseback as Pantycelyn traversed the Welsh countryside. Among the resulting influences on his *Christ's Kingdom* are the writings of Isaac Newton, as well as those of the Dutch inventor, mathematician and astronomer Christiaan Huygens, French anatomist and surgeon François Pourfour du Petit, the microscopist and inventor of the microscope Antonie van Leeuwenhoek, and the English clergyman and theologian William Derham (1657–1735).

Derham, who had given the Boyle Lectures of 1711 and 1712, published his *Physico-Theology: Or, a Demonstration of the Being and*

Attributes of God, from his Works of Creation in the same year that Newton added the *General Scholium* to the second edition of the *Principia* – 1713. Derham then followed *Physico-Theology* with his *Astro-Theology Or, a Demonstration of the Being and Attributes of God, from a Survey of the Heavens* in 1715. Some authors have regarded natural theology and physico-theology as essentially synonymous. However, Blair and Von Greyerz (2017) have suggested that physico-theology actually designates developments in natural theology that took place after 1650. These include the assumption that, as a rational being, God has a plan for everything and can be recognised by his goodness and benevolence. Furthermore, the physico-theological approach to nature was essentially utilitarian, thanks to God ensuring humanities welfare through the domination of nature. Physico-theological texts also exhibited a vocabulary that: highlights Gods excellence, omnipotence and goodness; describe arguments against deism and atheism; and provide an evidential base that rests on personal witnessing and experience.[27]

At least some of these criteria have already been seen in the earlier mentioned *A Relation of Apparitions of Spirits in the County of Monmouth and the Principality of Wales* (1779) by Edmund Jones. They include his arguments against deism and irreligion generally, and his numerous examples of personal witnessing and experience illustrated by accounts of apparitions of spirits. In another work, *A Geographical, Historical, and Religious Account of the Parish of Aberystruth* (1779), Jones also adopts a specific vocabulary when relating how the excellence of God is made apparent in his works of creation:

> it is reasonable to believe that some parts of [creation] must more visibly shew his creating excellencies than others, And of all others those formed into Mountains and Valleys claim the precedence in this respect. An instance of which is sufficiently shewed in the Parish of Aberystruth.[28]

Jones published his works in 1779, which shows that the sort of physico-theological ideas that also influenced the likes of Williams Pantycelyn were present in Wales throughout most of our long-eighteenth-century period. Nevertheless, despite the influence on Pantycelyn of physico-theology, and the other scientists mentioned earlier, he still says in his 'Preface' to *Christ's Kingdom* that the book he 'clung to chiefly was the Book of God'.

That an evangelical Methodist preacher should prioritise the Bible over works of science is no great surprise, and a similar subordination of science is seen among other evangelists too.[29] For example, Howell Harris noted several scientific events in his diary, but he usually couched them within a religious context, rather than seeing them as natural phenomena to be considered in their own right. In a diary entry of 1746 he declared, with millennial feeling, 'At midnight when I would see lightning in the sky or the Northern Lights, my soul would be ready to burst my body with joy being in hopes Christ was coming to judgment'. On visiting London, in 1748, Harris 'went to hear Brother [George] Whitfield preaching on the Great Eclipse', but he gives no description of the actual eclipse, beyond noting that it 'begins at 9 and ends at 12'.[30] On 18 May 1761, while at home in Trefeca in Breconshire, he records sitting up till 1 a.m. with his brother Joseph observing an eclipse of the moon, perhaps through the reflecting telescope that remains there today (see Figure 19). Yet Howell still described the sort of scientific interests enjoyed by his brother as mere 'head knowledge'. An attitude echoed in Williams Pantycelyn's 'But oh! it profits nothing, to know truth in the head, / Mere knowing may not influence, the heart is still not fed'.[31] In 1774 Williams Pantycelyn would write an essay titled *Aurora Borealis: or the LIGHT in the North, as a sign of the Success of the GOSPEL.*

Science was undoubtedly important to these men, but their subordination of science, as well as the consequences of doing so, led Evan Lloyd (1734–76) to satirise such attitudes in his 1766 poem *The Methodist.* Having first dealt with lawyers – 'For *Heav'n* they pray, as once for *Fees*' but now '*Religion* is the only Theme' – he turned to medicine; his lines reflecting the fact that before taking up their calling a number of Methodist preachers, including Williams Pantycelyn, had trained in the art of physic:

The *Physic-Tribe* their Art resign,
And lose the *Quack* in the *Divine*;
Galen lies on the Shelf unread,
A *Pray'r-Book* open in its stead;
Salvation now is all the *Cant*,
Salvation is the *only* Want.
'*Throw Physic to the Dogs*,' they cry,
'Twill never bring you to the Sky.[32]

We do not know whether Walter Williams or Williams Pantycelyn understood the mathematics on which Newton's *Principia Mathematica* was based, but another Welsh Newtonian and Dissenting minister certainly did. Though ill-health and various commitments led him to refuse a *c.*1762 request to edit a new edition of Newton's works, Richard Price taught, in the late 1780s, on Newton's *Principia*, and on fluxions (the newly invented calculus) at a Dissenter Academy in Hackney.[33]

Born at Tynton, a farmhouse in the village of Llangeinor near Bridgend, Price (1723–91) went to London in 1740. Educated in Wales in Dissenter academies at Pentwyn near Llannon in Carmarthenshire and Talgarth in Breconshire, and later at Moorfields in London, where one of his teachers had been Newton's friend John Eames FRS, Price settled in the village of Newington Green, just north of the City, in 1758. He lived and ministered to a congregation there through most of his working life, while making occasional trips home to Wales for sea-bathing and family visits. As a polymath who contributed to a number of sciences, he too subscribed to natural theology in terms reminiscent of Newton's *General Scholium*: 'It is impossible to survey the world without being assured, that the contrivance in it has proceeded from some contriver, the design in it from some *designing* cause, and the art it displays from some artist.'[34] Consequently, Price believed, the Creator must also be a rational and intelligent Being, since '[a]n unintelligent agent cannot produce order and regularity, and therefore wherever *these* appear, they *demonstrate* design and wisdom in the cause.'[35] Like Pantycelyn in his writings, Price used scientific imagery to enliven his sermons, some of it very up to the minute. In one he waxes lyrical about 'some discoveries in the heavens which have been lately made.' He then proceeds to illustrate them by taking his congregation on a verbal tour of a cosmos. One in which:

> our system moves towards other systems; [and] all the visible frame of sun, planets, stars, and milky-way forms one *cluster* of systems; and that, in the immense expanse of the heavens, there are myriads of these clusters which, to common glasses, appear like small white clouds, but to better glasses, appear to be assemblages of stars mixing their light.[36]

This particular sermon formed part of a collection Price published in 1786 (though dated to 1787), following a 1785 request for him to do so by his congregation. As an amateur astronomer, the source of his imagery is almost certainly a series of papers by William Herschel on the movement of the sun and the solar system (1783), and two papers on the Milky Way, nebulae, and 'On the construction of the heavens' read in June 1784 and February 1785 to the Royal Society, of which both men were Fellows. Herschel had also begun making observations of nebulae with his enormous 18.7-inch aperture reflector telescope in October 1783.[37] In other works, Price references Newton's *Principia*, including the Laws of Motion and the *General Scholium*, when outlining his thoughts on such fundamentals as the laws of nature, the first cause ('It follows certainly that matter can of itself do nothing; and that all the laws by which the order of the world is maintained, must be derived from a cause not material'), God's necessary existence, and the nature of that existence. Alongside his use of scientific imagery Price is not afraid to disagree in his sermons with the conclusions of others, nor is he afraid to admit to his own uncertainty and, sometimes, ignorance (as in the vexed question of matter being 'infinitely divisible').[38] He also poses questions of his own:

> For what is this earth to the whole solar system? How little do we know of the sun and of the worlds which move around him? What is light, and how does the sun dart it forth on all sides with

such inconceivable velocity? How are the planets furnished and peopled? What are the comets, and for what particular purpose were they created?[39]

In matters divine, the contrast between Price's attitude to science and that of Walter Williams and Williams Pantycelyn is notable, and might be explained by the denominational differences between the three preachers. Walter Williams certainly presents the more traditional Anglican view of natural and physico-theology – 'as a way to religious belief through reflection of the natural order'. The same can be said too of Williams Pantycleyn, whose Calvinistic Methodism still fell within the Anglican Church at this time. Pantycelyn's evangelism, however, allows for a more robust use of science with the aim, as Morus (2018) says, of showing 'his coreligionists in Wales's growing Methodist movement that God was on their side and that their success was assured'.[40] In Price there is more sense of science as a subject in its own right and not just as an adjunct to religion. Also, the questions that he often posed need answers from science rather than biblical analogy or revelation. His adoption of Arian Dissenting beliefs during his youth in Wales may well be the source of this outlook. In its rejection of the Holy Trinity, original sin and Calvinist predestination, and in its seeing Christ as not wholly of the same substance as God, Arian beliefs offered a somewhat more rationalist approach to religion. But, when Price writes that he sees every new vegetable or animal as 'a new production of Divine Power',[41] or he reflects that 'The vanishing of old stars, and the appearance of new ones, is probably owing to the destruction of old worlds, and the creation of new worlds … brought about under the direction of some superior power',[42] he retains, in common with Walter Williams, Williams Pantycelyn, Howell Harris and the Old Prophet Edmund Jones, a belief in the operation of divine Providence, and even miracles. As Price argues '[t]here was, undoubtedly, a time when this earth was reduced into its present habitable state and form' and '[t]his must have been a time of miracles, or of the exertion of supernatural power'.[43]

On the Probability of Miracles

It was Bertrand Russell who said that 'the pronouncements of science are made tentatively, on a basis of probability, and are regarded as liable to modification'.[44] Throughout the first half of the eighteenth century the *probability* of miracles occurring became a subject of much 'scientific' debate. For example, in 1727 Englishman Thomas Woolston published *Six Discourses on Miracles* in which he suggests that those performed by Christ did not actually occur. Instead, they were to be seen as allegorical, before quickly adding that in saying this he did not intend a denial of faith.[45] It was, though, the publication of David Hume's *Enquiries Concerning Human Understanding* (1748), with its (in)famous Section X *Of Miracles*, which brought the issue to a climax. The crux of the Scotsman's argument was that miracles were contrary to the laws of nature, and that witness testimony did not constitute proof of their occurrence. Here was a clear challenge to a fundamental Christian belief and an undoubted encouragement to scepticism and thereby deism and atheism.

In 1755, Welshman Henry Owen (1716–95) published *The Intent and Propriety of the Scripture Miracles*. He did not tackle Hume in the book, but between 1769 and 1771 Owen acted as a Boyle Lecturer. In the published version of his lectures (1773), which had the same title as his earlier work on miracles, he described Hume's argument as 'nothing more than down-right sophistry; the most fallacious reasoning ever'.[46] For Owen, God could 'as easily work a miracle, and "perform a new thing" [Numbers, 16.30] in the earth, as he may persevere in the old, common way, and keep things in their ordinary course'. He even saw gravity as 'no other than the constant agency of God';[47] thus providing a late example of Roger Cotes's concern, in his preface to the second *Principia* edition (1713), that '[s]ome say gravity is preternatural [beyond what is natural] and call it a perpetual miracle'.[48]

Born at the foot of Cader Idris in north Wales, Owen attended Ruthin School and Jesus College, Oxford, before writing a book on mathematics, practising as a medic and being ordained. By the time that his lectures appeared, he was at the church of St Olave in the City of

London, but he also had a long association as curate to Ralph Thoresby in Stoke Newington. Thoresby was a member of a 'supping club' formed with, among others, Richard Price, and it seems almost certain that Price and Owen would have known each other at some point. In his published lectures Owen's refutation of Hume relied on history and the need to consider the testimony of those in the past who saw miracles occur. In presenting his argument he primarily references Joseph Butler and *His Analogy of Religion Revealed*, but he also mentions Richard Price and his essay 'On the Nature of Historical Evidence and Miracles', the last essay in Price's *Four Dissertations* of 1767 (see Figure 6).[49]

Following a lengthy discussion of Hume's ideas, and in particular his argument that 'no testimony is sufficient to establish a miracle', Price refutes this last assertion and, in true Newtonian style, provides a long

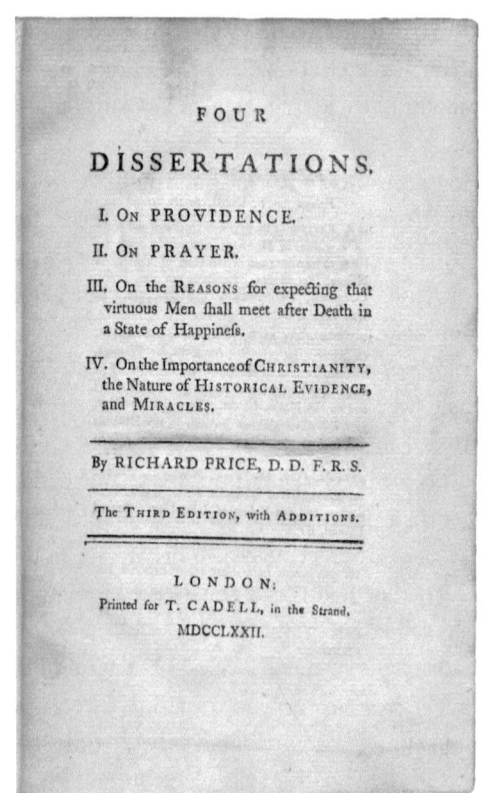

FIGURE 6 Title page to *Four Dissertations* by Richard Price (1767; London, 1772)

footnote in which he gives a mathematical calculation to prove his point. The basis of this calculation is Bayesian probability, in the revelation and development of which Price played a key role.

On the death of the Dissenting minister and mathematician Thomas Bayes (1702–61), Price found among his friend's papers a series of theorems concerning probability theory, or the 'doctrine of chances' as it was then called. He edited them, and added to the work, before its being read to the Royal Society by John Canton in December 1763. Tucker (2017) suggests that Price probably contributed about half of the finished paper, making in the process a 'distinct and substantial' contribution. Price also published a second paper in 1765 containing much new material. It is on the basis of Bayesian analysis, and a particular Bayes-Price probability theorem, that Price shows 'miracles may be possible and even far from improbable'. This represents the earliest known contribution to Bayesian Analysis (see Chapter 12).[50]

New Worlds, Shifting Foci

By the 1750s and 1760s the debate over miracles, coupled with continual scientific interrogation of creation and increasing biblical criticism, led once again to worries over the rise of scepticism. Science, though, began shifting its focus in the second half of the eighteenth century. It did so by moving away from seemingly endless discussion of its relationship with religion, and towards that of the wider society of which it was part. A society also changing as the more dangerous religious and political enthusiasms and sectarian debates of earlier decades became (somewhat) less divisive. Even the suspicion of science slowly gave way as the realisation gathered pace that its discoveries were not easily ignored, and that they could lead to direct improvements in everyday life. The development of probability theory proved to be a classic example. It became, and continues to be, of inestimable value in the development of pensions, annuities and life assurance, and insurance generally. As such, it helped provide not only greater financial security for people in old age but, by reducing risk, acted as a stimulus to wider travel and exploration and increasing investment at home and overseas. All destined to become subjects of ever greater importance in the long eighteenth century as the empire expanded and early nineteenth-century industrialisation approached.

Scientific discovery, and the new developments that it produced, also came to be seen as a fulfilment of the biblical assertion, in Genesis, of man's dominion over the natural world:

> male and female created he them. / And God blessed them, and God said unto them, Be fruitful, and multiply, and replenish the earth, and subdue it: and have dominion over the fish of the sea, and over the fowl of the air, and over every living thing that moveth upon the earth.[51]

But such dominance increasingly entailed moral and political dilemmas too, and science is, of course, amoral – it has no moral base whatsoever. At times this is its greatest strength, since, unlike religion, it is not subject to any prescribed text or revealed doctrine or dogma. At other times it is its Achilles Heel, since it must rely for its moral compass on scientists themselves and/or the society in which they operate, neither of which may provide such guidance. Scientists of Wales also contributed to long-eighteenth-century attempts to find acceptable resolutions to these moral dilemmas, most notably the attempt by Richard Price in his *A Review of the Principal Questions and Difficulties in Morals* (1758), and in his political writing on free will and civil liberties more generally.[52]

What none of the above suggests is a divorce between religion and science among the scientists of Wales in the long eighteenth century. How could there be when most of the works discussed above were written by clergymen, some of whom were still producing works near the close of the eighteenth century in the natural and physico-theology terms typical of the early eighteenth century. For example, the Old Prophet, Edmund Jones, besides publishing on spirits and apparitions in Aberystruth, wrote his manuscript 'A Spiritual Botanology' between 1771 and 1772 and gave to it the subtitle 'What of God Appears in the Herbs of the Earth; Together with Some of Their Natural Virtues and Uses'.[53] In 1794 Richard Price's nephew, Bridgend-born George

Cadogan Morgan, who trained as a Dissenting minister and later became a teacher, would unite religion and science in the 'Introduction' to his two-volume *Lectures on Electricity*. Not only were 'the power of gravity and the electric fluid … the agents of the same wise Omnipotence' but '[t]he faculties of the soul, like those of the body, become gigantic and mighty, by exertion only: and what delicious consciousness equals that of a mind triumphing in its own vigour?'[54]

However, the extent to which the vigour of various minds *had* changed the world since the writings of Thomas Vaughan and Edward Lhwyd a century and a half before can be seen in a paper George Morgan submitted to the Royal Society in January 1785. It was read to the Society by his uncle, Richard Price. Titled *Observations and Experiments on the Light of Bodies in a State of Combustion*, it is a work of pure science concerning electricity and photoluminescence. Even though Morgan and Price had trained and acted as ministers of religion, God and religion are entirely absent from the paper. There is also criticism of an element of the great Isaac Newton's work and, in a postscript, Price criticises part of his nephew's work.[55] The era of modern science had begun to appear.

With this chapter we come to the end of Part I of this book. In it, we have traced the foundations of modern science among the scientists of Wales. In the three chapters of Part II, that now follow, we will consider how scientists of Wales actually gained their knowledge of science in the long eighteenth century, and we begin with a topic in which religion continued to play a dominant role – education.

Notes

1. Nicholas Robinson, *A New Theory of Physick and Diseases, Founded on the Principles of the Newtonian Philosophy* (London, 1725), p. 6.
2. See A. de Morgan, 'On a Point Connected with the Dispute between Keil and Leibnitz about the Invention of Fluxions', *Philosophical Transactions of the Royal Society of London*, 136 (1846), 107–9.
3. de Morgan, 'On a Point Connected with the Dispute between Keil and Leibnitz'.
4. For all details, see *ODNB* entry for William Jones Sr; Paul Quarrie, 'The Scientific Library of the Earls of Macclesfield', *Notes and Records of the Royal Society of London*, 20 (2006), 5–24; de Morgan, 'On a Point Connected with the Dispute between Keil and Leibnitz'.

5. Nicholas J. Wade, *Perception and Illusion, Historical Perspectives* (New York, 2005), p. 96.

6. See Wade, *Perception and Illusion*, pp. 80–1.

7. An earlier, less detailed illustration of this crossing can be seen in Thomas Powell's, *Elementa Opticæ* (London, 1651); available online.

8. See Wade, *Perception and Illusion*, p.78.

9. Quoted in Wade, *Perception and Illusion*, p. 78, citing 'Harris, 1775, p. 100' – Joseph Harris, *A Treatise on Optics* (London, 1775).

10. Joseph Harris, *A Treatise on Optics* (London, 1775), p. 100. For a discussion of 'Newton's Aetherial Hypothesis, 1675', see A. Rupert Hall, *Isaac Newton Adventurer in Thought* (Oxford, 1992), pp. 135–8. It is notable that in quoting Newton's ideas on this subject from 1675, so before publication of the *Principia*, Hall notes that: 'The clockwork universe of Cartesian mechanical philosophy has made room for the animistic world of Paracelsus. It would be difficult not to see here ... the misty influence of recent reading in chemical and Platonic authors' (p. 137). Indeed, Newton's suggestion that 'Perhaps the whole frame of nature may be nothing but various contextures of some certain aethereal spirits, or vapours, condensed as it were by precipitation, much after the manner, that vapours are condensed into water, or exhalations into grosser substances, though not so easily condensable' suggests an interesting project would be a comparison of Newton's pre-*Principia* ideas with those expressed in the works of Thomas Vaughan.

11. Harris, *Treatise on Optics*, p. 100, in footnote.

12. Martin Griffiths, 'Joseph Harris of Trevecka Scientist, Artisan, Servant of the Crown', *The Antiquarian Astronomer*, 6 (2012), 19–33 (at 24).

13. George Makari, *Soul Machine: The Invention of the Modern Mind* (New York, 2015), p. 157.

14. Makari, *Soul Machine*, p. 159.

15. Evidence for Robinson's Welsh birth is taken from *ODNB* and from William Munk, see *https://history.rcplondon.ac.uk/inspiring-physicians/nicholas-robinson* (accessed 21 May 2024).

16. Nicholas Robinson, *A New System of the Spleen, Vapours, and Hypochondriack Melancholy* (London, 1729), p. 26.

17. Robinson, *A New System of the Spleen*, p. 28.

18. Makari, *Soul Machine*, p. 150.

19. Makari, *Soul Machine*, p. 153.

20. Robinson, *A New System of the Spleen*, p. 29.

21. Robinson, *A New System of the Spleen*, p. 29.

22. Thomas Knight, 'A Dissertation of the Operations of Chalybeat Medicines in Human Bodies', in *A Vindication of a Late Essay on the Transmutation of Blood* (London, 1731), 'Preface' (n.p.).

23. Thomas Knight, 'A Critical Dissertation upon the Manner of the Preparation of Mercurial Medicines, and their Operation on Human Bodies' (London, 1734).

24. Walter Williams, *Discourse Preached at St. Paul's Covent Garden, on St. David's Day, March 1, 1731*, (n.p., 1731), pp. 1–2.

25. Iwan Rhys Morus, 'On Science in a Small Country', *Physics Today*, 71 (December 2018), 42–8.

26. Eifion Evans, *Bread of Heaven: The Life and Work of William Williams, Pantycelyn* (Bridgend, 2010).

27. Ann Blair and Kaspar von Greyerz, 'Introduction', in Blair and Greyerz (eds), *Physico-Theology, Religion and Science in Europe, 1650–1750* (Baltimore, 2020), pp. 1–20 (at pp. 1–2 and 7–8). See also Kaspar von Greyerz, *European Physico-Theology (1650–c.1760): Celebrating Nature and Creation* (Oxford, 2022).

28. Edmund Jones, *A Geographical, Historical, and Religious Account of the Parish of Aberystruth* (Trevecka, 1779), 'Preface', p. v.

29. Evans, *Bread of Heaven*, see chapters 3, 15, 16 and 25.

30. Tom Beynon (ed. and transcriber), *Howell Harris's Visits to London* (Aberystwyth, 1960), p. 96 (northern lights), p. 205 (eclipse).

31. William Williams, *Bywyd a Marwolaeth Theomemphus o'i Enedigaeth i'w Fedd/The Life and Death of Theomemphus, from his Birth to his Grave* (London, 1764) cited in Evans, *Bread of Heaven*, p. 8.

32. Evan Lloyd, *The Methodist: A Poem* (London, 1766); available online.

33. See Carl Cone, *Torchbearer of Freedom* (Lexington, 1952), pp. 165–71. Price appears to have conducted his classes as a form of discussion group rather than on a formal teaching basis.

34. See D. D. Raphael (ed.), *A Review of the Principal Questions in Morals by Richard Price* (Oxford, 1974), p. 285.

35. Raphael, *Review of Morals*, p. 239.

36. Richard Price, *Sermons on the Security and Happiness of a Virtuous Course* (1787; Philadelphia, 1788), p. 174. Note that sermons in this American edition are in a different order to the 1787 London edition, the more controversial ones being relegated to the back.

37. See illustrative extracts in Michael J. Crowe, *Modern Theories of the Universe: From Herschel to Hubble* (New York, 1994), pp. 77–113.

38. For example, see his *Dissertation on the Being and Attributes of the Deity* written before 1758, but published in 1787 in *A Review of the Principal Questions in Morals* (1758; London, 1787), and his *Four Dissertations* (London, 1767).

39. Richard Price, 'Sermon IX: On the ignorance of man, and the proper improvement of it', in William Morgan (ed.) *Sermons on Various Subjects, by the late Dr. Richard Price, D.D. F.R.S.* (London, 1816), p. 167.

40. Morus, *Science in a Small Country*, p. 44.

41. Richard Price, *Four Dissertations* (1767; London, 1772), pp. 74.

42. Price, *Four Dissertations*, pp. 435–6.

43. Price, *Four Dissertations*, pp. 435–6.

44. Bertrand Russell, *A History of Western Philosophy* (1946; London 1989), p. 480.

45. Thomas Woolston, *Six Discourses on Miracles* (London, 1729); see also Judd Bernstein, 'Hume and Campbell the Miracles Debate and its Eighteenth-Century Background' (unpublished MA thesis, McGill University, March 1977), available online.

46. Henry Owen, *The Intent and Propriety of the Scripture Miracles*, 2 vols (London, 1773), I, p. 67.

47. Owen, *Intent and Propriety*, I, p. 53 and 52.

48. I. Bernard Cohen and Anne Whitman (trans.), *Isaac Newton: The Principia, Mathematical Principles of Natural Philosophy, A New Translation* (Los Angeles CA and London, 1999), p. 392.

49. Price, *Four Dissertations*, pp. 362–464.

50. See John V. Tucker, 'Richard Price and the History of Science', *THSC* (2017), 69–86 for a full discussion of Price's rebuttal of Hume, as well as his work on Bayes, probability and other science matters.

51. Genesis, 1.27–8.

52. For details see Paul Frame, *Liberty's Apostle: Richard Price his Life and Times* (Cardiff, 2015).

53. See Adam N. Coward (ed.), *A Spiritual Botanology*, S. Lucilius Verus (Edmund Jones) (Newport, 2017).

54. George Cadogan Morgan, *Lectures on Electricity*, 2 vols (Norwich, 1794), I, pp. xv and xliv.

55. George Cadogan Morgan, 'Observations on the Light of Bodies in a State of Combustion', *Philosophical Transactions of the Royal Society*, 75 (1785), 190–212.

PART II

GAINING A KNOWLEDGE OF SCIENCE

6

LEARNING ABOUT SCIENCE

A wretched people [the Welsh], they know little,
have seen less, and they cannot be taught.[1]

Morris Kyffin (*c.*1555–98; trans. Meic Stephens)

So wrote the Oswestry-born, Welsh and English-language poet, and soldier, Morris Kyffin. Whatever we make of his blanket statement, it is certainly true that, even with the founding or re-founding of grammar schools at St Asaph (n.d.), Brecon (1541), Abergavenny (1543), Bangor (1557), Presteigne (1565), Ruthin (1574) and Carmarthen (1576),[2] educational provision in Wales, up to the end of the sixteenth century, remained sparse and favoured the gentry above all. However, educational provision began to widen post-1600 and in this chapter we explore the place of science within that development.

Places to Learn

Between 1600 and the start of our period in 1650, sixteen schools endowed by wealthy or otherwise prominent benefactors were established. These operated alongside an informal home education and fee-paying private tuition sector, one that continued throughout the long eighteenth century (1650–1820) and, in some form or other, exists to this day. Within our long century another twenty-two endowed schools appeared between 1650 and 1700, supplemented by sixty or so free schools created under the terms of the parliamentary *Act for the Better Propagation of the Gospel in Wales* (1650). Although these free schools represent the State's first intervention into education in Wales, two-thirds of them disappeared after the restoration of the monarchy under Charles II in 1660. Into the breach stepped the Welsh Trust (*fl.*1674–81), whose educational and Bible publication efforts Robert Boyle had supported. In the ensuing eighteenth century a further eighty-seven schools were endowed, accompanied between 1699 and 1740 by the 157 charity schools of the Society for Promoting Christian Knowledge (SPCK), whose work the mathematician William Jones Sr aided.[3] These were succeeded in turn by Griffith Jones's famous circulating schools, of which more than 3,000 were established between 1737 and 1761, including one in Morris Kyffin's Oswestry. In the period of their operation, the circulating schools are estimated to have reached approximately 35 per cent of the Welsh population, and given basic Welsh-language literacy to some 200,000 of the 300,000 children taught, and to 100,000 of the 150,000 adults.[4] The schools were, and are, justly famous and, besides the eighteenth-century enquiries concerning them made by Catherine the Great of Russia, Griffith Jones's circulating schools were suggested as a model for others to follow by UNESCO as recently as 1955.[5]

Conspicuous by its absence in long-eighteenth-century Wales is a university, though this did not result from any want of trying to establish one. Owain Glyndŵr (*c.*1354–1415/17), during his rebellion against Henry IV, sent his famous Pennal Letter to Charles VI of France in 1406. In the letter Glyndŵr outlined his desire for universities to be established in north and in south Wales. A further attempt was made in the Puritan 1650s when Aberystwyth, Cardigan and Machynlleth were considered as possible sites, along with Shrewsbury: as 'a place of some name' lying 'within the verge of England ... that your sons may learn English'. As usual, funding was an issue and the proposers also expected their efforts to be opposed by Oxford and Cambridge. In fact, it was Cromwell's death and the restoration of the monarchy in 1660 that ultimately swept aside both hope and opportunity.[6] What replaced them was a continuation of the long-established brain drain of Welsh scholars to the two universities in England, as well as those in Scotland, Ireland and the continent, with some colleges and institutions maintaining links to particular schools in Wales in order to facilitate this movement.

Founded principally at the instigation of Welsh lawyer and clergy-man Hugh Price (1495?–1574), and numbering John Aubrey's great-grandfather, William Aubrey (c.1529–95) of Brecon, as a founding Fellow, the 'Welsh College', as Jesus College in Oxford was sometimes known, maintained close links with Cowbridge Grammar School. The school had been left to the college in the will of Jesus College alumnus and benefactor Sir Leoline Jenkins (1625–85), who clearly intended it to 'be an appendage which would have the useful function of supplying the College with students'.[7] In Cambridge, St John's College maintained similar links to a number of Welsh schools, as did the London-based Worshipful Company of Haberdashers to the grammar school at Monmouth (founded 1614).[8] Much later, in the first decade of the nineteenth century, Englishman Charles Greville (1749–1809) included 'The College of King George III' in his planned new town of Milford Haven. An early form of technical college rather than a true university, it would specialise in 'engineering, mathematics, navigation, mechanics and, of course, the sciences'. Building on the chosen site began in 1805 to plans drawn up by the Swansea-based architect William Jernigan, but Greville's death just before the planned opening in April 1809 led to the project's abandonment.[9]

Despite the absence of a university within Wales,[10] a university education was available for Welsh Anglicans at Oxford and Cambridge. However, the unwillingness of the growing body of Dissenters in long-eighteenth-century Wales to subscribe to the terms of the *Act of Uniformity* (1662) and to the Thirty-nine Articles of the established Anglican Church, which was required at Oxford on matriculation and at Cambridge on graduation, gave rise to alternative sources of higher education. Many enrolled at Scottish and continental universities, while others chose to attend the university-like Dissenter academies, a num-ber of which appeared in Wales and England after 1660.

In broad terms these were the principal educational establishments available in Wales in our period, and there is no intention to relate their historical development in any greater detail since that information can be found elsewhere.[11] Instead, our concern will be with their provision of a science education, as determined from the limited evidence of surviving curricula, and the recollections and reflections of some of the scientists of Wales who attended them.

Science and the Curricula

Subjects taught at the various free and charity schools in Wales reflected the overriding concern of their sponsors, whether private individuals, such as Griffith Jones, or charitable societies like the Welsh Trust and the SPCK. While their good intentions are not in doubt, and they are justly celebrated for substantially increasing literacy rates among the Welsh population over the course of the long eighteenth century, their principal concern remained religion, not science. Griffith Jones's circulating schools centred on learning to read the Bible and other religious literature in Welsh, while a circular from the SPCK in 1705 noted how 'a Christian and useful education for the children of the poor is absolutely necessary to their piety, virtue and honest livelihood'.[12] With this 'honest livelihood' in mind, some of the charity schools taught basic writing and arithmetic skills. For boys, there were also a few trade schools in areas where particular skills were needed, such as the teaching of navigation in coastal towns like Aberystwyth, or in river ports such as Carmarthen. At Neath, an early example of a 'works' charity school was established by the Shropshire-born industrialist Sir Humphrey Mackworth (1657–1727), one of the founders of the SPCK. As a member of the Company of Mines Adventurers, he obtained £40 from the company to educate forty children of miners, and other workers, at schools established in Glamorganshire and Cardiganshire.[13] Girls, who received short shrift from most educational establishments, generally received an elementary education coupled, sometimes, with knitting and spinning.[14] Their further education, if they received one at all, took place largely at home and, for those who could afford it, through private tuition. As Caroline Williams (1893) notes of the eighteenth-century Morgan family of Bridgend, even though its scientific members, George Cadogan Morgan and William Morgan, 'had very liberal notions as to female education' generally, and of their own daughters in particular, a 'superior education did not at that time open out a career for women'.[15]

Most of the endowed schools in Wales (those supplied by their benefactors with funds and/or land for their running and upkeep) were formed as grammar schools. Beyond the basics of reading, writing and arithmetic their syllabus centred on a classical education; the order of the day being the learning of Greek, Latin and sometimes Hebrew, together with reading classical authors such as Ovid, Homer, Pliny and Horace in their original languages. Other subjects might include English, French and Geography, for which, in 1732, the headmaster at Cowbridge Grammar requested 'a pair of ordinary Globes for ye use of ye School'.[16] Light relief came with dancing and drawing lessons, both of which were available at Carmarthen Grammar school in 1783 for 10s. 6d a quarter. Lectures in experimental philosophy were available at a guinea.[17]

In the Dissenter academies religion and the training of students for the ministry remained the primary objective, but from the mid-eighteenth century the teaching of science played an increasing role. Science, though, had formed part of the curriculum in some academies from very early in their development. In 1690, a Carmarthenshire-born ex-student of Carmarthen Grammar, James Owen (1654–1706), together with Presbyterian minister and diarist Philip Henry (1631–96),[18] founded an academy in the border town of Oswestry. The day there began at 6 a.m. in summer and 7 a.m. in winter. At 9 a.m., following prayers and various readings, lectures were given in logick, metaphysick, physics, geometry, astronomy, and more religious concerns.[19] As a format for learning it resembles that recorded nearly fifty years later by Thomas Morgan during his time at Carmarthen Academy in the 1740s. He too was up before 6 a.m., and at prayers in 'ye lecture room at 7'. There followed a lecture in Hebrew at 10 a.m., and one from Keill's Astronomy at 3 p.m. But for Morgan, neither a weekly lecture in the Greek Testament, nor work on conic sections, trigonometry, fluxions, astronomy, natural philosophy and the law of nature, could entirely dull the 'several evil inclinations arising from ye flesh', as he confessed to his diary.[20] By 1773 the same Carmarthen academy could boast the acquisition 'of the most curious Philosophical, Optical, and Mathematical Instruments and Machines':

The Telescopes and Microscopes are good. The Air-pump, Quadrants, etc. are of the best but the Orery, which consists of Planetarium[,] Tellurian, and Lenarium – The new constructed Electrical Machine, with the insulated Cushion and Electrometor – The set of Magnets – The concave and convex Mirrors, with the universal Dial, are certainly the most curious things of art – they exceed description – their elegance can scarcely be conceived – and their usefulness are too well known to be descanted upon.[21]

The instruments 'were *all* made under the direction of the judicious, and learned Dr. [Richard] Price, author of the Treatise on Reversionary Payments, etc.' Price had himself attended an academy (1735–38)[22] at Pentwyn, in the parish of Llan-non in Carmarthenshire, where one of his tutors, Samuel Jones (*fl.*1715–?1767/8), is said to have given lectures on science with a skeleton perched on his knee.[23]

Among Saints and Sinners

So, what did our budding scientists think of their education in Wales and elsewhere? Education as a subject in its own right elicited powerful and eloquent responses from several eighteenth-century Welsh scientists whose words still resonate today. 'On the bent given to our minds as they open and expand depends their subsequent fate, and on the general management of education depends the honour and dignity of our species',[24] declared Richard Price in a 1787 discourse to supporters of a new Dissenter college at Hackney in London. He would teach there for a time on life assurance and public finance, the higher mathematics and fluxions, as well as other discoveries made by Newton. In his *Observations on the American Revolution and the Means of Making it a Benefit to the World* (1784), Price advised the American Revolutionaries that:

The end of education is to direct the powers of the mind in unfolding themselves and to assist them in gaining their just bent and force. And, in order to this, its business should be to teach how to think, rather than what to think, or to lead into the best ways of searching for truth, rather than to instruct in truth itself.[25]

His nephew, George Cadogan Morgan, saw the issue more specifically through the lens of science; in particular, the new and wondrous one of electricity – 'There is scarcely any science, whose light may not be increased by [studying] it'[26] – and through classical mathematics: 'It may be safely asserted, that there is nothing, in the whole circle of science, unattainable by a man conversant with the elements of geometry'.[27]

Like his uncle, Morgan makes no specific comment regarding his education in Wales, but we might presume it coloured, at least to some degree, the opinions that he expressed in *Lectures on Electricity* (1794) and *Directions for the Use of a Scientific Table* (1796). In *Lectures on Electricity*, which formed the basis of his teaching at the new Dissenter academy in Hackney after he succeeded his uncle there,[28] he suggests that 'Labour and patience, and an ardent devotion of the whole mind to the object of its pursuit, are the essential properties of those who successfully attempt either to learn or to improve a science'.[29] Undoubtedly many in Wales followed this course, but such single-mindedness was not always evident among the pupil and student bodies. In his paper on Welsh non-conformist academies H. P. Roberts (1930) describes how 'a perusal of the Presbyterian Board minutes establishes the fact that the students of Carmarthen were more turbulent and disobedient than those of other (English) Academies with which the Board had to do'. And at Rhydygroes, near Carmarthen, the health of one tutor, Robert Gentleman of Shrewsbury, 'gave way chiefly because of "excessive vexation proceeding from the disobedient and turbulent behaviour of the students"'.[30]

George Morgan saw such waywardness as a part of growing up. Writing in *Directions for the Use of a Scientific Table* he suggests that:

No minds are more eager to wander than those of children. This is the kind disposition of nature; for without the relief of variety, the intellectual principle is actually killed, or it is frozen into a pigmy: besides, the restraints of children most commonly form associations in their minds, very unfavourable to knowledge; and hence, the moment they quit school, they are eager to sell their books, and to throw off their habits of slavery.[31]

Consequently, he rails against children being 'chained to their books' and, despite being a one-time classics scholar himself, taught through dead classical languages:

> Is there no knowledge but that which is acquired through the medium of the dead languages? Is the temple of science open only to the Greeks, or to those who understand them when they speak? Are there no interpreters, of whom we can avail ourselves; or has knowledge no language but theirs?[32]

On the other side of the coin, the challenge of being a teacher was recognised by Morgan's uncle, Richard Price:

> The principal part of education is certainly not teaching the languages or the sciences, but directing the passions and forming the mind and temper; and this is a work that requires greater abilities and more wisdom than the generality of even Scholars and Philosophers possess.[33]

For the alchemist Thomas Vaughan, who had criticised the outdated ideas taught to him at Oxford in the late seventeenth century by the scholastic method, a related concern was the authority imparted by the teaching of so-called 'established' truths: for these, when 'coming from our superiors carry with them the awe of the Tutor, and this breeds in us an opinion of their certainty; so that an university-man, cannot in all his life time, attain to so much reason and confidence, as to look beyond his lesson'.[34] A problem reflected in the education the deist David Williams received in the eighteenth century. Having first gone to a school near Caerphilly, he later attended Carmarthen Academy in order to fulfil his father's dying wish that his son become a Dissenting minister; this despite Williams's own 'indisposition to the dissenting ministry'.[35] Given such reluctance, and his later attraction to deism, it was perhaps inevitable that Williams was a severe critic of his education in Wales and of its science deficit in particular. As he writes in his three-volume *Lectures on Education: Read to a society for Promoting reasonable and humane improvements in the discipline and instruction of youth* (1789):

> I had undergone the usual forms of a liberal education; and had felt its absurdities. Of the facts of natural history, natural philosophy, astronomy, mechanics, chemistry, and anatomy, I had obtained a scanty, precarious knowledge, and they had been introduced to me as matters of scientific faith.

Williams later found that when induced by 'curiosity' or 'fashion' to examine the facts taught to him, he discovered 'it would require many years to correct the inaccuracies of an ostentatious but bad education'.[36] Williams left Carmarthen Academy in 1757 and moved to England in 1758, later declaring in his *Essays on Public Worship* (1773), 'I was educated among the Saints; and I now live, thank God, among Sinners'.[37]

In another pithy comment on education Richard Price declared '[s]eminaries of learning' to be 'the springs of society which, as they flow foul or pure, diffuse through successive generations depravity and misery, or, on the contrary, virtue and happiness'.[38] Both his nephew George Cadogan Morgan, and David Williams, sought to ensure those 'springs' flowed as pure as possible. At different points in their careers both men became teachers, both established schools of their own in London, and both wrote on education. In *Lectures on Education*, Williams declared his determination that:

> the children under my care should be informed of facts, and instructed to compare them, before they were initiated in the art of reasoning on ideas or principles. – At this period I required all the support I could derive from the superlative abilities of Lord Bacon.[39]

Just as Francis Bacon had proposed for an adult audience, so Williams believed 'children should be made acquainted with natural objects, their common relations, and general uses, by actual observation or experiment'.[40] In order to achieve this, and in preference to established methods that fostered memory at the expense of far more valuable cognitive skills, he had his pupils draw from nature and living models so as to 'quicken their powers of natural observation'.[41] Williams also studied the works of continental educational reformers such as Johann Basedow[42] whose ideas, according to Dybikowski (1993), Williams wanted to discipline 'to a Baconian model of scientific enquiry' adapted 'to English conditions'.[43] This concern with continental ideas is also reflected in his *A Treatise on Education* (1774) and its extended title, *In which the general method pursued in the public institutions of Europe: and particularly those of England; that of Milton, Locke, Rousseau, and Helvetius are considered; and a more practical and useful one proposed.* To make 'practical and useful' improvements demanded new educational materials, the existing ones being considered wholly inadequate. In response, Williams used the available scientific journals. By a process of simplification, he converted the *Philosophical Transactions* of the Royal Society and productions of the French and Prussian academies into usable school books. As a result, and on seeing his pupils 'drinking with avidity, at genuine and suitable fountains of knowledge', he could not but regret 'the years I had misemployed, or the irritable loss I had sustained, in not having the operations of my fancy, judgment and reason, founded on early and accurate acquaintance with natural or experimental truths'.[44]

Morgan and Williams also shared a compassion for those Morgan describes as having 'been flogged and cuffed, poisoned by bad air, and depressed by the most relaxing or debilitating confinement for six or eight years, in our common schools'.[45] Williams went so far as to reject using the physical punishments common at the time, a course that aroused suspicion and even led to rumours of a secret punishment chamber being built below the academy that he established at Lawrence Street in London in 1773.[46] To improve discipline and student behaviour he actually instituted a form of student government, as a form of social contract, together with a system of 'reciprocal assistance' in which students were allowed to tutor each other in class, while remaining under the general supervision of a teacher.[47] Today, working together in pairs or small groups is commonplace in schools and universities, especially in technical subjects. Lastly, Williams believed in time being given for the bud that is a young mind to open properly, and with the subjects taught being suited to the level of preparation and ability of the students, a proposal indirectly reflected in the *'Andrometer'* drawn up, as he approached the age of twenty-eight in 1774, by London-born William Jones Jr, the son of Anglesey mathematician William Jones Sr. Intended as nothing more than a piece of whimsy, it would allow, he said, every man to measure his 'merit' and 'to what degree he has risen in arts, sciences and ornamental qualifications' by looking for his age in a vertical scale running from year zero at base to seventy at top. He must then compare it with the various intellectual accomplishments listed alongside and to be achieved by a referenced age. It is interesting to see where Jones placed the attainment of science on his scale. After Latin at age twelve, Greek at thirteen, French and Italian at fourteen, Rhetorick and Declamation at seventeen, History and Law at eighteen, we arrive at 'Logick and Mathematicks' at nineteen, with 'Science improved' being held off till age forty-two; further 'mathematical works' follow at fifty-five. This perhaps suggests that for Jones Jr, science was to be seen as a mainly adult pursuit, despite the youthful mathematical prowess developed by his father. The scale ends between sixty-five and seventy with the 'Perfection of Earthly happiness' and a 'Preparation for Eternity'.[48]

Degrees of Difference

The author of the *Andrometer*, William Jones Jr, was the son of the deist mathematician from Anglesey, William Jones Sr, who died in 1749 when his son was but three years old. Consequently, Jones Jr received his early education from his mother, Mary Jones (née Nix, 1705–80). She had learned 'the Jonesian specialisms of algebra, trigonometry, and navigation' with help from her husband. Having achieved a degree of prosperity, Jones Sr managed to leave to his son and to his daughter £1,000 each, along with funds for 'their respective maintenance, tuition and education'.[49] The daughter, Mary, though considered 'to have pursued a track of learning similar to that which distinguished her brother', faced the strictures described earlier by Caroline Williams. Nevertheless, she and her close friend, Anne Welch, who studied mathematics, music and seven languages, became '*avant la lettre* feminists, rejecting the dubious delights of domesticity for the more vibrantly "masculine" public sphere'.[50] William Jones Jr went on to study at Harrow and later Oxford. He became a lawyer on the Carmarthen circuit, and finally a judge in Bengal, where he founded the Royal Asiatic Society and achieved lasting fame as a philologist with a companion interest in the sciences (see Volume 2).

Money has always guaranteed good schooling and a university education, though not always the best uptake or use of that education. Money combined with gentry or aristocratic position also meant an interest in science could take place in a well-stocked private library, and amid the clutter of expensive scientific apparatus. Such was the good fortune of barrister John Lloyd (1749–1815, FRS 1774, FLS 1799, FSA), who inherited estates at Wigfair, near St Asaph, and at Hafodunnos in Denbighshire. As a Fellow of the Royal Society, the Linnean Society (FLS) and the Society of Antiquaries (FSA), the diversity of his interests is evident in the scale and nature of his library. When sold following his death in 1815 the sale catalogue of 1816 described it as 'one of the BEST GENTLEMAN'S COLLECTIONS in the kingdom'. It encompassed, among other things, volumes on poetry, classics, divinity, philosophy, mathematics, physic, chemistry, natural history, minerals, astronomy, politics and Welsh history, together with a large collection of clocks, telescopes, scientific instruments and material derived from the Pacific voyages of James Cook; the last having been presented to Lloyd by Joseph Banks.[51] The collection took two weeks to sell and the importance of its contents can be gauged from a much later sale of just one volume that had once been in Lloyd's library: *The Recuyell of the Histories of Troye* was the first book printed in the English language, and the first book from the press of William Caxton, who had translated it from the French of Raoul Lefèvre. The ex-Lloyd copy sold at Sotheby's in London in 2014 for £1,082,500.

Some among the less well-heeled middling-sort – ministers and teachers, for example – tried to emulate the gentry's facilities for the pursuit of science, as can be seen from this description of George Cadogan Morgan's 'study':

> His schoolroom was his study [George became a teacher after having given up the dissenter ministry for which he originally trained], an immense apartment, not less than sixty feet in length, and twenty-five feet wide, and proportionably high, with book shelves requiring a step ladder to get access to all the valuable works in his extensive library. Considerable space was appropriated to his philosophical apparatus, specimens for the study of natural history, anatomy and mineralogy, occupied the shelves of large glass cases, while a broad table supported electrical machines, jars and other experimental models, as well as globes and a grand telescope, manufactured by Dolland, which had been a legacy from Dr. Price.[52]

A telling difference between the study of George Cadogan Morgan and the library of John Lloyd is that the former also acted as a work-a-day schoolroom. What Morgan lacked, like most of the middling folk mentioned in this chapter, was gentlemanly leisure time. Most of the scientists of Wales mentioned thus far had been, or still were, practising doctors, ministers of religion or teachers whose passion for science needed support from the daily grind of earning a living. Furthermore, following their initial education in Wales, and the absence there of a university or major urban intellectual centre, many of them inevitably gravitated to establishments in England to take their education further; with many subsequently choosing to stay, live and work there. Before attending Tenter Alley Academy in London, near the modern Barbican, Richard Price had been first educated at home, then at schools in south Wales, and later at Pentwyn Academy in Carmarthenshire. His nephew George Cadogan Morgan had been head boy at Cowbridge Grammar School and entered Jesus College Oxford to study classics. He did not finish his degree. Instead, and for reasons that remain uncertain, he left to attend a Dissenter academy at Hoxton, before moving on to employment as a minister in Norwich and later in London. David Williams attended a school near Caerphilly, and then Carmarthen Academy, before working as a Dissenting minister in Frome and Exeter, and later moving to London. This clearly represents a well-established route to advancement for our middling class of scientists from Wales, but it also contrasts, sometimes markedly, with the route taken by the even less well off; those among the 'gwerin' who harboured scientific interests and aspirations, not least among them William Jones Sr.

The epithet usually applied to this second route is 'self-education', which the artist Hugh Hughes (1790–1863), of Pwll-y-gwichiad in Llandudno, is said to have regarded as superior to a university education and a virtue of his nation and class.[53] Certainly a degree of self-education must have featured in the early lives of scientists from poorer backgrounds in Wales. Yet, despite their relative poverty, most received some aid along the way from the charity schools and/or interested patrons. Early reading probably helped William Jones Sr to develop his mathematical skills, but it was at the charity school that he attended in Llanfechell on Anglesey where they were first properly recognised. Having been brought to the attention of the local aristocrat, Viscount Buckley, Jones was recommended to a London merchant. On-the-job learning followed, as well as time spent on a man-of-war teaching mathematics and navigation. The skills garnered in navigation (on which he published) no doubt fostering his later passion for the then vexed question of determining longitude at sea (see Chapter 15). Jones would also become a private tutor and scientific adviser to members of the aristocracy, while no doubt continuing to self-educate through books from one of the finest mathematical libraries in Britain, his own.[54]

A similar path was taken by Joseph Harris, the brother of evangelical preacher Howell Harris. As a blacksmith's apprentice in Breconshire, Joseph seems to have developed an early interest in practical or artisan science in the manner of 'a tinkerer, enjoying making small instruments for varied purposes'. Somewhat surprisingly, from the records available, this practical attitude, coupled with an enquiring mind, does not seem to have led to any formal education; a marked contrast to his brother

Howell who attended grammar school (with Joseph's financial support) and, briefly, Oxford University. As with William Jones Sr, a leg up for Joseph came from a patron, in his case the local MP Roger Jones. He introduced Joseph to the London Society of Instrument Makers and, via them, to astronomer and Royal Society Fellow Edmond Halley. Like Jones Sr, Joseph undertook long sea voyages and became skilled in aspects of navigation and astronomy (see Volume 2).[55] Like Jones too, he used his scientific skills to make significant contributions to eighteenth-century society, while at the same time accumulating an extensive library and a wealth of scientific instruments.

Finally, and late on in our study period, we have the examples of John Evans (1796–1861) and Griffith Davies (1788–1855, FRS 1831). As the son of a Cardiganshire weaver, John Evans had been expected to follow in his father's footsteps. Instead, he chose to walk to London as a seventeen-year-old. There he became a door-to-door book salesman and, with help from David Davies, 'a fellow-countryman and musical instrument tuner', he attended the London school run by Griffith Davies. Hailing from the parish of Llandwrog, in Caernarfonshire, Davies's own 'early educational advantages were scant', and he worked as a farm labourer and quarryman before discovering an aptitude for mathematics; this was after his savings had allowed him to attend school in Caernarfon, aged seventeen. Arriving in London in 1809 he eventually opened a school teaching 'mathematics and allied subjects'. Davies remained in London, eventually becoming actuary to the Guardian Assurance Company from where he made significant contributions to new methods in actuarial science. Meanwhile, John Evans, having made good progress in mathematics at Davies's school, returned home to Wales and established his *Mathematical and Commercial School* in Aberystwyth *c*.1817–20. There he taught mathematics and navigation and continued to develop his skills in astronomy and the making 'of sundials and mechanical instruments'.[56]

Such examples suggest a determination on the part of those from poorer backgrounds to gain a more secular education than the charity schools offered. A desire perhaps fostered during their early self-education and, if lucky, exposure to the stimulation of mathematics through arithmetic at school; the latter often providing a route to employment through its links to navigation and surveying. William Davis (1771–1807), for example, began life as a land surveyor in his home area of Gresford in Denbighshire, but became a London-based mathematician. He was a member of the Spitalfields Mathematical Society, the editor of *The Gentleman's Mathematical Companion*, and a major publisher of mathematical works. The last including a new edition of Andrew Motte's 1729 first English translation of Isaac Newton's *Principia*, 'carefully revised and corrected by W. Davis'. It appeared in 1803 and was reissued in 1819, after Davis's death.[57] John Evans (1770–1851) of Rhuddlan, another who had been expected to follow in his father's footsteps by becoming a Parish clerk, set up his own school when he was sixteen. He later became a land surveyor, played the cello in church, wrote on and composed music, and published two *Cyfrifydd Parod* or 'ready-reckoners' used in the buying and selling of grain and the measuring of timber and haystacks.[58]

We can see, therefore, that science, despite significant disparities in facilities and opportunities in Wales, could form part of a person's education there in the long eighteenth century. Though the diversity of science subjects taught might be restricted, the stimulus for some interested individuals to develop their knowledge further remained, even though this often meant moving away from home and oftentimes out of Wales.

Notes

1. Quoted in Meic Stephens (compiler and ed.), *A Most Peculiar People: Quotations about Wales and the Welsh* (Cardiff, 1992), p. 17.

2. Malcolm Seaborne, *Schools in Wales 1500–1900: A Social and Architectural History* (Denbigh, 1992), p. 22, Table 1.

3. In 1745, the Anglesey-born mathematician William Jones Sr, though based in England throughout his adult life, involved himself in the production of a Bible subsidised by the SPCK for the use in the circulating schools of Griffith Jones. In a letter to William Jones of 14 May, Richard Morris noted: 'the progress I have already made in collating and correcting ye Orthography of the Welsh Bible, which I submit to your consideration and beg the favour of your sentiments upon.' However, persons of 'great learning' had charged that the Bible was a translation from the 'English and not the original Hebrew and Greek'. Morris believed this to

be untrue but 'would be proud' of Jones's 'opinion upon ye whole matter, for which I will make bold to call in a few days'. The Bibles were printed at a cost of £6,000 between 1746 and 1752. See Michael J. Franklin, *Orientalist Jones* (Oxford, 2011), p. 48; AML-I, pp. 152–3.

4. All dates and figures taken from Seaborne, *Schools in Wales*, chapters 1–3; W. T. R. Pryce, 'The Diffusion of the "Welch" Circulating Schools in Eighteenth-Century Wales', *THSC*, 25/4 (2011), 486–519 (see 491 for numbers taught).

5. See entry for 'Circulating Schools', in John Davies et al., *Encyclopaedia of Wales* (Cardiff, 2008), p. 145.

6. See Geoffrey F. Nuttall, 'The Correspondence of John Lewis, Glasrug, with Richard Baxter and with Dr. John Ellis, Dolgelley', *Merionethshire Historical and Records Society*, 2/2 (1954), 120–34; E. L. Ellis, *The University College of Wales, Aberystwyth 1872–1972* (Cardiff, 1972), pp. 1–4.

7. Iolo Davies, *'A Certaine Schoole': A History of the Grammar School at Cowbridge Glamorgan* (Cowbridge, 1967), see chapter 1, quote at p. 23.

8. Seaborne, *Schools in Wales*, pp. 65, 72–3.

9. His brother Robert is said to have tried to re-establish it in 1812 but this too came to naught. For all details see Martin Rowland, 'The Doomed Dome', *Pembrokeshire Life* (May 2006), p. 24. See also Greville's *DWB* entry.

10. Although *The Principles of Collegiate Education Discussed and Elucidated in a Description of Gnoll College, Vale of Neath, South Wales* (London, 1857) would outline an education in mathematics, mechanics, physics, chemistry and natural history, a full university would not be established in Wales until 1872, at Aberystwyth, through the combined efforts of a few prominent proposers and financial contributions from the people of Wales, and largely in the face of initial government indifference.

11. For example, Gareth Elwyn Jones and Gordon W. Roderick, *A History of Education in Wales* (Cardiff, 2003) and Seaborne, *Schools in Wales (1500–1900)*.

12. Joseph Downing, *An Account of the Methods Whereby the Charity-schools have been erected and managed* (London, 1705), p. 1, quoted in Seaborne, *Schools in Wales*, p. 45.

13. Seaborne, *Schools in Wales*, p. 47; *DWB* entry; and Eiluned Rees, 'An Introductory Survey of 18th Century Welsh Libraries', *Bulletin of the Welsh Bibliographical Society*, 10 (1971), 197–259, at 203–4.

14. One such charity school was established by Mrs Johnes, the wife of Thomas Johnes of Hafod, at which the girls learnt to read and spin, see Rees, 'Survey of 18th Century Welsh Libraries', 204.

15. Caroline E. Williams, *A Welsh Family from the Beginning of the 18th Century* (London, 1893), p. 39. See also Nicola Bruton Bennetts, *William Morgan: Eighteenth Century Actuary, Mathematician and Radical* (Cardiff, 2020) for the education of George Cadogan Morgan, William Morgan and their family, including the daughters.

16. Davies, *A Certaine Schoole*, p. 34.

17. Martin Evans, *An Early History of Queen Elizabeth Grammar School Carmarthen 1576–1800* (Carmarthen, n.d.), p. 84.

18. See *DWB* entries. Philip Henry was born in Whitehall where his father, who came from Briton Ferry near Swansea, became 'Keeper of the Orchard' to Charles I. Philip Henry and James Owen held a debate with William Lloyd (1627–1717), then Bishop of St Asaph and later of Worcester. Lloyd, as a friend of William Whiston, was one of those who worried that Whiston was becoming a 'subverter of souls' with his writing and questioning of religious orthodoxy (see Chapter 4 of this volume, and Larry Stewart, *The Rise of Public Science*, (Cambridge, 1992), pp. 78–9).

19. R. Brinley Jones, '"Grace rather than Law": Aspects of the History of the Education of Dissenters', *THSC*, New Series, 7 (2001), 83–95 (at 88).

20. NLW MS 5456A Henllan 5, Diary of Thomas Morgan.

21. See Gwyn Walters, 'Richard Price and the Carmarthen Academy', *The Price-Priestley Newsletter*, 4 (1980), 69, citing *The Cambrian Magazine; or Useful Repository, of Science and Entertainment*, 1 (June 1773), 29–30.

22. See Roland Thomas, *Richard Price: Philosopher and Apostle of Liberty* (London, 1924), pp. 10–11.

23. Geraint H. Jenkins, *The Foundations of Modern Wales 1642–1780* (Oxford, 1987), p. 315.

24. Richard Price, 'The Evidence for a Future Period of Improvement in the State of Mankind (1787)', in D. O. Thomas (ed.), *Price, Political Writings* (Cambridge, 1991), p. 166.

25. Richard Price, 'Observations on the Importance of the American Revolution (1785)', in D. O. Thomas (ed.), *Price, Political Writings* (Cambridge, 1991), p. 137.

26. George Cadogan Morgan, *Lectures on Electricity*, 2 vols (Norwich, 1794), I, pp. x–xi.

27. George Cadogan Morgan, *Directions for the use of a Scientific Table, in the Collection and Application of Scientific Knowledge* (London, 1796), p. 12. Note, in D. O. Thomas's, 'George Cadogan Morgan (1754–98)', *The Price-Priestley Newsletter*, 3 (1979), 53–70, there is another quote attributed to Morgan's *Directions*, namely: 'Men of science must preside in our schools: and the elements of geometry must become the first grammar that is taught.' However, the quote has not been found in the 1796 first edition. It may come from the preface to the 1826 edition used by Thomas, or the *Monthly Magazine* reference he also gives (see Thomas, 68 n. 3 and 6).

28. For Morgan's teaching at Hackney Academy, along with that of Richard Price (whom Morgan succeeded), and the involvement of Montgomeryshire-born Abraham Rees as one of the academies founders, see Chapter 8 in Olive Lewis, 'The Teaching of Science in English Dissenter Academies 1662–1800' (unpublished MPhil thesis, The Open University, 1989), available online.

29. Morgan, *Lectures on Electricity*, I, p. i.

30. H. P. Roberts, 'Nonconformist Academies in Wales (1662–1862)', *THSC* (1928–9; 1930), 1–98 (at 88 9 and 25). For school-based examples, see Davies *'A Certaine School'*, pp. 35–7.

31. Morgan, *Directions for the use*, p. 14.

32. Morgan, *Directions for the use*, pp. 12, 13, 14.

33. W. Bernard Peach and D. O. Thomas (eds), *The Correspondence of Richard Price*, 3 vols (Durham NC and Cardiff, 1983–94), I (1983), p. 98, RP to Earl of Shelburne, 22 May 1771.

34. *WTV(E)*, p. 535.

35. Quoted in J. Dybikowski, *On Burning Ground: An Examination of the Ideas, Projects and Life of David Williams* (Oxford, 1993), p. 20.

36. David Williams, *Lectures on Education. Read to a Society for Promoting Reasonable and Humane Improvements in the Discipline and Instruction of Youth*, 3 vols (London, 1789), II, pp. 315–16.

37. David Williams, *Essays on Public Worship, Patriotism, and Projects of Reformation* (London, 1773), p. 25.

38. Price, 'The Evidence for a Future Period of Improvement', p. 56.

39. Williams, *Lectures on Education*, II, p. 316.

40. Williams, *Lectures on Education*, I, p. 164.

41. Dybikowski, *On Burning Ground*, p. 121

42. Johann Basedow (1724–90), German educational reformer who founded the *Philanthropium* a progressive though short-lived school. He was one of the first to introduce popular illustrated textbooks for children and was strongly influenced by Rousseau. For Williams and Rousseau see Dybikowski, *On Burning Ground*, pp. 129–33.

43. See Dybikowski, *On Burning Ground*, pp. 121 and 125.

44. Williams, *Lectures on Education*, II, pp. 316–17.

45. Morgan, *Directions for the Use*, p. 12.

46. See Dybikowski, *On Burning Ground*, p. 118.

47. Dybikowski, *On Burning Ground*, pp. 118, 121.

48. Garland Cannon (ed.), *The Letters of Sir William Jones*, 2 vols (Oxford, 1970), I, pp. 174–5 and Plate 3.

49. Michael J. Franklin, *Orientalist Jones* (Oxford, 2011), pp. 51–2.

50. Franklin, *Orientalist Jones*, pp. 56–7.

51. See NLW MS 12500B – *Bibliotheca Llwydiana: a catalogue of the entire library … of John Lloyd, Esq. LL.D. … by Mr Broster of Chester*, 1816. Also, Gwynfryn Walters, 'Bibliotheca Llwydiana: Notes on the Sale Catalogue (1816) of John Lloyd's Library', *NLWJ*, 10/2 (1957), 185–204.

52. The 'Autobiography of Richard Price Morgan, Senior', in Mary-Ann Constantine and Paul Frame (eds), *Travels in Revolutionary France and A Journey Across America* (Cardiff, 2012), pp. 141–2.

53. See Ian McCalman (ed.), *An Oxford Companion to the Romantic Age: British Culture 1776–1832* (Oxford, 1999), p. 549. Between 1816 and 1832, Hughes pioneered the illustration of popular Welsh-language texts including some that related science information; see, for example, illustrations in Peter Lord, *Hugh Hughes Arlunydd Gwlad* (Llandysul, 1995), pp. 197, 198.

54. Jones left his library to the Earl of Macclesfield at Shirburn, where it remained until sold, along with the wider Macclesfield Library, at auction in six sales between March 2004 and November 2005. See Paul Quarrie, 'The Scientific Library of the

Earls of Macclesfield', *Notes and Records: The Royal Society Journal of the History of Science*, 20 (2006), 5–24.

55. See Martin Griffiths, 'Joseph Harris of Tevecka, Scientist, Artisan, Servant of the Crown', *The Antiquarian Astronomer*, 6 (2012), 19–33 (quote at 19); and *DWB* entry for Harris.

56. See respective *DWB* entries.

57. Other than being 'British' the nationality of Davis is not usually recorded (in *ODNB* for example), and he currently has no entry in *DWB*. However, details of his Welsh birth and work there as a surveyor, together with his later career in London, can be found in *The Literary Panorama, and National Register*, 2 (1807), pp. 1357–8. For the *Principia* details and quote, see I. Bernard Cohen, 'A Guide to Newton's Principia', in I. Bernard Cohen and Anne Whitman (eds), *The Principia: A New Translation* (Los Angeles CA and London, 1999), p. 26.

58. For details and quotations see *DWB* entries for each person, but see note 57 here for William Davis.

7

THE LANGUAGES OF SCIENCE

Among all the countries of the world, there is no nation more
lacking in love and more hostile to its own language than the
Welsh, although our language, because of its antiquity and
richness, deserves as much respect as any other language.[1]

John Edwards, *Madruddyn y Difinyddeaeth
Diweddaraf* (1652; trans. Meic Stephens)

In the previous chapter we considered the relationship between education and science. Here, we are concerned with the languages of science
itself. Wales in our period was a monoglot Welsh-speaking country with
bilingual and English-only speakers restricted largely to the eastern
border area and the southern coastal fringes.[2] Then, as now, encroach-
ment by the English language and an associated decline in the use of
Welsh formed subjects of much debate. Many people were extremely
concerned at the trend while others, though not exactly favouring it, saw
English usage as a means to a better life and an inevitable consequence
of modernity. Although the Welsh language had long and mutually
sustaining links with religion, the provision of a university and sizeable
towns might well have provided additional linguistic bastions while
simultaneously acting as centres for the development of the sciences.
With their absence it was inevitable that some with scientific interests
felt impelled to try and improve their lot outside Wales where English,
in these Isles at least, would be mandatory for many purposes. Griffith
Davies (1788–1855), for example, declared his purpose in going to
London was to improve his English as well as his mathematics. While
many did the same and then returned to Wales, others, like Davies,

did not. Although essential to Welsh culture and nationhood (and to the history of wider Britain), the Welsh language came to be seen as a barrier to progress in a dominantly English-speaking and increasingly modernising and imperialistic Britain. Some even saw bilingualism or multilingualism as detrimental to learning itself; one of them being the deist David Williams (1738–1816) who we first met in Chapter 4.

In a Land of Babel?

Williams knew the Welsh-language dialect of his local area of Caerphilly and admired the linguistic agility of the language: 'Its prose is copious, expressive, and susceptible of great force and sublimity. Its system of Poetry is extensive, accurately divided into classes; and its varieties of versification seem to have no limits.' Yet, at the same time, he still believed that for Welsh speakers 'the perpetual business of translation occupies their minds and their time; and grave errors occasioned by annexing wrong ideas to terms founded on unknown customs, give them the appearance of folly, stupidity, and inferiority'.[3] Of course, his argument applies equally to any first-language English speaker faced with a Welsh, French, German, Thai or Maltese speaker. As Walter Davies pointed out in *The Cambrian Register* of 1795, the Welsh 'commit, it is true, just as many blunders in speaking English, as the English themselves would do in speaking French, or Irish, before they are taught; and no more'.[4] Nevertheless, David Williams felt their language would exclude the Welsh 'from all the speculations of industry' – an argument still heard today. He even went so far as to suggest in his *History of Monmouthshire* (1796) that 'Enquiries concerning the origin of moral evil should have been directed to the tower of Babel, as well as to the garden of Eden'. And, if given a free hand in his London school, which he was not, he would have limited a child's learning to one language.[5]

Sometimes used as an illustration of declining Welsh-language use in the long eighteenth century is the visit in 1782 by William Jones Jr to the Court of Louis XVI of France. Having been entertained by Jones speaking the languages of the places that he had visited, Louis declared him to be 'an extraordinary man!' To which a court minion added: 'he

understands almost every language in the world but his own'. 'Then
of what country is he?' asked the king; 'he is, please your Majesty, a
Welchman!'[6] In fact Jones Jr divided the twenty-eight languages of
which he had knowledge into eight that he had 'studied critically',
these were English, Latin, French, Italian, Greek, Arabic, Persian and
Sanskrit; eight 'studied less perfectly, but all intelligible with a dictionary', among them Hindi, Turkish, Spanish and Bengali; and twelve
'studied less perfectly, but all attainable', including Tibetan, Russian,
Ethiopic, Coptic, Swedish, Dutch, Chinese and Welsh.[7] While few
in Wales could equal the language skills of Jones Jr a sort of linguistic
Babel did exist among its scientists. But it was a useful one rather than
the *confusion of tongues* with which the word Babel is normally associated, and which so disturbed not only David Williams but the likes of
John Wesley, who deplored 'the confusion of tongues' when preaching
in Wales.[8]

The Languages of Science

The scientists of Wales used one or more of three principal languages
in the long eighteenth century – Latin, English and Welsh. Even
though some of the earliest texts on mathematics to appear in the
English language had been written by Tenby-born, London-based
Robert Recorde by the mid-1500s (see Chapter 1), Latin had long
been the universal language of international science. As a *lingua franca*
Latin ensured a wide European readership for works such as Edward
Lhwyd's *Lithophylacii Britannici Ichnographia* ('Map of the British Stone
Cabinet', 1699). In the year of its London publication a pirated copy
also appeared in Leipzig and, according to Emery (1971), the work was
known to scientists as far afield as France, Italy, Holland and Germany.
It also appeared in Russia, where a copy once owned by Robert Areskin
(d.1718), a physician to Tsar Peter the Great, has been found in the
Academy of Science in St Petersburg. It quite possibly came to Russia
via the German doctor and naturalist David Krieg who had it from
Hans Sloane, a notable sponsor of Lhwyd's volume, along with Isaac
Newton and others.[9]

Just as importantly, Latin facilitated international communication through correspondence: the naturalist and traveller Thomas Pennant (1726–98) of Flintshire using it to write to the non-English-speaking Swedish naturalist Carl Linnaeus (1707–78), founder of the binomial classification of genus and species. For younger Welshmen, knowledge of Latin allowed them to seek an advanced education in universities such as Leiden, Franeker (in Friesland), Basel, and Altdorf (near Nuremberg in Germany). These places were generally more welcoming of Dissenters and Nonconformists than Oxford and Cambridge.[10] They were also places where Latin-based lectures were given by such key figures in European science as Herman Boerhaave in Leiden: 'one of the greatest scientific teachers of the century and one of the first outside Britain to incorporate Newton's work into his lectures' (see also Volume 2).[11]

Nevertheless, the trend towards using national languages (English in Britain) for the presentation of scientific data and results increased throughout Europe in the long eighteenth century. So, while Bassett Jones of Glamorgan wrote his 1648 alchemical treatise *Lapis Chymicus Philosophorum Examini Subjectus* in Latin, he followed it in the early 1650s with a work of 2,917 lines of English poetry and 7,000 words of English prose titled *Lithochymicus, or A Discourse of a Chymic Stone præsented to the University at Oxford.* The work, which contains an occasional use of Welsh, Latin and Greek, remained as an unpublished manuscript until 1995 when it was described by its editor, Robert M. Schuler, as 'the most ambitious alchemical poem of the Renaissance'. Thomas Vaughan's alchemical works, also written in the 1650s, contain a smattering of Latin and his Welsh mother tongue but are otherwise wholly in English. Alongside his religious tracts the Revd Thomas Powell of Breconshire published *Elementa Opticæ: Nova, Facili, & Compendiosâ Methodo Explicata* ('Elements of Optics: a New, Easy, and Compendious Method Explained') in Latin in 1651. It also contains Latin panegyrics addressed to him by Thomas Vaughan and his brother Henry, both of whom Powell may have tutored. However, in 1660 Powell's *Humane Industry: or, a History of most Manual Arts, Deducing the Original, Progress, and Improvement of them* appeared in English. As a discourse on scientific inventions and events through the ages, together with what he sees as their religious connections, he restricted the use of Latin (and Greek) to titles and quotations or chapter and section headings. A further encouragement to the use of national languages came with the founding of journals linked to national science societies. Among the earliest of these journals was the *Philosophical Transactions of the Royal Society*, which first appeared in 1665. Papers in Latin occasionally appeared in the journal but articles in English dominated from the start, with one relatively early paper being Edward Lhwyd's *An Account of a sort of paper made from Linum Asbestinum found in Wales* (1685). In the absence of a similar provision in Wales many of the nation's scientists contributed to the journal in English.

The trend towards English usage continued into the early eighteenth century. The mathematician William Jones Sr used it for his 1702 work *A New Compendium of the Whole Art of Practical Navigation* and, despite its part Latin title, for his historically important *Synopsis Palmariorum Matheseos, or, A New Introduction to the Mathematics* (1706). For his later *Analysis per Quantitatum series, fluxiones ac differentias* of 1711, however, Jones turned to Latin. This work included material from the papers of the mathematician John Collins (1625–83, FRS 1667) that Jones had acquired in 1708. Among them was a draft, made by Collins, of an unpublished work by Isaac Newton – *De Analysi* (1669). Jones published the work, with Newton's agreement, together with other early mathematical essays by Newton, including *De quadratura, Enumeratio linearum tertii ordinis* and *Methodus differentialis*.[12] Jones also included a brief history of Newton's mathematical development. Another fuller account of Newton's mathematics written in English by Jones remains unpublished.

Newton, of course, had used Latin for the first edition of his *Principia Mathematica* in 1687. The fact that it would be forty-two years before the first English translation was made, by the Gresham College lecturer Andrew Motte in 1729, again indicates the continuing ability of the scientific community to access works in Latin. Nevertheless, by the mid-eighteenth century English prose had come to dominate scientific writing in Britain, and Jurgen Leonhardt (2013) suggests that the years between 1760 and 1840 represent the closing of the Latin millennium inaugurated in Europe 'by Charlemagne between 800 and 814'.[13]

Rivalry between Latin and English usage in Wales is certainly evident in the 1650s. In July 1652, in mid-Wales, 'the news that filleth all mouths' was a 'Dispute ... at New-Chappel in Montgomery' between Dr George Griffith (1601–66), later appointed Bishop of St Asaph, and the 'stout-hearted and iron-lunged Puritan' Divine, Vavasor Powell (1617–70). Their very public dispute centred on matters of church doctrine, but these need not detain us. Of more interest are the opening salvos of the dispute and the attitudes expressed towards language and the social groupings present at the time.

Dr Griffith, who is our not wholly unbiased source, wanted to undertake the discussion in Latin, but he realised that this would cause a problem for some of those present because:

> the company met together, consisted of three sorts of men; many professed Scholars that expected satisfaction in the way of Learning: Divers Gentlemen of rank and quality, who though not professed Artists, yet no strangers to the Sciences and wayes of Disputation; and their expectation also he conceived might be the same: The third sort unlearned women and illiterate men, that knew not how to judge of such matters.[14]

Consequently, the unlearned and illiterate were asked for their patience while the dispute continued in Latin, in order to be '*more Academico*'. What transpired would be declared to them later 'in the way and language they were capable of'. To which iron-lunged 'Powell and his company immediately began to cry out, *English*', while 'the better sort, if not the greater number, *Latin*'.[15] The importance attached to Latin is no surprise in a gathering of scholars and gentry, given their education in grammar schools and universities where classical language learning prevailed. Nor is Powell's demand for 'English' a surprise; although a Welshman he preached and wrote his religious texts in English, and Montgomery lies close to the border with England where bilingual (Welsh/English) and monoglot English speakers were most prevalent.

The missing language of course is Welsh, which we can only assume most of those present would have known regardless, perhaps, in this period at least, of their social position.

If the Welsh language is represented at all in the Latin and English publications already noted in this chapter, it makes but brief appearances. For example, some of the text illustrations in Bassett Jones's mammoth *Lithochymicus* bear Welsh phrases, poems and mottoes. Not least on the hand-drawn title page depicting Jones's coat of arms and its motto *Duw ar fy Rhan* ('God on my Side'). It also includes reference to his fellow countryman '*Merlin*' who, as Robert Schuler points out, has been central to the 'British myth' since Tudor times.[16] Other scientists of Wales used the Welsh language in their more private correspondence. In north Wales we have the letters written by the four Morris brothers of Anglesey – Lewis (1701–65), Richard (1703–79), William (1705–63) and John (1706–40) – between 1728 and 1786. Filling four substantial volumes today, they contain a wealth of general and scientific information written in Welsh and English, with both languages often appearing in the same letter, sometimes the same paragraph, and occasionally the same sentence.[17] In south Wales we have three equally substantial volumes of letters in Welsh and English sent or received by Iolo Morganwg (Edward Williams) between 1770 and 1826.[18] Others who do not appear to have made use of Welsh in their public or private endeavours might still possess the language. As with the eighteenth-century father of modern actuarial science, William Morgan; he happily entertained his family by turning 'a Welsh song into elegant English verse on the spur of the moment'.[19]

To find works of a scientific nature written wholly in Welsh is less easy. In the medical sphere, for example, Alun Withey notes in his seminal *Physick and the family* (2011) that 'in Wales, and in Welsh, the number of specifically medical books printed in the period 1600–1733 is somewhat easier to gauge precisely than for England; there were none'. The first full medical text in the language being *Llyfr Meddyginiaeth, ir anafys ar chlwfus* ('a book of medicine for the wounded and sick') published by William Bevan of Llanfaes in Anglesey in 1733. 'Priced cheaply' and 'distributed from the author's own home' the book appears to have been based on a now lost earlier work by Thomas Ab Robert Shiffery of Llanberis, in Caernarfonshire.[20]

Almanacs

From 1680 onwards almanacs were the most commonly available Welsh-language publications to impart a degree of scientific knowledge. They did so largely through astronomical data relating to tides, phases of the moon and planets, and often with associated links to astrology; the last being reflected in the titles given to the almanacs themselves. For example, *Newyddion oddi wrth y Sêr* ('News from the Stars') published in 1680 by 'the sweating astrologer' Thomas Jones (1648?–1713), who described himself as 'a lover of Learning, and a student in Astrology, and Autodidactus';[21] or that of John Harris in the late eighteenth century bearing the Latin title *Vox Stellarum & Planetarum* ('Voice of the Stars and Planets'), even though its actual content was in Welsh (see Figure 7).

The first published discussion of arithmetic in Welsh may have been a now lost work of *c.*1716 by another almanac publisher, John Roderick (or Sion Rhydderch, 1673–1735). Further contributions came with *Cyfarwyddiadau i Fesurwyr* (*c.*1700) by Thomas Durston and *Cyfrifydd Parod* by John Evans, both being forms of ready reckoner. In 1768 *Arithmetic: mewn Trefn Hawdd ac Eglur* ('Arithmetic: in easy and clear order') by John Roberts (1731–1806) appeared. *Annerch i Ieuengctyd Cymru* ('Address to the Youth of Wales', 1795) by one-time weaver, sailor, schoolmaster and customs officer, John Thomas (1757–1835) of Denbighshire, covered arithmetic, algebra and logarithms, as well as natural philosophy and astronomy. Both Roberts and Thomas were involved in almanac production. Broader compendia discussing modern scientific ideas, though in very general terms and always with a religious subtext, appeared as *Hanes y Byd a'r Amseroedd* ('History of the World and Times', 1718) by Simon Thomas, and *Golwg ar y Byd* ('A View of the World', 1725) by the Revd Dafydd Lewys; the latter volume beginning with the essence and certainty of God before covering everything from anatomy to astronomy and what we would today call geology, biology, botany and zoology.

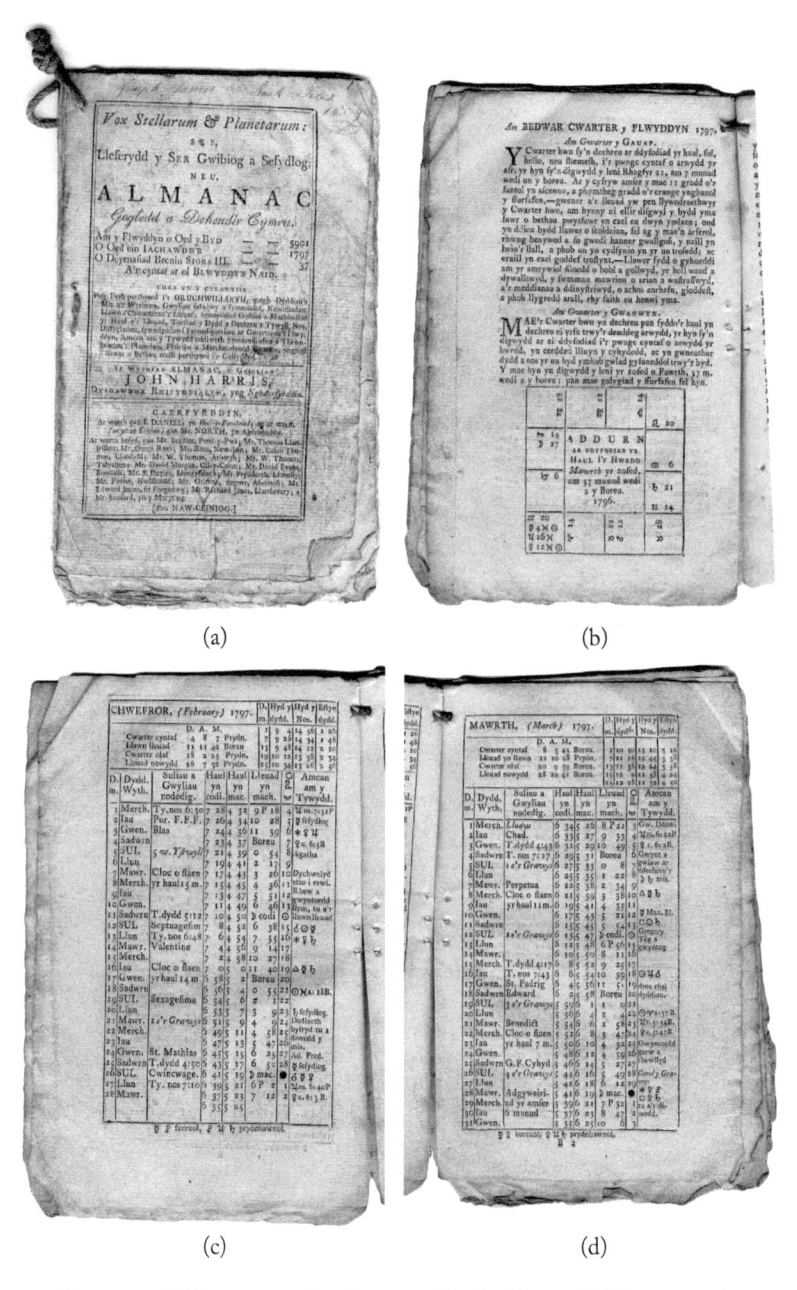

FIGURE 7 Title page to the almanac *Vox Stellarum & Planetarum* by
John Harris (Caerfyrddin, 1797) (a) together with an astrological
horoscope for the spring quarter of 1797 (b), and pages of astronomical
and weather data for February and March 1797 (c and d).

In the early nineteenth century yet another almanac maker and printer, Robert Roberts (1777–1836), published his *Daearyddiaeth* (Geography, 1816). As a 500-page compendium describing the countries of the world, together with an initial and detailed section on astronomy, the book echoed similar content published in *Speculum Terrarum & Coelorum: neu Ddrych Y Ddaear A'r Ffurfaven* ('Glass, World and the Firmament: or the Mirror of the Earth and Heavens', 1804) by Mathew Williams (1732–1819). As another land surveyor and almanacker,[22] Williams, in his own fifty-six-page discussion of astronomy, includes:

an outline of the theories of Ptolemy and Tycho Brahe;

URANOLOGY, sef Eglurhad o'r Copernican a'r Newtonian Hypothesis, neu System yr Haul, (Uranology, an explanation of the Copernican and the Newtonian hypothesis, or the Solar System);

HELIOGRAPHY, neu, Fynegiad am yr Haul (Heliography, or, an expression about the sun; in which he references Newton's *Opticks*);

Planetography;

Cometography;

Selenography, neu Hanes philosophyddaidd am y Lleuad (Selenography, or a philosophical History of the Moon);

Aerology, neu Natur ac Effeithiau'r AER (Aerology, or the nature and effects of the air).

Of note too are bilingual publications such as *Pharmacopoeia, Or Medical Admonitions, English and Welsh*, by Nathaniel Williams (1793), and those incorporating just a few pages of English text within an otherwise entirely Welsh-language publication. An example of the latter is the almanac published in 1797 by John Harris. It contains a few pages at the end listing Welsh and English books available to purchase at the shop of John Daniel, Bookseller and Stationer in Market Street, Carmarthen. It also has a single page of English text inside the front cover informing 'the Public' of Harris's recent move from Kidwelly to

Carmarthen where, aside from publishing his almanacs, he taught various branches of mathematics:

Arithmetic, vulgar and decimal.

Mensuration [the mathematics of determining lengths, areas and volumes] of Superficies [surfaces] and solids.

Gauging by pen and sliding rule.

Book-keeping according to the Italian method [double entry book-keeping].

Practical and theoretical surveying.

Trigonometry, plain and spheric.

Navigation logorithmetically by scale and compasses, geometrically and by inspection, together with the newest methods of determining the latitude and longitude at sea.*

The elements of astronomy, geometry, and algebra.

Also, geography, the use of the globes, maps, charts, quadrants, and other mathematical instruments, both in theory and practice, on reasonable terms.[23]

[*Presumably the successful time-keeper method championed by John Harrison in the 1760's and/or Nevil Maskelyne's Lunar tables, which continued to be used until the early nineteenth century.]

The advert further illustrates the extent to which a 'scientific' education in late-eighteenth-century Wales centred on mathematical learning through arithmetic and geometry.[24] The advert also suggests that Harris believed a good proportion of the readers of his Welsh-language publication could understand and read English.

In a slightly different vein there is, at the very start of the 500 pages of Robert Roberts's *Daearyddiaeth* ('Geography'), a single page bearing two columns of words. One in Welsh and one in English with the Welsh words listed alphabetically to the left and their English meaning given to the right. They suggest Roberts too believed readers of his otherwise Welsh-language publication knew English but would require help interpreting less familiar Welsh terms. Some of the Welsh words on the list were constructs by Roberts himself; 'Tân-dyniad' for electricity, and 'Moel danllyd' for volcano, being examples. Their inclusion in the volume represents their first iteration and in creating them Roberts was, like scientists everywhere in the long eighteenth century, involving himself in the development of a fourth language – the technical language of science itself.

Scope of Language

One of the fullest mid-seventeenth-century engagements with language in relation to the new science came in *Herm'aelogium; or an Essay at the Rationality of the Art of Speaking: as a Supplement to Lillie's Grammar* (1659)[25] by the Glamorgan alchemist, Bassett Jones. Lily's grammar formed the standard Latin grammar used in education in England and Wales since Tudor times. Yet, as Robert M. Schuler (2013) notes in his introduction to Jones's alchemical poem *Lithochymicus*, 'For the most part, Jones's linguistic work, if noticed at all by historians of grammar, has been dismissed as being either "unoriginal or absurdly fanciful"'.[26] But as Schuler generously opines, the linguistic work reveals:

> an energetic mind at work, synthesizing material from classical philology, Neoplatonic philosophy, English literature, science, magic and alchemy. Jones cites numerous Greek and Latin authors for grammatical and philosophical analysis; he quotes three of Chaucer's poems, Raleigh's *History of the World*, poems by Jonson, Donne and Davenant; Francis Bacon is on every other page; the opinions of Sir Thomas Browne and Sir Kenelm Digby [a friend of Jones and a founding member of the Royal Society] are discussed; and Hermes, Plato, Proclus, Cicero of the *Somnium Scipionis* and *De Natura Deorum* rub shoulders with Campanella, the Paracelsian Gerhard Dorn, and the professor of chemistry William Davisson.[27]

At one point in the text of *Herm'aelogium*, which is an admittedly difficult read, Jones describes himself in 1658 as 'scribbling this at my paternal hermitage in Glamorganshire'. This Welsh context is confirmed by an occasional use of the Welsh language and, on one occasion, the illustration of a grammatical point by reference to the aquatic fauna of the River Dulais in Neath. As a supplement to Lily, though, *Herm'aelogium* resulted from a conversation between Jones and some 'select companions ... relating to the Grammatical part of [Francis Bacon's] the *Advancement of Learning*'. For Bacon, a key to attaining knowledge that could be used for humanity's improvement was the avoidance of prejudices in the mind that might bias observations, and so distort the methods of enquiry that he advocated. One such prejudice was the misuse or misinterpretation of words – what Bacon in *Novum Organum* calls the idols of the marketplace – whereby 'the great and solemn disputes of learned men often terminate in controversies about words and names'.[28] According to Schuler (2013), Bassett Jones wanted to explain what Bacon called 'the lost rationality of Latin syntax':

> that is, to present a new philosophy of language based on the universal principle that syntax in any language is rooted in the rationality of its speakers, not in the forms of its words. He is thus eschewing the tradition of 'literary grammar' and aiming toward the kind of 'philosophical grammar' called for by Bacon in the *Advancement of Learning*, whereby 'the analogie between words and things, or reason' is to be established.[29]

In this regard *Herm'aelogium* is part of the wider conversation concerning language seen in other grammatical works of the time, such as Francis Lodwick's *The Ground-work or Foundation Laid ... for the framing of a New Perfect Language* (1652), John Wallis's *Grammatica Linguae Anglicanae* (1653), and John Wilkins's *Essay Towards a Real Character and a Philosophical Language* (1668), a work sponsored by the Royal Society.

Of course, the need for new words to *accurately* describe developments in the new science applied equally to Welsh, English and Latin. For those scientists of Wales who wrote in English it was relatively easy to adopt newly created words in that language, and in Chapter 4 we saw Thomas Vaughan credited with a very early use of the words 'telescope' (1650) and 'mineralize' (1655). As Hawke (2018) has noted, the root of such 'technical terminology' in English is based 'extensively on Latin and Greek'. According to the *OED*, for example, the word 'astronomer' was probably formed in English by derivation or borrowing from Latin. Numerous Welsh words also derive from Latin originals, but for technical words a simpler and more easily understood methodology was often adopted. Thus, the Welsh equivalent of 'astronomer' – *seryddwr* (first recorded in 1816) – is formed by taking the Welsh word for 'star' – *sêr* – and adding 'the suffix *-ydd* denoting an agent'. Similarly, *trydan* ('electricity'), which was probably developed *c*.1803 to translate 'electric fluid' derives from 'the intensive prefix *try-*' and *tân* ('fire').[30] In a similar vein, when Mathew Williams noted himself as the author of *Speculum Terrarum & Coelorum* in 1804 he did so as a 'Daear-Fesurwr' (literally an 'Earth-measurer'; a 'surveyor'). Such a method encouraged considerable inventiveness among Welsh-language writers; even though some constructs were destined to fall by the wayside, as was the case with some of the Robert Roberts examples above. A 1755 letter written by Lewis Morris in London, to his brother William in Wales, captures the excitement this creative process could produce *and* the importance of new words being made available in Welsh:

> You must make your cregyn ['shells'] Welsh names if they have none, there is no if in the case. You must give them names in Welsh. I'll send you a catalogue of ye English names of some sales here, which are all foolish whims, and it is an easy matter to invent new names, and I warrant you they will be as well received as Latin or Greek names. Tell them they are old Celtic names, that is enough. They'll sound as well as German or Indian names, and better.[31]

As Geraint H. Jenkins has written of Lewis Morris: 'His manuscripts are littered with Welsh words which he coined for a variety of shells, seeds, plants, ores, tools and machines, and some of his most memorable letters were devoted to the correct use and spelling of words related to science, mathematics and technology'.[32] In essence, the concerns of Lewis Morris echo those of Bassett Jones and Francis Bacon. The linguistic Babel was within a language, not in the multiplicity of languages – as Morris explained in a letter of 1760 to schoolmaster and poet Edward Richard (1714–77):

> You see I interlard my letters with Welsh, while men of learning adorn theirs with Greek and Latin quotations. But this is the highest pitch of my learning, except I through [i.e., throw] in a dish of Geometry and Algebra, which perhaps would be fitter for me than to meddle with any language. The art of writing and speaking any language seems to me a bottomless pit. I see no end of it. Custom has so high a hand over it, that it is extream uncertain, and the whims of mankind in setting such arbitrary marks on our ideas hath made a sad jumble of things, and I think the confusion of Babel is acted over and over every day.[33]

Other Welsh words minted in the long eighteenth century include: *awyrgylch* ('atmosphere', 1770), *dyfeisgar* ('inventive'; 1771) and *daearyddiaeth* ('geography'; 1793), as well as others adapted from English. The Welsh for 'botany', a subject under much investigation in Wales in this period, appears as *botani* (later becoming *botaneg*) in a 1741 letter between two of Lewis Morris's brothers – the Wales-based botanist William (1705–63) and Richard (1703–79) the London-based founder, with Lewis, of the Cymmrodorion Society. In 1813 cleric Hugh Davies would produce his *Welsh Botanology; A Systematic Catalogue of the Native Plants of Anglesey in Latin, English, and Welsh*. In writing to Sir James Edward Smith, the president and founder of the Linnean Society, to which Davies had been elected a fellow in 1790, he wonders whether Smith had seen *Welsh Botanology* and if, or when, Smith does:

> I Hope you at least bestow so much time on it, as to read the Preface, and then tell me if it be inconsistent with your design to add the British, or Welsh, name to each subject in your work

---

[Smith intended a work on British Flora – *Flora Brittanica*], as I can without vanity say, that they are now more accurately ascertained than ever before.[34]

New words continued to appear in Latin too, and were adopted by those scientists of Wales who still wrote in the language. For example, when discussing the problems of refraction caused by 'the sphere of exhalations, spirits and vapours that surrounds the earth globe from all sides and is expired by it', Thomas Powell, in his *Elementa Opticæ* of 1651, uses the Latin word 'Atmosphærum'. The word had been created in 1608 by Willebrord Snellius for his Latin translation of cosmological works by the eminent Dutch-Flemish mathematician and physicist Simon Stevin (1548–1620), whose work Powell references.[35] Interestingly, the *OED* notes the first use in English of the word 'atmosphere' as being in 1638, but in relation to the atmosphere surrounding 'any of the heavenly bodies'. The earliest *OED* entry for the 'mass of aeriform fluid surrounding the earth', the sense in which Powell uses the word in 1651, dates to 1677.

The creation of new words or the evolution of existing ones is just one side of the technical language of science. The other is the language of symbols, particularly in relation to mathematics, including geometry, both of which developed rapidly in this period. Robert Recorde of Tenby, in his sixteenth-century publications, had already made a significant contribution to this language through the introduction of the equals sign to mathematics, and, probably, the first use in an English language publication of the symbols for addition and subtraction. In the long eighteenth century William Jones Sr (see also Chapters 4 and 5), in his *Synopsis Palmariorum Matheseos, or, A New Introduction to the Mathematics* (1707) contributed further with the introduction of the symbol π to denote the ratio of a circle's circumference to its diameter. He also involved himself in editing and publishing Isaac Newton's work in *Analysis per Quantitatum series, fluxiones ac differentias*, and so may well be responsible for the wider use of the dot symbolisation developed by Newton for the calculus.

These, then, were the languages of science in Wales in the long eighteenth century. However, before leaving the subject we should note one other area where the Welsh language could play a part, particularly when combined with knowledge of English and other modern and classical languages. This was in the attempt to obtain employment within a scientific milieu. For example, there is the case of Methusalem Bowen (dates unknown) who contended with a number of other candidates for the position of 'Clerk to the [Royal] Society' in 1723. The post became vacant after its previous incumbent, Alban Thomas (1686–1771), made a rapid exit from it in 1722. Originally from Carmarthenshire, Thomas had been librarian of the Ashmolean Museum in Oxford and, since 1713, an assistant secretary to the Royal Society. In 1718 he published *A List of the Fellows of the Royal Society* and, in 1719, took a medical degree at Aberdeen University. The latter was just four years after James Francis Edward Stuart, the Old Pretender and son of James II, had returned to Scotland at the head of a Jacobite army aimed at reclaiming for the exiled Stuarts the English throne – it being currently occupied by the Hanoverian, George I. Possibly as a consequence of his time in Scotland, a suspicion of Jacobite sympathies seems to have dogged Thomas and this may have precipitated his hasty departure from the Royal Society and London.

Several eminent names challenged Methusalem Bowen for Thomas's position. They included Francis Hauksbee Jr (1688–1763), whose uncle and namesake had been a very prominent experimenter at the Royal Society and a close associate of Newton. The nephew now came 'recommended by diverse members of the Society'. Another candidate was Philip Henry Zollman (d.1748). A correspondent of Leibnitz, his language skills included Latin, French, Dutch, Italian, Spanish and his native German, as well as English. At a meeting of the Society's council on 4 April 1723, chaired by Isaac Newton as president, with Hans Sloane, Edmond Halley, Martin Foulkes and James Jurin among the council members present, Bowen presented himself along with the other candidates for consideration. He came recommended by 'Judge Vaughan [possibly John Vaughan (1633–1722) of Derllys in Carmarthenshire], Sir John Phillips [probably John *Philipps* (1666?–1737) of Picton, Pembrokeshire, a friend to Newton and Hans Sloane],[36] Mr Chamberlain and My Lady Williams'. With his £500 security duly noted in the council minutes, Bowen, like the other candidates, faced examination in handwriting, 'skill in Natural knowledge or History', 'conversation in Library's and Books' and 'knowledge of Languages'. Betraying no sign of a 'cultural cringe' he declared his language skills to be in 'Latin, Welsh and English'. Also called into the meeting, though seemingly not as one of the formal applicants, was Cardiganshire-born Moses Williams (1685–1742, FRS 1719). Williams had earlier been an assistant to Edward Lhwyd at the Ashmolean Museum in Oxford and he made major contributions to Welsh manuscript preservation and antiquarianism. Having already taken 'care of the Society's House', and 'been conversant in the Society's repository [its museum and archive] in Mr [Alban] Thomas's time', he now offered himself up as clerk and Library Keeper to the Society and expressed his willingness to serve 'in all the employments which Mr [Alban] Thomas held'. On examination he professed an ability to provide 'any security the society should demand' and his knowledge of Greek, Latin, French and Welsh.[37]

Methusalem Bowen's application proved unsuccessful, and he does not seem to have been heard from again. Zollman was disqualified from the position on account of his not being a naturalised Briton. Nevertheless, his language skills proved too good to lose and he subsequently became an assistant to the secretary managing the Society's foreign correspondence. Moses Williams joined a committee charged with inspecting the state of the Society's library and repository, but their actual care was assigned to Francis Hauksbee Jr. Unable or unwilling to resume his life in London, Alban Thomas returned to Wales to live and practise medicine in Cardiganshire, to which, we might hope, he brought some of the scientific knowledge that he surely gleaned from the Royal Society, its meetings, and its members. Among the latter is Hans Sloane, with whom Thomas corresponded from west Wales.[38] This brings us to the theme of the next chapter: by what processes did scientific knowledge arrive in and spread through Wales in our period?

Notes

1. Meic Stephens (compiler and ed.), *A Most Peculiar People* (Cardiff, 1992), p. 21.
2. See maps for *c.*1750 in W. T. R. Pryce, 'The Diffusion of the "Welch" Circulating Schools in Eighteenth-Century Wales', *THSC*, 25/4 (2011), 500–1.
3. J. Dybikowski, *On Burning Ground: An Examination of the Ideas, Projects and Life of David Williams* (Oxford, 1993), p. 122 n. 44.

4. *CR*-I (1796), p. 280.

5. See Dybikowski, *On Burning Ground*, pp. 121–2.

6. See Michael J. Franklin, *Orientalist Jones* (Oxford, 2011), pp. 96–7 and n. 5.

7. Lord Teignmouth, *Memoirs of the Life and Writings and Correspondence of Sir William Jones* (London, 1804), p. 376.

8. For Wesley, see Geraint H. Jenkins, Richard Suggett and Eryn M. White, 'The Welsh Language in Early Modern Wales', in Geraint H. Jenkins (ed.), *The Welsh Language before the Industrial Revolution* (Cardiff, 1997), p. 89.

9. See F. V. Emery, *Edward Lhwyd FRS 1660–1709* (Cardiff, 1971), p. 41. For Russian occurrences, see N. N. Ivanova, 'An Unpublished Copy of Edward Lhuyd's Catalogue of British Fossils', *NLWJ*, 26/2 (1989), 123–8 (at 127).

10. Dissenters could not take degrees at either university. There were also social and class considerations, as when six practising Methodist students were expelled from St Edmund Hall, Oxford, in 1768. Three of them, it was declared in a gathering of the student and academic body, were students who 'should not have been admitted, because they were "bred to trades"'. Among them was Thomas Jones, a barber, who may well be the Jones resident in north Wales with whom Lady Huntingdon corresponded. See Howard D. Weinbrot, *Literature, Religion, and the Evolution of Culture 1660–1778* (Baltimore, 2013), pp. 185–6; John R. Tyson and Boyd S. Schlenther (eds), *In the Midst of Early Methodism: Lady Huntingdon and her Correspondence* (Plymouth, 2006), pp. 135–6, Letter 178 to T. Jones, 3 August 1772, and p. 154, n. 23.

11. *CHS*-4, p. 296.

12. For details see *ODNB* entries for Jones, and Collins; Paul Quarrie, 'The Scientific Library of the Earls of Macclesfield', *Notes and Records: The Royal Society Journal of the History of Science*, 20 (2006), 5–24 (at 11–15); A. Rupert Hall, *Isaac Newton Adventurer in Thought* (Oxford, 1992), pp. 83, 357–8.

13. Jurgen Leonhardt, with Kenneth Kronenberg (trans.), *Latin: Story of a World Language* (Cambridge MA, 2013), p. 245.

14. George Griffith, *A Welsh Narrative, corrected, and taught to speak true English, and some Latine ... Containing a narration of the disputation between Dr Griffith and Mr Vavasor Powell, neer New Chappell in Montgomery-shire, July 23, 1652* (London, 1652), p. 2.

15. Griffith, *A Welsh Narrative.*

16. Robert M. Schuler, *Alchemical Poetry 1575–1700: From Previously Unpublished Manuscripts* (1995; Abingdon, 2013), p. 221.

17. See *ML*-I and II; AML-I and II.

18. Geraint H. Jenkins, Ffion Mair Jones and David Ceri Jones (eds), *The Correspondence of Iolo Morganwg*, 3 vols (Cardiff, 2007).

19. Caroline E. Williams, *A Welsh Family from the Beginning of the 18th Century* (London, 1893), p. 137.

20. Alun Withey, *Physick and the Family: Health, Medicine and Care in Wales, 1600–1750* (Manchester, 2011), p. 64.

21. For an excellent biographical study of Jones see Geraint H. Jenkins, '"The Sweating Astrologer": Thomas Jones the Almanacer', in R. R. Davies, Ralph A. Griffiths,

Ieuan Gwynedd Jones and Kenneth O. Morgan (eds), *Welsh Society and Nationhood, Historical Essays Presented to Glanmor Williams* (Cardiff, 1984), pp. 161–77; also *DWB* entry for Jones. For the quote, see Jenkins, 'Jones the Almanacer', p. 163, citing T. Jones, *An Astrological Speculation of the late Prodigy* (1681), title page.

22. He published *Britannus Merlinus Liberatus* from 1777 to *c*.1814.

23. John Harris, *Vox Stellarum & Planetarum* (Caerfyrddin, 1797), p. 2.

24. During his tour of Wales in 1819 Michael Faraday would be 'much amused by a School-master's card' written in a similar vein, though with its added proviso that 'The strictest attention will be paid to the Morals and improvement of the pupils' by the master concerned. See Dafydd Tomos, *Michael Faraday in Wales including Faraday's Journal of his Tour through Wales in 1819* (Denbigh, n.d.), pp. 52–3.

25. According to Schuler, *Alchemical Poetry*, p. 210 n. 20, '"Herm'aelogium" seems to allude to Hermes Trismegistus (whose achievements included the invention of writing as well as alchemy and other arts) and *logos*, as "unifying principle in the world", "word", "wisdom", "word of God" etc'.

26. Schuler, *Alchemical Poetry*, p. 209. See also Joseph L. Subbiondo, 'Neo-Aristotelian Grammar in Seventeenth Century England: Bassett Jones' Theory of Rational Syntax', *Historiographia Linguistica*, 17 (1990), 87–98.

27. Schuler, *Alchemical Poetry*, p. 209.

28. A. Wolf, *A History of Science Technology and Philosophy in the 16th and 17th Centuries*, 2 vols (1935; London, 1962), II, p. 635.

29. Schuler, *Alchemical Poetry*, p. 209.

30. See Andrew Hawke, 'Coping with an Expanding Vocabulary: The Lexicographical Contribution to Welsh', *International Journal of Lexicography*, 31/2 (2018), 229–48 (at 240).

31. *ML*-I, p. 347.

32. Geraint H. Jenkins 'The Cultural Uses of the Welsh Language 1660–1800', in Jenkins (ed.), *Welsh Language before the Industrial Revolution*, p. 387.

33. AML-II, p. 482.

34. Linnean Society Collections, GB-110/JES/COR/4/34: James Edward Smith Correspondence, Hugh Davies to Smith, 26 August 1820.

35. Simon Stevin, 'Van de eertclootsche damphoogde', Book 4 in *Vant Eertclootschrift* (1605), the term 'atmosphere' was introduced in the Latin translation *Hypomnemata mathematica* by Willebrord Snellius (1608). I am grateful to Ineke Loots for her translations and this information.

36. See W. Moses Williams, *The Friends of Griffith Jones* (London, 1939), p. 21; 'Address to the Electors of a Borough', *Notes and Queries*, I (3rd series), p. 244. Both Vaughan and Phillips (Philipps) were friends of Griffith Jones and supported the Circulating School movement.

37. RSC CMO/2/309 (alternate CMO/2/265) *Minutes of Council Meeting, 4 April 1723*.

38. See *https://dralun.wordpress.com/2013/02/28/a-welsh-doctor-sir-hans-sloane-and-the-disappearing-catheter/* (accessed 22 May 2024).

8

SPREADING THE WORD

Books are not written, produced, or read in a vacuum; they
reflect underlying changes in society which, in turn, shape
and govern the needs and aspirations of people. It may be
true that normal, day-to-day social functions remained
predominantly oral, but literate middling sorts now expected
people to do rather more than listen and hear.[1]

Geraint H. Jenkins, *The Eighteenth Century*
in *A Nation and Its Books* (1998)

The scientific developments of our period stem from multiple points
of origin – the France and Netherlands of Descartes, the England
of Isaac Newton, and the Sweden of Linnaeus, to give a few famous
examples. Given these multiple sources, any attempt to track the spread
of scientific ideas as they move into and through Wales is fraught with
difficulty. Consequently, we concentrate here on the processes by which
scientific knowledge was transmitted to, and within, Wales in the long
eighteenth century; namely, by oral and literary processes.

Oral Processes

In Chapter 6 we touched on two oral methods of imparting scientific
knowledge, with our first – the teaching of subjects bearing on the sciences – illustrated by reference to surviving school curricula, and the
attitude of scientists from Wales to the education they received. Far
more difficult to quantify and illustrate is the extent to which teachers
themselves were influenced by the new developments in science. For

example, between 1721 and 1763 a French-speaking Channel Islander, Daniel Durel (1694–1766), was master at Cowbridge Grammar School. A little later, in 1768, the first teacher appointed to the newly established Dissenter college at Trefeca, in Breconshire, was Swiss born Jean Guillaume de la Fléchère (John William Fletcher, c.1729–85). That Durel and Fletcher had some interest in natural philosophy is suggested, in the case of Durel, by his purchase of books such as Varenius's *Geography*, with notes by Sir Isaac Newton, and volumes on physics, mathematics and astronomy. Fletcher's interest is revealed in a letter that he wrote on taking up his appointment at Trefeca. It concerned the curriculum that he wanted to see introduced, which would include 'Grammar, Logic, Rhetoric, and Ecclesiastical History, and a little Natural Philosophy and Geography, with a great deal of practical Divinity'.[2] Given the continental influences on both men, we might wonder what effect the ideas of Descartes and Leibnitz, let alone Newton, had on them and their teaching of a student body which, in the case of Durel, in Cowbridge, came not only from Wales but also England, France, the Channel Islands and Jamaica.[3]

Given the religiosity of Wales in the long eighteenth century, the second method of oral transmission – the pulpit oratory of itinerant evangelical preachers and their use of scientific imagery – had wide societal influence. Sermons were long and often contained ample references to matters of the day, some even including scientific imagery, as we saw in relation to the London-based preaching of Richard Price (Chapter 5).

To these we can now add the secular science preached by itinerant lecturers armed not just with oral imagery but with miraculous explanatory models. Given his name, we might expect 'Mr William Griffis' to have occasionally strayed beyond his 1750s stomping ground in the Midlands, Bath, Bristol and Salisbury to preach a little science in Wales. Little or nothing is known of his origins, or his life in general, other than his willingness, for a guinea, to lecture on 'Newtonian natural philosophy, mechanics, hydrostatics and hydraulics, pneumatics' and 'their application to improving trade in the arts and professions'. Lectures he illustrated by 'a very large and noble set of instruments' numbering some '40 models'; among them were a planetarium, a cometarium, a

Ptolemaic sphere and an improved orrery.[4] For evidence of a similar evangelism more conclusively within the boundaries of Wales we have Thomas Morgan's diary entry for 12 March 1743:

> At 11 o'clock went, with ye rest of my fellow students to attend Mr Domkey's Lecture, who read experimental Lectures in Natural Philosophy, in this town [Carmarthen] – the first experiment he made was to show ye pressure of ye air, by some Glass images/ gauges going up & down in a glass filled with Water, by ye pressing of a bladder yt was on ye orifice of ye Glass. 2. The way to find ye Specifick gravity of liquids & solids by a Hydrostatical Balance. 3. The divining tool & its uses. 4. The spouting fountain.[5]

Made while Thomas was a student in Carmarthen, the entry also notes lectures on air pumps over three or four days, and several experiments in gravitation and attraction and mechanical powers. In another example, Taliesin Williams (1787–1847) wrote in the first week of August 1813 telling his father, Iolo Morganwg (Edward Williams, 1747–1826), of attending 'lectures on light, vision and colours' in Neath. They were given by the Dissenting minister from London, Jeremiah Learnoult Garrett (1764–c.1806), who stayed in the town for some days.[6] Along with giving his lectures, Garrett exhibited:

> an optical museum, containing [ma]ny other very curious things: two solar microscopes, transparent [op]aque. The microscopes were fixed in a window. The thing to be [mag]nified was, I believe, inserted inside of the lens and, after [da]rkening the room, the object was reflected on a large sheet the opposite side, wonder-fully magnified.[7]

By this means, Taliesin informed his father, cheese mites appeared as large as lobsters, the process of crystallisation could be seen, and small quarter-inch pieces of various woods appeared 'magnified to about 8 ft'. The last allowing Taliesin to see 'the various modifications of the pores of different woods' and to draw some conclusions about them: 'Tis this, perhaps, that constitutes the difference in the nature and durability of timber, for the primary particles may be the same in all, only differently arranged in different woods.' An insight immediately followed by a lack of confidence: 'I am very probably talking nonsense, so do not show this letter to any one.'[8]

Some nine days later Taliesin's father, accompanied by his two daughters, went to see Garrett perform in Cowbridge. Though Peggy and Nancy 'were highly pleased' with the event, Iolo was less impressed: 'I found every thing infinitely inferior to what I had many times seen of the kind in London and elsewhere', he informed Taliesin. Nevertheless, he still felt Mr Garrett's lectures and exhibition may:

> produce some good effects in places remote from the habitations of real science and profound knowledge. They will excite attention and curiosity and study, and of course put some out of their present high roads of ignorance and lead them into the paths of true knowledge, of genuine philosophy.[9]

They certainly inspired Iolo's son to a greater interest in science. As Taliesin records: 'Previously to attending the lectures, I read the numbers of the scientific dialogues that treat on optics lent me by Mr Davis [a teacher in Neath], and which proved of great assistance to me.' Garrett's lectures and exhibition also encouraged him to 'give greater time & attention to the subjects than I otherwise should have done'.[10]

Itinerant lecturers were not the only visitors coming to Wales with scientific knowledge to be discussed and exchanged and enhanced by those they met in the course of their journeys. As early as 1655 Sir Owen Wynne (1592–1660), of Gwydir Castle in the Conwy Valley, wanted to know, in a list of queries directed at those working in the lead mines of Cardiganshire, 'if there be amongst the miners at the lead works any outlandish man, (as a Dutchman or High Dutchman,) that hath skill in mines and in how to find out and discover the lead mines in the ground where lead ore is found in several places thereof'.[11] Wynne, like other north Wales landowners, also consulted German experts in mining and minerals as they were the leading exponents in the field at this time, and had been for a considerable time.[12] Among the most frequent visitors to Wales in the long eighteenth century were those who sought to hire expert mountain guides and knowledgeable local plant collectors for their botanical expeditions.[13] They include such luminaries in the history of botany as John Ray, Daniel Solander, J. J. Dillenius, Samuel Brewer (who is said to have climbed Snowdon thirteen times) and Joseph Banks, to name but a few.[14] Welshmen toured the country too, and both Edward Lhwyd and Thomas Pennant entered into extensive preparations by sending out questionnaires to numerous local experts and contacts before beginning their respective journeys. Joseph Harris, the London-based scientifically involved brother of evangelical preacher Howell Harris, toured his home-country in 1746 and 1748.[15] Others came to gain geological insights, with rock man Arthur Aikin publishing a journal of his visit in 1797.[16] Visitors such as Henrik Kalmeter and Reinhold Rücker Angerstein from Sweden wanted to see industrial processes at work in the early iron and copper works[17] and others, such as Edmond Halley, came to undertake scientific experiments (see Chapter 9); all activities likely to have stirred curiosity and discussion among at least some of the general populace.

Another source of oral scientific knowledge were those students who chose to return to Wales following completion of their studies in the likes of Oxford, Cambridge, Edinburgh, Glasgow, Dublin and the continental universities (students at the last are discussed in Volume 2). Taking medical degrees was particularly popular and the acquired learning, when back home in Wales, would have informed their daily work as doctors and, quite possibly, answers to the queries of their patients.

Finally, we must not forget the everyday story of country-folk. When the gentry, the scholars and the 'unlearned' and the 'illiterate' gathered in Montgomery to hear Vavasor Powell and George Griffiths dispute in July 1652, their numbers were so great the event had to be held out of doors. Such was the pulling power of religion and the spoken word in mid-seventeenth-century Wales. But for those present that day, and for many in later times, practical science, as much as religion, formed the conversational backbone of their everyday lives. Discussion of the latest methods of plant, crop, and livestock breeding would have been of interest to the gentry, the tenant farmer and the farm labourer alike. Similarly, planting times and weather patterns linked to astrological predictions based on the astronomical data provided in the increasingly available (and cheap) almanacs would also be part of everyday gossip.

Sadly, its ephemeral nature makes oral discussion and its possible consequences almost impossible to quantify. It might be the consideration of a moment, or it might be held in the mind and memory of the hearer to be passed on from one person or generation to the next, but ultimately it only achieves true permanence when written down.

Print Processes

Each year between 1763 and 1767 Thomas Beynon, from the parish of Llansadwrn in Carmarthenshire, wrote into his commonplace book a list of the books that he had read. They total 409 volumes over the four years, with between sixty and 106 read each year. As in many such lists, there is a plethora of religious, philosophical, geographical, travel and history books in English, with a few volumes in Welsh. Of particular interest here are the four or five science-based volumes that Beynon read each year (see Table 1 for some examples).

TABLE I Selection of books read by Thomas Beynon

Among the books read by Thomas Beynon
Derham's *Astro-theology* and *Physico-theology*
Boyle's *Experiments on Air*
Fonning's Arithmetic
Watts on ye Globes
Keill's Euclid
Bradley's new improvements in planting
Whiston's *History of ye Mathematics*
Dr Woodward's *Natural History of ye Earth*
Clare's *Motion of Fluids*
Pointer on the Weather
Rowning's volumes on astronomy, optics, mechanicks and hydrostatics, and pneumatics

From internal evidence in the commonplace book, Thomas (1956) suggests Beynon 'was consciously following one of the many schemes for self-education that were such a feature of the mid-eighteenth century publishing world'. He also suggests Beynon, who was ordained as a deacon in 1768, may have purchased the books.[18] Certainly this is possible. Yet thirty years later, in 1797, the range of science-based volumes for sale at John Daniel's bookshop in Carmarthen, whose catalogue John Harris published in his almanac for 1797, remained limited. Of 177 English-language volumes listed, 107 are classed as 'Divinity'. Of the remaining seventy, in 'Law, History, Miscellanies, &c', only ten titles relate to the sciences, including:

> Lord Bacon's *Letters and Memoirs* 4s., Pennant's *Tour in Wales*, calf gilt, £1 10s., Holmes's Geography and Astronomy, *maps*, 4s., Hamilton Moore's Navigation 7s. and 8s., Guthrie's Geography 10s 6d., Taplin's Farriery, calf lettered, 16s., Moxon's Astronomy and Geography 3s., Buchan's Domestic Medicine 8s., Ferfuson's Mechanical Exercises 4s., and Select Essays from the Encyclopedy at 2s. 6d.[19]

Of course, Beynon could have purchased his books from London or other centres of production and had them shipped into Wales; just as the SPCK did with its religious publications. But another scenario is that Beynon simply borrowed many of the books he read. Such borrowing certainly took place. As early in our period as 1651, David Pennant wrote to the bibliophile and collector of 'chymick bookes' Owen Wynne at Gwydir, asking to borrow Jean Baptiste van Helmont's *A Ternary of Paradoxes. The magnetic cure of wounds, nativity of tartar in wine, image of God in man.*[20] In mid-January 1778 young Daniel Walters, son of the Vicar of Llandough in Glamorgan, called at 'Henry Llewellyn's for a little book I lent him'. Perhaps it was Shelvocke's *Voyage round the World* that Daniel spent the next day reading before concluding: 'very entertaining, but I believe he now and then advances Falsities'.[21]

Libraries were another source of borrowed books. Schools sometimes had their own libraries, including the Cowbridge Grammar attended by young Daniel Walters, and for which the books mentioned earlier as being bought by the school's master, Daniel Durel, were perhaps destined. At Carmarthen Grammar the original library had been destroyed in the Civil War. It was restocked by the Company of Merchant Taylor's of London and by donations from ex-students, though the nature of any science content is currently unknown.[22] It is a similar picture for the Dissenting academies. Despite the belief of an anonymous correspondent to *The Cambrian Magazine* in 1773 that scientific instruments presented to Carmarthen academy would form part of an existing 'most valuable, and voluminous library', we again have little or no knowledge of that library's scientific content.[23]

Of the other types of library established in Wales in the long eighteenth century,[24] the major diocesan libraries, such as the one operating in Cowbridge between 1711 and 1848, and the many parochial libraries established by the SPCK, were overwhelmingly religious in content.[25] Of the circulating and subscription libraries and various book clubs established, their nature varies. Newspapers, journals and periodicals, which might contain news of the latest scientific developments, were certainly found in some of them. At Mrs and Miss Oakey's circulating library in Wind Street, Swansea, newspapers and periodicals could be accessed in the reading room for a guinea a year, 10s. 6d per half year, 6s. per quarter, or 2s. 6d per month. Of the content of such libraries in general, however, one author of an 1803 guidebook concluded that, 'Though the scholar would be disappointed, those who seek only for amusement would be sufficiently gratified';[26] a conclusion also reached slightly earlier by Iolo Morganwg when attempting to establish his own 'little circulating library at Cowbridge'. Writing to the 'Clergy of the town and neighbourhood' near the end of the eighteenth century, he pragmatically notes how his library would be made useful:

> by storing it with such a select assortment of books as might be deemed something superior to the trash too often found in circulating libraries, which however, such is the depravity of public taste every where, cannot well be dispensed with.[27]

Perhaps to help remedy this situation, he proposed, as part of his plan for writing 'a superb history of Glamorgan', the establishment of 'a respectable county library'. It would not only include his own donated manuscripts and books, but also 'ancient and modern history, civil, ecclesiastical and natural, containing the best authors in each department – history, botany, mineralogy, zoology, to which also should be added agriculture and biography'.[28]

As Kaufman (1963) notes, a major difficulty in ascertaining the content of the various book clubs and circulating libraries in Wales is the scarcity of their published catalogues.[29] However, two do survive for the Pembroke Society. Established as a subscription or book club by 1741, the catalogues are dated 1776 and 1791 and were published in Hereford and Pembroke respectively. Among science volumes the club members could enjoy were:

> three volumes of Bacon's works, Boerhaave's Chemistry, three volumes of Boyle's works by Shaw, eleven abridged volumes of the *Philosophical Transactions of the Royal Society*, Sprat's history of that Society, Pemberton's view of Newton's philosophy, volumes on Husbandry by Randal and DuHamel, two volumes of the Medical Museum, Mills *On the Disorders of Cattle*, and five volumes of Pennant's *Zoology*, as well as his *Synopsis of Quadrupeds* and the *Tour through Scotland*.

In a paper on the Pembroke Society, Walters (1967) notes how such content 'runs counter to the general tendencies of book clubs' anywhere in this period.[30]

Other resources available were the private libraries established by the gentry, not all of whom were willing to lend. In 1758 John Owen (d.1759) wrote to Evan Evans (Ieuan Fardd, 1731–88) concerning the library of Mr Vaughan at Hengwrt: 'mae'n debygol, mal yr ydych yn dywedyd, mai y Llygod fydd fwyaf eu croesaw ymysg ei lyfrau' ('it appears, as you say, that the mice will be the most welcome amongst his books').[31] For those who did lend there were other dangers. Owen Wynne of Gwydir needed recourse to law to obtain the return of some thirty books that he had lent to his wife's deceased uncle, John Williams (1582–1650), Archbishop of York.[32] In the main, though, most gentry seem to have been willing to lend beyond close family and friends, and an obvious borrowing source for young Thomas Beynon would have been the library of the Vaughan family, at Golden Grove in Carmarthenshire.

Elsewhere in Wales there were substantial collections of scientific material in Gwydir Castle. As a 'man of books' with an interest in mining, Sir Owen Wynne had many volumes and manuscripts on alchemy, chemistry and metallurgical operations, though it seems his collection did not long survive his death in 1662.[33] There was also the magnificent late eighteenth-century library of Thomas Johnes of Hafod (1748–1816), in Cardiganshire. Sadly, a fire in March 1807 appears to have destroyed a substantial part of it, including many manuscripts by Edward Lhwyd. Even though Johnes had begun rebuilding the library by November,[34] we have no surviving catalogue of its complete contents, science or otherwise. Yet we can assume some science presence among the mass of antiquarian manuscripts and books, given the agricultural and sylvan interests of Thomas Johnes, and the botanical studies undertaken by his daughter Mariamne (1784–1811).[35] Both Thomas and Mariamne were correspondents of Sir James Edward Smith, President and founder of the Linnean Society, who visited Hafod on a number of occasions.[36] Catalogues for two other important book collections do survive – those kept by Lewis Morris at Penbryn, the house of his wife, at Goginan, just outside Aberystwyth, and the substantial library of John Lloyd of Wigfair.

Various book lists relating to the collection of Lewis Morris are preserved. In one, titled the 'books and manuscripts lockd up in my chest upon our being drove out of Cardiganshire', he records at least fourteen books of science among the fifty-one titles. They include Love's *Surveying*, Hawksbee's *Mechanics*, various mathematical works, Newton's *Optics*, Ray's *Synopsis*, Keill's *Astronomy* and Henry Pemberton's *View of Sir Isaac Newton's Philosophy*.[37] In a list of the collection held at Penbryn, and probably made just before, or just after, Morris's death in 1765, we find more than 500 titles with some eighty or ninety bearing on science.

As with most such collections, the volumes reflect their owner's varied scientific interests. Given Morris's surveying of the Welsh coastline, there are works on navigation, coastal piloting and surveying. As a local dispenser of medical treatments, we find works on the letting of blood and anatomy. His position in Cardiganshire as 'deputy steward of the crown manors' led him into mining disputes with local landowners, as well as prospecting for lead on his own account; hence, various works on alchemy, chemistry, metallurgy and mining. Other interests are represented in books on fossils, mathematics, astronomy, botany and natural history. Of publications by scientists from Wales we find William Jones Sr's treatise on navigation, and his *Analysis per Quantitatum series, fluxiones ac differentias* of 1711; *The Natural History of Barbados* by Griffith Hughes (see Volume 2); and Edward Lhwyd's *Archaeologia Britannica*. Also included are dictionaries, such as Minshew's *Dictionary of Eleven Languages* (1617) and a German dictionary, together with a Dutch Grammar, and occasional manuscripts bearing titles such as 'Mechanicks a MS by Mr L[ewis] M[orris]', 'Improvement in Mechanicks by L.M.' and 'Machines MS by L.M.'.[38]

Such polymath interests were a characteristic of the time and are nowhere more in evidence than in the library belonging to John Lloyd. In a catalogue of the libraries at his estates of Wigfair and Hafodunos, created for an auction to be held at Wigfair in January 1816, following Lloyd's death in 1815, the auctioneer, Mr Broster of Chester, noted how Lloyd had collected 'such subjects as tended to illustrate the universal principles of science and philosophy'. The importance of the collection is emphasised by his making the catalogue available in 'London, Oxford, Cambridge, Bath, Manchester, Halifax, Birmingham, Leeds, Liverpool, Shrewsbury, Dublin, Edinburgh, Wrexham, Holywell, Ruthin, St Asaph, Denbigh, Chester' and Wigfair. Of the eleven classes of books and equipment sold over thirteen days the science books, in Class IV, came to sale on days five, six and seven. They numbered some 461 volumes, encompassing 'Philosophy, Mathematics, Arts and Sciences, Physic and Chemistry, Natural History and Minerals, Agriculture and Gardening'. Among them were a seven-part edition of Newton's *Principia*, edited and published by Denbighshire-born,

London-based mathematician and publisher, William Davis; George Cadogan Morgan's *Lectures on Electricity*, as well as volumes by Pennant; many of Robert Boyle's works; Erasmus Darwin's *Zoonomia*; Raspe's account of German volcanoes; Hutton's *Theory of the Earth* (1802); and a sixty-nine volume set of the *Philosophical Transactions* (Lloyd had been made FRS in 1774). The volumes were part of the 'uncommon confused heaps' in which Mr Broster found them at Lloyd's two estates.

But books were not the only scientific content on sale. Class X contained 'Astronomical, philosophical, chemical, and optical instruments, with phosphorous, quicksilver, and minor articles used in experiments'. Class XI contained minerals, fossils and various petrifactions. Beyond sauce and soup ladles at £22 9s. 7½d, a richly embossed tea kettle and stand (£36 8s.), a large venison dish (£27 12s. 4d) and a Caxton printing of 'The Recuyele of the Historyes of Troye's' at £126, the highest prices paid at the auction were for telescopes made by Herschel and Peter Dolland – they sold for £99 15s. and £84, respectively.[39]

Such is the comprehensive nature of the sales catalogue it is ripe for detailed interrogation. What follows here is the result of an initial and very cursory analysis from which two preliminary inferences are drawn. The first is that the science volumes sold in Class IV actually fall into some twenty or so subcategories, so that the abundance of volumes in each subcategory might give us an indication of Lloyd's principal interests. The resulting analysis shows the greatest numbers of volumes relate to mathematics (forty-eight) followed by astronomy (thirty-seven), botany (thirty-four), medicine (thirty-one), agriculture (thirty), chemistry (twenty-three), zoology (twenty-one) and physics (twenty-one). Categories containing the least number of volumes are optics (three), marine navigation (two), electricity (two), magnetism (one), engineering (one), and clocks (one). The last is something of a surprise given Lloyd's significant collection of timepieces. It perhaps suggests that abundance is not necessarily an indicator of preference.

Our second and somewhat more significant inference comes from a comparison between the number of science (Class IV in the sale catalogue) and religious/theological titles (Class III in the sale catalogue). When the total number of scientific titles and religious titles published pre-1700, between 1700 and 1750, and between 1750 and Lloyd's demise in 1815 are plotted, an interesting change in the nature of the libraries emphasis is seen (see Figure 8). In both the pre-1700 and 1700–50 categories the number of divinity titles trumps science, but when the 1750 to 1815 volumes are added the reverse is observed. The number of science volumes *far* exceeds the divinity content, with the latter showing a very marked decline from its pre-1700 position. Perhaps what this illustrates is the point made by Geraint H. Jenkins in the opening quotation of this chapter, that books 'reflect underlying changes in society, which, in turn, shape and govern the needs and aspirations of people'. If so, then a significant change in the needs and aspirations of John Lloyd seems to have taken place in the 1750s.[40]

As Jenkins (1998) has also written, 'There is no doubt that the middle years of the eighteenth century witnessed a new interest in ways and means of developing the considerable economic and strategic resources

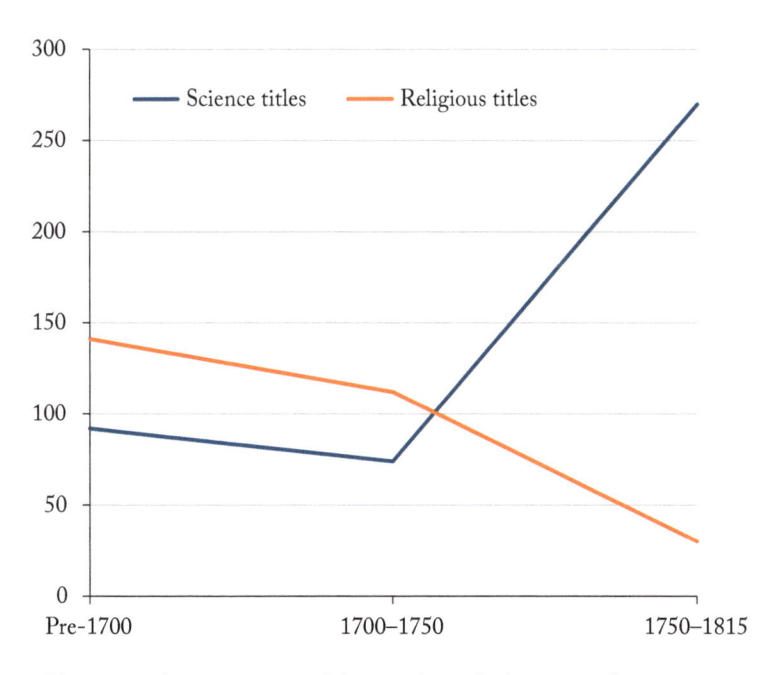

FIGURE 8 A comparison of the number of religious and science texts contained in John Lloyd's library based on their date of publication.

of Wales. Much greater vitality and an enthusiasm for change are detectable'.[41] Tying into this post-1750 'vitality and enthusiasm' are two short pamphlets published in 1755: *Proposals for Enriching the Principality of Wales: Humbly submitted to the Consideration of his Countrymen* by 'C.B', writing under the pseudonym *Giraldus Cambrensis* (see Figure 25) and the *Constitutions of the Honourable Society of Cymmrodorion in London*. The first tract aimed for the improvement and benefit of Wales through agriculture and commerce, with the betterment of landed estates to be achieved by improved land and livestock management through claying and marling, turnip-sowing, fattening of cattle, and the sowing of clover and rye-grass seeds. Whether inspired by such publications, or simply as a general consequence of the changing times, it is notable that the first agricultural society in Wales, and one of the earliest in Britain, was formed in Breconshire in 1755, the same year as *Proposals for Enriching the Principality* appeared. Later, in 1800, Thomas Johnes of Hafod published *A Cardiganshire Landlord's Advice to his Tenants* containing his own ideas on husbandry and farming methods; a Welsh-language version was published the same year, in London, by Samuel Rousseau, cousin to the philosopher Jean-Jacques Rousseau. Also published at the private press established by Johnes at Hafod was *The Rules and Premiums of the Society for the Encouragement of Agriculture and Industry in the County of Cardigan* (1804).[42] In 1816 the Anglesey Agricultural Society ordered the translation into Welsh of two essays on agriculture. Printed in Caernarfon they appeared as *Garddwriaeth ymarferol* ('Practical Horticulture') with sponsorship from the Anglesey and Caernarfonshire Agricultural Societies and the local gentry.[43] In 1817 the Breconshire Agricultural Society introduced plans 'for [agricultural] experimentation and the purchase of books for a circulating library'. By 1818, every county in Wales possessed an agricultural society; though how effective the proposed new developments and experiments proved to be is debatable.[44] Even in 1817 Iolo Morganwg suggested that 'the landed gentleman of Glamorgan will never be able to make the most of their estates until they become geologically acquainted with them'. Soils, he argued, will differ from one another depending on the subsurface rock from which they derive. Some will be 'light', others 'strong and heavy, &c. and this in various degrees'. So, a 'manure that is proper for one soil may not be so for another. What ameliorates one deteriorates the other, but our farming clodhoppers apply the same manure indiscriminately to all soils'. His solution to help resolve such ignorance was 'the formation of a Glamorgan Geological Society', though the proposal was not taken up in his time.[45]

Whatever its true effectiveness at the time, this new mid-century 'vitality and enthusiasm' also found its way into the *Constitutions of the Honourable Society of Cymmrodorion*. The Society was founded in London in 1751 largely through the efforts of Richard and Lewis Morris, with its Constitutions being agreed at the *London-Stone Tavern*, in Cannon Street, in 1753, and at the *Half-Moon Tavern*, in Cheapside, in 1755, the year of their publication (see Figure 9).

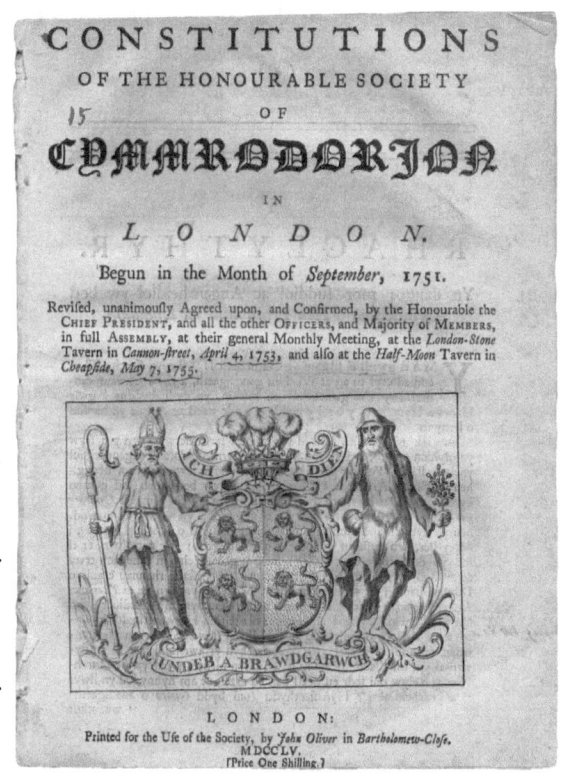

FIGURE 9
Title page of
the *Constitutions
of the Honourable
Society of
Cymmrodorion
in London.
Begun in
the month of
September, 1751*
(London, 1755)

The *Constitutions* declared that the Council of the Society should be composed not only of 'Gentlemen of Learning' in the British language, history, poetry, genealogies and antiquities of the Ancient Britons, but those 'acquainted with the present state of Wales, with respect to Learning, Trade, Manufactures, Fisheries, Mine-works, Husbandry etc.' Furthermore, although antiquarian and cultural interests were to be the principal remit of the society, it would also concern itself with Natural Philosophy, including: 'Plants found in some parts of Wales, not hitherto described by any Botanists', as well as the country's fossils, fish, birds, beasts, insects and medicinal waters. Donations to the Society of 'books, manuscripts, medals, fossils, ores, shells, or any other curious productions of art or nature' would be recorded as to their donor's name.[46] Yet, despite such good intentions, the cultural and antiquarian sides of the *Cymmrodorion* would eventually squeeze out its earlier scientific aspirations. It did, though, provide an important scientific legacy when it oversaw the publication of Thomas Pennant's self-funded *British Zoology* in 1766. Pennant had become a corresponding member of the Society in 1751 and a full member in 1761 and his book was 'sold for the benefit of the British [i.e., Welsh] Charity-School on Clerkenwell-Green' (today the Marx Memorial Library). Publication, however, left Pennant seriously out of pocket and the school does not seem to have actually benefited much from the endeavour.[47]

Two slightly different print processes also have a bearing on the spread of scientific knowledge in Wales in our period – the translation of foreign-language books into Welsh and English, and the publication of dictionaries and encyclopaedias. With regard to the translation of foreign language works, we have already noted *Reflections upon Monsieur Des Cartes's Discourse* (1654), translated from a French manuscript by John Davies of Kidwelly – 'the most active translator from the French in his age'. In 1665, in partnership with another translator, George Havers, he produced (again from a French original) *Another Collection of Philosophical Conferences of the French Virtuosi, upon questions of all sorts; for the Improving of Natural Knowledge. Made in the Assembly of the Beau Esprits at Paris, by the most ingenious persons of that nation.* A discussion of this volume and its significance will be found in Volume 2.[48] In 1759 the

Wrexham-born weaver, miner, book binder, 'tallow-chandler' and itinerant preacher John Evans (1723–1817) translated John Wesley's *Primitive Physick* into Welsh. A popular handbook of largely homeopathic remedies it was guaranteed a wide readership in the country. In 1800 Swansea publishers Voss & Morris printed an English translation of Gmelin's edition of Linnaeus's *Systema Naturæ* ('*A General System of Nature*'). The translation was the work of the Swansea-based, English natural historian and conchologist, William Turton (1762–1835), his work being the first translation into English of the complete *Systema Naturæ* to appear (see Figure 16). In 1814, a Welsh translation of a work by P. Medalon, a French surgeon, appeared as *Meddyg Anifeiliad: Neu Pob Dyn yn Feddyg i'w Anifail ei Hun* ('The Veterinarian: everyman his own animal doctor'; see Figure 10). We shall return to the subject of translation in Chapter 14.

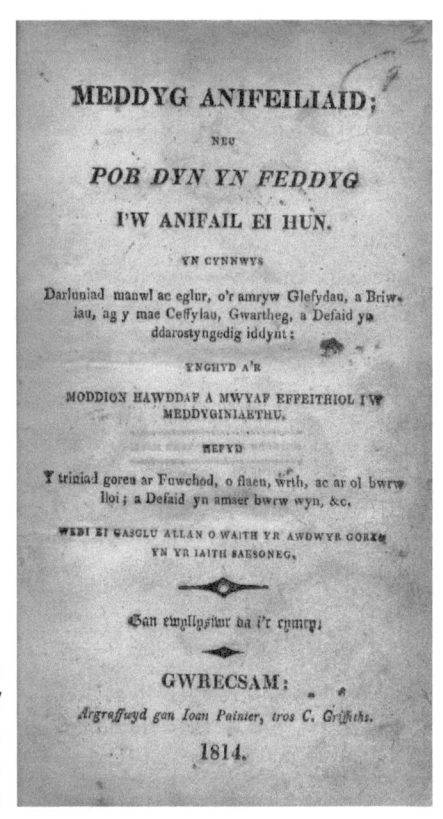

FIGURE 10 Title page to *Meddyg Anifeiliad* ('Animal Doctor'), based on a French original (Gwrecsam, 1814)

With regard to dictionaries and encyclopaedias, a number of the former were produced in the long eighteenth century, among the earliest being the dictionary of alchemical terms and procedures that Bassett Jones added to his unpublished manuscript *Lithochymicus* of *c.*1650. It came complete with hand-drawn illustrations with the purpose of providing 'For the better understanding as well of this discourse as of other chymic writings'. Among later efforts we find an important English-Welsh dictionary serialised between 1770 and 1794 by Revd John Walters, father of the young Daniel Walters mentioned earlier. The publication history of this and other such dictionaries can be found elsewhere;[49] their importance here being as an aid to introducing newly minted technical words of science to a wider Welsh and/or English-speaking populace. Following his discussion of astronomy in *Daearyddiaeth* (Geography) of 1816, Robert Roberts included for the benefit of his readers 'Eglurhad o'r prif enwau y sydd yn arferedig mewn Seryddiaeth' ('An explanation of the main names that are common in Astronomy'). It includes explanations of Welsh and English terms used in the subject.

Finally, our look at print processes and their part in spreading scientific knowledge in Wales concludes with two monumental encyclopaedias. Both were the product of the Wales-born, London-based Dissenting minister Abraham Rees (1743–1825, FRS 1786). Not only did Rees edit, substantially revise and add to a late eighteenth-century re-publication of Ephraim Chamber's 1728 *Cyclopaedia* (for which Rees would be made FRS), he followed it up between 1800 and 1820 by editing his own *Cyclopaedia: or, universal dictionary of arts, sciences, and literature*. This vast undertaking resulted in a truly wonderful forty-five-volume, illustrated, English-language compendium of eighteenth-century scientific and technological knowledge. It is considered more fully in Chapter 16.

As with the *Constitutions of the Cymmrodorion*, the *Proposals for Enriching the Principality*, and Pennant's *British Zoology* most works by the scientists of Wales in the long eighteenth century were published in London. Prior to the lapsing of the Printing Act, in 1695, books in Wales and England had to be published, with a few exceptions, in

London, Oxford or Cambridge. Commercial printing of books did not begin in Wales until 1718, though from then on numerous presses would be established throughout the country. A fact of particular importance to the publication of Welsh-language books related to science. Like Thomas Johnes at Hafod, Lewis Morris established for a short time his own printing press at home in Holyhead and he instinctively understood the importance of such printing:

Yr Argraffwâsg [*medd y doethion*] yw Canwŷll y Bŷd a Rhyddid Plant PRYDAIN. Pam i ninnau [a fûom wŷr Glewion gŷnt! os oes côel arnom] na cheisiwn bêth o'r Goleuni?

(The printing-press [so say the wise] is the Candle of the World and the Freedom of BRITAIN'S children. Why do not we [who once were valiant men! If we can be believed] seek some of the Light?)[50]

Nevertheless, the fact that most works by scientists of Wales were printed outside the country can darken the scene by fostering a sense that little science was being done *within* the nation in this period. The perception is compounded by the difficulty of knowing where works were actually being written. Few of them contain the sort of overt declaration of being 'scribbled' in Wales that we see in Bassett Jones's *Herm'aelogium*. More often, we are left wondering where they were 'scribbled'. For example, Thomas Powell's *Elementa Opticae* was published in London, which he undoubtedly visited often, but his career was largely spent as a clergyman in Wales. Hugh Davies's *Welsh Botanology* appeared in London in 1813, yet its content alone makes it likely that the writing took place in Wales. Equally difficult to know is what works published by scientists of Wales were actually being read in Wales. That some were is clear from their presence in libraries and collections already discussed; but how far did such works filter into the consciousness of those beyond the gentry and middling classes? Given the current state of research into the history of science in Wales it is impossible to take this further, other than to say that publication outside

Wales in the long eighteenth century, and even the slow uptake of some ideas, should not be construed as reflecting a lack of scientific writing or activity within the country. Indeed, there is still much treasure to be found among the archival collections available to today's researcher, and particularly in the surviving correspondence between scientists of Wales and others within and beyond Wales.

Correspondence and Networking Processes

Among the most important resources available to the historian of the long eighteenth century are letters. Today this treasure is more readily available than ever before, thanks to the searchable digital archives that are increasingly appearing online. At Early Modern Letters Online (EMLO),[51] some 700 letters sent, and more than 3,000 letters received by Edward Lhwyd are available digitally. Of the 2,000 or so extant letters of Thomas Pennant, many are also accessible through EMLO, and the *Curious Travellers* portal;[52] the latter having been established as part of a major study of Pennant undertaken at the Centre for Advanced Welsh and Celtic Studies in Aberystwyth (CAWCS). As the result of another CAWCS study, on Iolo Morganwg, the three published volumes of his correspondence contain some 1,230 letters. Another 705 letters occur in the four volumes of correspondence relating to the Morris Brothers of Anglesey. The Linnean Society of London has digitised some eighty-four letters of Thomas Johnes, sixteen by his daughter Mariamne and thirty-five by Hugh Davies, author of *Welsh Botanology*. In the NLW we have the correspondence to and from John Lloyd of Wigfair. It includes numerous letters from Joseph Banks discussing everything from gossip about mutual scientific acquaintances to the first balloon flights in France and Banks's sometimes imperious demands for botanical specimens. The digitised collection of Samuel Hartlib, at the University of Sheffield, and the Hans Sloane correspondence, in the British Library, add a further wealth of correspondence from Welsh alchemists, plant collectors, weather watchers, doctors and plain ordinary folk acting as observers of the world around them. Such abundance is clear evidence of the degree of networking through correspondence undertaken by scientists of Wales,[53] and while some of the characters involved, both at home and abroad, appear in subsequent chapters, a truly comprehensive discussion requires a volume of its own.

A more manageable source of networking by the scientists of Wales is provided by the subscriber lists included in their published works. A perhaps underused resource, the lists not only reveal the names of those interested enough to pay towards the cost of a publication, but often their locale and in some cases their occupation.[54] By the time his important and pioneering *Plans of Harbours, Bars, Bays and Roads in St George's Channel* finally saw the light of day in 1748 (see Chapter 9), Lewis Morris had managed to network a healthy 1,240 subscribers. Of that number some 400 or so were from Wales, approximately 150 from Ireland, 550 from England, twenty-five from Scotland, one from the American colonies, and one from Barbados. Of the 114 subscribers to William Davis's *Complete Treatise of Surveying* (1798) at least sixty were from England, twelve from Wales, and one each from Ireland and the United States. Among the occupations noted were mathematician, optician, schoolmaster, drawing master, merchant, land surveyor and bookseller. Of the 940 subscribers to Robert Roberts's 1816 Welsh-language volume *Daearyddiaeth* ('Geography') some eighty-three were from England, including the president of the Royal Society, Joseph Banks, eleven from Ireland, and the rest from Wales. Among the forty-three different occupations recorded were: surgeon, engraver, architect, schoolmaster, sieve maker, wheelwright, sawyer, stone cutter, glazier, joiner, smith and weaver. All of which suggests a widespread interest in science-based volumes from all but the very lowliest in society by the late eighteenth/early nineteenth century.

Evidence contained in Henry Pemberton's *A View of Sir Isaac Newton's Philosophy* (1728) also indicates that the Welsh subscribed to works of science beyond the country's borders. Already mentioned as being contained within the Pembroke Society catalogues, and the Lewis Morris library, Pemberton's volume is an attempt at a simplification of Newton's works in order 'to encourage such young gentlemen as have a turn for the mathematical sciences, to pursue those studies more chearfully [*sic*]'. It begins, though, with a very long list of subscribers. Aside from aristocrats, and the likes of mathematician Abraham de Moivre, many cannot be identified beyond their name, but a number of Welsh contributors can be discerned. They include the thirty-seven-year-old Swansea-born, London-based apothecary, FRS and future committee member of the Royal Society, Silvanus Bevan (1691–1765). Also present are a number of clergymen from Wales, including the Bishop of St David's who, at the time Pemberton published, was Englishman Richard Smalbroke. As Bishop between 1724 and 1731 he is said to have had enough Welsh to officiate for his congregation. Recorded too is the 'Rev. Mr Richards, Rector of Llanvyllin, Montgomeryshire' and now identified as Thomas Richards, who published *The Happiness of a good Christian after Death, at the funeral of Lady Price* in 1727.[55] In all probability the 'Rev. Samuel Price' is the Llangeinor-born uncle of mathematician, philosopher and political pamphleteer, Richard Price (1723–91). Samuel had moved to London in 1703 as an assistant to the hymnist Isaac Watts. Among the gentry names are Marmaduke Gwynne (1691–1769) of Llanelwedd in Radnorshire, and his brother Roderick. In 1737 Marmaduke, as a local magistrate armed with a copy of the Riot Act, would set out to arrest the dangerous evangelical preacher Howell Harris, only to find himself converted by Howell's preaching and inviting him back home for tea.[56] How far this later conversion to Methodism influenced Marmaduke's earlier interest in Newton's philosophy is not currently known. Nor do we know the effect on another subscriber, 'Sir John Pryse, Newton Hill, Montgomeryshire'. This is probably the 5th Baronet of Newton (d.1761) who married three times. Before she would agree to the marriage, his third wife, Eleanor Jones, insisted on the removal of the embalmed bodies of her prospective husband's two previous wives from their place either side of his bed.[57] Other subscribers to Pemberton's work include John Campbell of Stackpole Court in Pembrokeshire, together with Mrs Campbell; a William Phillips of Swanzey [*sic*]; Rowland Dawkin of Glamorganshire, who is probably associated with Kilvrough Manor in Gower; and Morgan Morgan Esq., to whom the concurrent names Charles Morgan and Francis Morgan may be related.

A convivial aspect of networking, for most participants at least, came through membership of scientific and philosophical societies. The development of agricultural societies throughout Wales has already been considered. Apart from these, and the London-based Cymmrodorion, whose early scientific energy faded as the eighteenth century progressed, there were few truly scientific societies established in Wales until the early nineteenth century. One significant formation was the 1807 Cyfarthfa Philosophical Society in Merthyr Tydfil, dubbed by historian Gwyn Alf Williams as the Welsh equivalent of the Birmingham Lunar Society of Josiah Wedgwood, Matthew Boulton, Erasmus Darwin and Joseph Priestley.[58] With their guinea subscription used to buy instruments, and its members being known for their inventiveness, the Merthyr society also formed part of radical Merthyr politics, with secret readings of Voltaire, Thomas Paine, and d'Holbach, held on Aberdare Mountain.[59] In the main, though, for scientists of Wales who lived and worked within Wales, as well as those who chose to settle in London, the Royal Society remained the principal society of the long eighteenth century. Fortunately, it too provides us with an underused networking resource: the surviving certificates of election to fellowship of the Society.

Certificates do not always exist for early Welsh fellows of the Society, such as Edward Lhwyd (FRS in 1708); nor even for Silvanus Bevan, who became FRS in 1725. Where early certificates do exist, they simply note the applicant for fellowship as 'recommended' by those whose names were appended; the appended names usually being written in a single (secretarial) hand, rather than by each individual concerned. An existing example is that of the Leiden graduate in medicine 'Roger Jones M.D. of Abergele', he became FRS in 1736 and was 'recommended' by the likes of astronomer Edmond Halley and the future founder of the British Museum and President of the Royal Society, Hans Sloane.

Later in the eighteenth century the language of the certificates changed. Each proposer of a nominated fellow now had to individually sign the certificate of recommendation *and* declare their personal knowledge of the applicant. The change opens up for scrutiny the social and scientific milieu of the scientists of Wales over the course of our period of interest. To take just two examples: among those who sponsored the mathematician, moral philosopher and civil libertarian Richard Price for fellowship, in 1765, were his close friends the American polymath, and future founding father of the United States, Benjamin Franklin, and the English electrical experimenter John Canton. In turn, Price, during his twenty-six years as an FRS, sponsored not only his nephew, the Bridgend-born William Morgan (FRS 1790), father of modern actuarial science and winner of the Royal Society's Copley Medal, its highest honour, in 1789,[60] but such other luminaries as the chemist and discoverer of oxygen, Joseph Priestley (FRS 1766); the inventor of the steam engine, James Watt (FRS 1785, Price mistakenly signing his certificate twice); the encyclopaedist Abraham Rees (FRS 1786, following postponement of an earlier attempt in 1778); the writer on astronomy, electricity and horology, Jean-Baptiste Le Roy (FRS 1773); and, just weeks before Price died, the Chargé d'Affaires of Portugal, and Member of the Academy of Sciences in Lisbon, Cypriano Ribeiro Freire (FRS 1791). When John Lloyd of Wigfair came to be sponsored by the likes of lawyer, antiquary and naturalist Daines Barrington, the surgeon and secretary of the Royal Society Charles Blagden, and botanists Daniel Solander and Joseph Banks, Lloyd went on to add his own signature to the certificates of explorer James Cook (FRS, 1775); engineers Matthew Bolton (FRS, 1785) and John Rennie (FRS, 1797); anatomist John Hunter (FRS, 1785); the astronomer William Herschel (FRS, 1781); the scientific instrument maker Jesse Ramsden (FRS 1786); and James Lind (FRS, 1777), cousin to the curer of scurvy.

* * *

With the close of this chapter we have come to the end of Part II of this book. It has concentrated on the manner in which the scientists of Wales gained their scientific knowledge. We now turn, in Part III, to the practical investigations and studies undertaken by these scientists, and the significant contributions they made to scientific and social developments within the Isles in the long eighteenth century.

Notes

1. Geraint H. Jenkins, 'The Eighteenth Century', in Philip Henry Jones and Eiluned Rees (eds), *A Nation and its Books: A History of the Book in Wales* (Aberystwyth, 1998), p. 119.

2. For Durel, see Iolo Davies *'A Certaine School': A History of the Grammar School at Cowbridge Glamorgan* (Cowbridge, 1967), p. 35. For Fletcher, see his biographical entry at, Stephen Orchard and Isabel Rivers, 'John William Fletcher (1729–1785)', *Dissenting Academies Online: Database and Encyclopedia*, https://qmul.ac.uk/sed/religionandliterature/dissenting-academies (accessed 28 May 2024).

3. Davies, *'A Certaine School'*, pp. 37–8.

4. For details, see Hsiang-Fu Huang, 'Theatres, Toys and Teaching Aids: Astronomy Lecturing and Orreries in the Herschel's Time', in Silvia De Bianchi (ed.), *The Harmony of the Sphere* (Newcastle upon Tyne, 2013), pp. 132–55, at p. 144 and n. 43; Martin Beech, *The Wayward Comet: A Descriptive History of Cometary Orbits, Kepler's Problem and the Cometarium* (Boca Raton FL, 2016), p. 141; A. E. Musson and Eric Robinson, *Science and Technology in the Industrial Revolution* (1969; London, 1994), p. 164 and n. 4. See also Larry Stewart, *The Rise of Public Science* (Cambridge, 1992), p. 128 and n. 82.

5. NLW MS 5456A

6. *CIM*-III, p. 172 n. 8.

7. *CIM*-III, pp. 165–6.

8. *CIM*-III, p. 166.

9. *CIM*-III, pp. 172–3.

10. *CIM*-III, p. 166.

11. *CWP*, Item 2071, p. 339.

12. See Owen Morris, *The 'Chymick Bookes' of Sir Owen Wynne of Gwydir* (Cambridge, 1997), pp. 2–3.

13. See Dewi Jones, *The Botanists and Guides of Snowdonia* (Llanrwst, 1996).

14. For a fuller listing and description, see *https://sublimewales.wordpress.com/introduction/types-of-tourist/natural-historians/botanists/* (accessed 22 May 2024).

15. See Joseph Harris, *Journal of Two Visits to Wales 1746 and 1748*, in the Ynysfor Collection, NLW (currently un-catalogued); also, Peter Wakin and Joan Day, 'Joseph Harris in Bristol, 1748', *Bristol Industrial Archaeological Society Journal*, 15 (1982), 8–12.

16. Arthur Aikin, *Journal of a Tour through Wales and part of Shropshire: with Observations in Mineralogy and other Branches of Natural History* (London, 1797).

17. See listings, *https://sublimewales.wordpress.com/introduction/types-of-tourist/early-18th-century-travellers/* (accessed 22 May 2024); also W. Linnard, 'A Swedish Visitor to Flintshire in 1760', *Flintshire Historical Society Journal*, 30 (1981–2).

18. R. George Thomas, 'The Complete Reading List of a Carmarthenshire Student, 1763–7', *NLWJ*, 9/3 (1956), 354–64 (at 356).

19. John Harris, *Vox Stellarum & Planetarum* (Caerfyrddin, 1797); see appended 'John Daniel Book Catalogue', pp. 1–7.

20. Given the English title, and date of the request, this probably relates to the 1650 translation by Walter Charleton, see *CWP*, Item 1974, p. 326.

21. See Davies, *'A Certaine School'*, Appendix IV: The Diary of Daniel Walters (31 March 1777–3 May 1778), pp. 362–74 (at p. 372).

22. Martin Evans, *An Early History Queen Elizabeth Grammar School, Carmarthen* (Carmarthen, n.d.), pp. 85–7.

23. See Gwyn Walters, 'Richard Price and the Carmarthen Academy', *The Price-Priestley Newsletter*, 4 (1980), 69, citing *The Cambrian Magazine; or Useful Repository, of Science and Entertainment*, 1 (June 1773), 29–30.

24. For a full discussion of these, see Eiluned Rees, 'An Introductory Survey of 18th Century Welsh Libraries', *Journal of Welsh Bibliographical Society*, 10/4 (1971), 197–258.

25. See Ewart Lewis, 'Cowbridge Diocesan Library 1711–1848', *Journal of the Historical Society of the Church in Wales*, 4 (1954), 36–44; 7 (1957), 80–91.

26. See Paul Kaufman, 'Community Lending Libraries in Eighteenth-Century Ireland and Wales', *The Library Quarterly: Information, Community, Policy*, 33/4 (1963), 299–312.

27. *CIM*-I, pp. 795–6.

28. *CIM*-II, pp. 143, 146.

29. Kaufman, 'Community Lending Libraries'.

30. See G. Walters, 'The Eighteenth-Century "Pembroke Society"', *WHR*, 3/3 (1967), 291–8; the digitised catalogue for 1776 is available at the NLW website.

31. Quoted in Rees, 'Introductory Survey of 18th Century Welsh Libraries', 197–258; also AML-I, p. 364.

32. Born in Conwy, Williams succeeded Francis Bacon as Lord Keeper of the Privy Seal in 1621. For details of the book controversy see *CWP*, Items 1943, 1959, 1962.

33. See Morris, *The 'Chymick Bookes' of Sir Owen Wynne*, p. 7.

34. *CIM*-II, p. 838.

35. See *DWB*, *ODNB* and R. J. Moore-Colyer, 'Thomas Johnes of Hafod (1748–1816): Translator and Bibliophile', *WHR*, 15/1, (1990), 399–415.

36. Smith visited Hafod as a doctor to treat Mariamne, and as her friend and tutor in all things botanical.

37. See AML-II, Appendix C, pp. 790–1.

38. See AML-II, Appendix E, pp. 794–807.

39. See *Bibliotheca Llwydiana: A Catalogue of the entire Library ... of John Lloyd Esq., LLD ... by Mr. Broster, Chester (1815)*, NLW MS 12500B. Also, Gwynfryn Walters, '"Bibliotheca Llwydiana": Notes of the Sale Catalogue (1816) of John Lloyd's Library', *NLWJ*, 10/2 (1957), 185–204.

40. For discussion of a proposed and somewhat controversial suggestion of a mid-eighteenth century 'revolution in the reading' of novels, see Robert Chartier, with Arthur Goldhammer (trans.), *Inscription and Erasure; Literature and Written Culture from the Eleventh to the Eighteenth Century* (2005; Philadelphia, 2007), p. 113.

41. Geraint H. Jenkins, *The Foundations of Modern Wales* (Oxford, 1987), p. 280.

42. See Ifano Jones, *Printing and Printers in Wales and Monmouthshire* (Cardiff, 1925), p. 182.

43. Eiluned Rees, *The Welsh Book-Trade Before 1820* (Aberystwyth, 1988), p. lviii.

44. R. J. Colyer, 'Early Agricultural Societies in South Wales', *WHR*, 12/4 (1985), 567–81, at 570–1, and 572 for quote.

45. *CIM*-III, pp. 440–1.

46. Cymmrodorion, *Constitutions of the Honourable Society of Cymmrodorion in London*, (London, 1755), pp. 17, 21, 35.

47. See C. Stephen Briggs, 'Thomas Pennant: Some Working Practices of an Archaeological Writer in Late Eighteenth-Century Britain', in Mary-Ann Constantine and Nigel Leask (eds), *Enlightenment Travel and British Identities: Thomas Pennant's Tours in Scotland and Wales* (2017; London, 2019), p. 42.

48. See Joseph E. Tucker, 'John Davies of Kidwelly (1627?–1693), Translator from the French: with an Annotated Bibliography of his Translations', *The Papers of the Bibliographical Society of America*, 44/2 (1950), 119–52; David Hook, 'John Davies of Kidwelly, a Neglected Literary Figure of the Seventeenth Century', *The Antiquarian*, 11 (1975), *www.kidwellyhistory.co.uk/Articles/JohnDavies.htm* (accessed 22 May 2024).

49. For Bassett Jones, see 'Lithochymicus', in Robert M. Schuler (ed.), *Alchemical Poetry 1575–1700* (1995; Abingdon, 2013) pp. 327–56; for Revd Walters, see Brian Ll. James, 'The Cowbridge Printers', in Stewart Williams (ed.), *Glamorgan Historian*, vol. 4 (Cowbridge, 1967), pp. 231–44 (at pp. 232–4); M. T. Burdett-Jones, 'Early Welsh Dictionaries', in Jones and Rees (eds), *A Nation and Its Books*, pp. 75–81; Andrew Hawke, 'Coping with an Expanding Vocabulary: The Lexicographical Contribution to Welsh', *International Journal of Lexicography*, 31/2 (2018), 229–48.

50. Quoted and translated in Rees, *The Welsh Book-Trade*, p. xxiv.

51. At *http://emlo.bodleian.ox.ac.uk* (accessed 28 May 2024).

52. At *https://curioustravellers.ac.uk* (accessed 22 May 2024).

53. See, for example, the place of William Jones Sr among the contacts of Abraham de Moivre in D. R. Bellhouse, E. M. Renouf, R. Raut and M.A. Bauer, 'De Moivre's Knowledge Community: An Analysis of the Subscription List to the *Miscellanea Analytica*', *Notes and Records: The Royal Society Journal of the History of Science*, 63 (2009), 137–62. See, in particular, Figure 2 on page 145.

54. One of the early papers to deal with such lists is Eiluned Rees, 'Pre-1820 Welsh Subscription Lists', *Journal of the Welsh Bibliographical Society*, 11 (1973–4), 85–119.

55. See Robert Watt, *Bibliotheca Britannica; or a General Index to British & Foreign Literature*, 4 vols (Edinburgh, 1824), II, p. 802.

56. For a fascinating biography of Marmaduke see Jason Evans, 'Marmaduke Gwynne (1691–1769): A Methodist Squire', *NLWJ*, 35/4 (2013), 248–70.

57. *DWB*.

58. Gwyn A. Williams, 'The Merthyr of Dic Penderyn', in Glanmor Williams (ed.) *Merthyr Politics: The Making of a Working-Class Tradition* (Cardiff, 1966), p. 15. Also, Gwyn A. Williams, *The Merthyr Rising* (London, 1978), pp. 46, 73, 80.

59. Williams, *The Merthyr Rising*, p. 46.

60. For the 'political' arguments that are a background to Morgan's nomination, and a list of his other sponsors, which included Astronomer Royal, Nevil Maskelyne, physicist Henry Cavendish, and James Edward Smith of the Linnean Society, see Nicola Bruton Bennetts, *William Morgan: Eighteenth-Century Actuary, Mathematician and Radical* (Cardiff, 2020), pp. 64–6.

PART III

PRACTICAL SCIENCE

THE SCIENCE OF THE EARTH

Nay, the paradise of a philosopher, devoted to the study of nature, is
not confined to earth, or sea, or air; to the third or fourth heavens;
but includes those distant regions which the light of a day, that
shone thousands of years ago, has not yet reached! He feels it
everywhere; he enjoys it at all times. Its pleasures burst forth from
every visible object; they spring from the ground he treads upon. The
atmosphere teems with them. They are to be found in the mass and
in the lowest depths of the ocean. They flourish in the desert and
amongst precipices and rocks. They are present amongst the horrors
of volcanoes and earthquakes; nor do they forsake the mind when
looking forward to the destruction of worlds and their systems.[1]

George Cadogan Morgan, *Lectures on Electricity* (1794)

In this and the next two chapters, we follow the scientists of Wales
as they turn to the practical study of the earth, of nature and of the
heavens. We see too how such studies presented an increasing challenge
to older world views and systems, and how advances in knowledge and
understanding led to greater awareness of the natural world and the
bounty that it contains, in both scientific and economic terms.

The Sea and the Land: Determining Boundaries

In 1905 the English mathematician and physicist Lewis Fry Richardson
(1881–1953) took up a post at Aberystwyth University. In the course
of a varied career that soon took him beyond Wales, he developed what
came to be called the 'Richardson effect'; crudely put, the realisation that
the measured length of a boundary changes as the unit of measurement

changes. Thus, the coastline of Cardigan Bay, on which Aberystwyth is situated, appears much shorter when measured by a ruler one mile long than when all its inlets and bays are measured by a ruler of twelve inches. Following a mention of Richardson's mathematical and statistical work on this subject in a paper by Polish-born, French mathematician Benoit Mandelbrot – *How Long is the Coast of Britain?* (1967) – 'The Richardson effect' is now seen as one of the foundations of the modern science of fractals.[2]

Indirectly, the effect also reflects the changing cartographic representation of the Welsh coastline as we move from the smoothed-out one of Humphrey Llwyd's 1568 *Cambriae Typus* (Figure 11), published in 1573 as part of *Theatrum Orbis Terrarum* by the Dutch cartographer and geographer Abraham Ortelius, to the more indented coastlines seen in the 1729 map of south Wales by Emanuel Bowen (1694–1767), and the 1797 map of north Wales by Llanymynech-born cartographer John

FIGURE 11 *Cambriae Typus* by Humphrey Llwyd and Ortelius (1573). By permission of Llyfrgell Genedlaethol Cymru/National Library of Wales.

Evans (1723–95; Figure 12). Delving deeper (in the manner of fractal geometry), some of the most striking and accurate coastal maps produced in the long eighteenth century were those of Holyhead Customs Officer, Lewis Morris of Anglesey.

Like many in Wales with an education grounded in arithmetic and geometry, Morris began his working life as a land surveyor; in his case, on the Bodorgan Estate of landowner Owen Meyrick of Anglesey. Following his appointment to the Customs Service in Holyhead in 1729, Morris proposed to the Treasury and Admiralty in London a hydrographical survey of the Welsh coast. His proposal received the barest of encouragement despite the inadequacy of British coastal charts of the time, the principal ones then in use being Greenville Collins's *Great Britain's Coastal Pilot* published in 1693. Nevertheless, on 4 July 1737

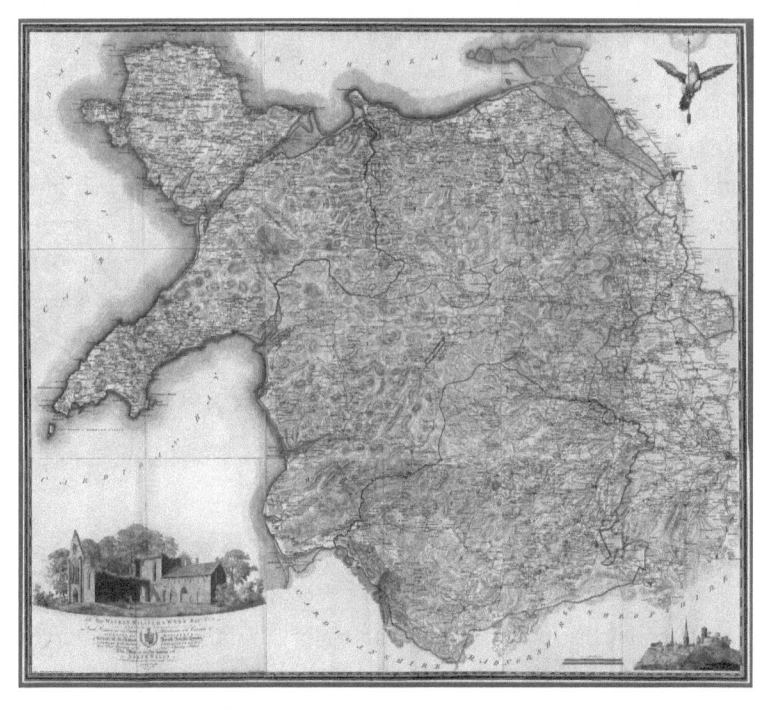

FIGURE 12 Map of north Wales by John Evans (begun before 1776, published 1795). By permission of Llyfrgell Genedlaethol Cymru/National Library of Wales.

Morris began his coastal survey at the Great Orme, near Llandudno, on the north Wales coast. Through various travails – natural in the shape of wild weather and sometimes dangerous waters; man-made in the continuing lack of aid from the Admiralty and his Custom House employers; and scientific from the nature and difficulty of using land and marine surveying equipment, some of which he constructed himself – the survey continued until 1739, at which point his employers refused to grant him further leave of absence. Having eventually been allowed to proceed in 1742, the survey finally came to a complete halt at Tenby in 1744 due to an outbreak of war with France.

Having examined the survey maps produced by Morris, the Admiralty recommended their publication. These included not only his large-scale map of the entire surveyed coastline, but also a series of smaller hydrographical plans of individual harbours and bays. The latter Morris had made for his own use: 'to refresh my memory in case of Storms or other sudden Disasters, when on my survey'. Twenty-five of these smaller maps appeared in *Plans of Harbours, Bars, Bays and Roads in St George's Channel* published in 1748. Not only did they detail specific Welsh harbours and bays, but also their near-shore hydrographical features of crucial importance to inshore sailors. As such, *Plans and Harbours* represented a major improvement (at least for the Welsh coast) over the inaccurate inshore maps then in use. By 1761 the first edition of *Plans and Harbours* had sold out, and in 1801, Lewis's son William would prepare a new and expanded edition.[3]

But what was the actual relationship between practical science and the land encompassed by sea to the west, north and south, and by a long land-border to the east that was politically defined?

The Lay of the Land

Towards the close of her magnificent history of the Ordnance Survey, Rachel Hewitt (2010) suggests 'thinkers of the Enlightenment and Romantic periods often entertained conflicting interpretations of the significance of maps – one as emblems of reason and the other as images of imagination'.[4] Through their scientific accuracy, the already mentioned maps of Emanuel Bowen and John Evans can certainly be seen as 'emblems of reason', from their detailing of the coastline and of the developing Welsh road network, to the positioning of towns, villages and the houses of the gentry, many of whom sponsored such works in order to have their family seat included on the map produced. The level of detail achieved by both Bowen and Evans would not be surpassed until the advent of the one-inch Ordnance Survey maps of the nineteenth century.[5] The coastal maps of Lewis Morris similarly express reason through precision and accuracy, despite the difficulty of mapping a coastline having sometimes dangerous hydrographical features.

To see these same maps as 'images of imagination' we need only look at how they represent the interior physical landscape of Wales. As a coastal surveyor Lewis Morris concerns himself solely with landscape features immediately adjacent to the coast, and while these appear accurately positioned – since they sometimes bear landmarks presumably intended for the taking of bearings at sea – they remain highly imaginative representations (see Figure 13).

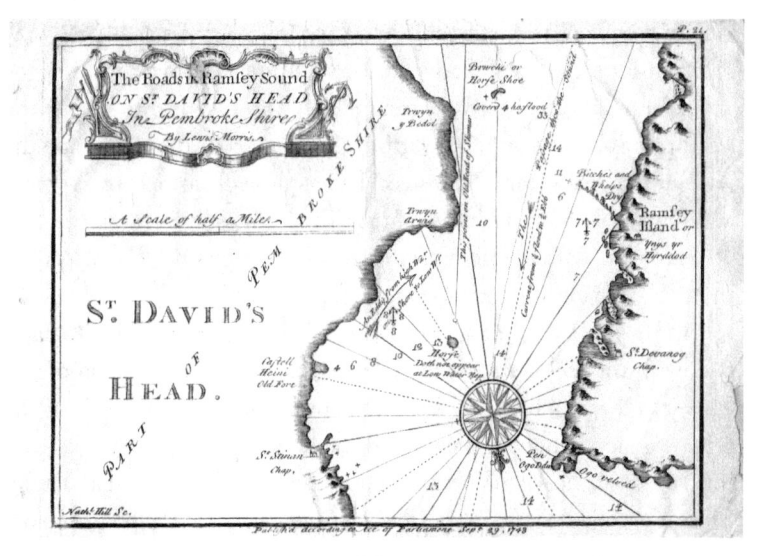

FIGURE 13 Lewis Morris, 'The Roads in Ramsey Sound', Pembrokeshire, from his *Plans of Harbours, Bars, Bays and Roads in St. George's Channel* (London, 1748)

By contrast, and in an era before the advent of contours on British maps (they only appear in the mid nineteenth century), Bowen and Evans attempted a more realistic portrayal of the physical landscape through the technique of 'hachuring' – a series of small lines drawn to represent the steepest slopes (see Figure 12). However, without accompanying measured heights this too is an essentially imaginative representation of landscape, one that could result in some surprises for adventurous eighteenth-century hillwalkers using such maps – as when historian William Hutton suddenly and unexpectedly found himself on a north Wales 'precipice two hundred feet high', the shock only compensated for by his view of 'a most beautiful valley, nearly one mile wide and four long'.[6]

For reason to inform these imaginative landscapes measured heights were clearly needed and, in the development of these, Yr Wyddfa (Snowdon), the highest peak in Wales and England at 1,085 metres (3,560 feet), played a significant role. The first person to try and accurately establish the mountain's height was John Caswell (1654/55–1712). A Somerset-born mathematician, who later became Savilian Professor of Astronomy at Oxford, he was employed in conducting a survey of England and Wales with senior partner John Adams (c.1670–1738) of Shropshire.[7] In 1682 Caswell used traditional geometric and trigonometric methods (based on triangulation and angles) to record the height of Snowdon as 1,240 yards (3,720 feet).[8] At the same time the Astronomer Royal, John Flamsteed, had requested Caswell's senior partner, John Adams, to have the famous Torricelli experiment performed on Snowdon, with the necessary equipment supplied by Flamsteed; in this case a Torricelli mercury barometer.[9]

In Italy in 1643, Evangelista Torricelli, a one-time pupil of Galileo, had, with the aid of an assistant,[10] inverted the open end of an otherwise sealed tube full of mercury in a bowl filled with the same. If the tube were long enough, they found the level in the mercury fell, leaving a space at the top. Whether or not this space constituted a vacuum became a subject of much controversy, since many natural philosophers, including Thomas Vaughan and René Descartes, believed a vacuum in nature to be impossible. Having set up the now vertical tube, Torricelli noted over the next few days minor fluctuations in the level of the mercury. Believing 'we live submerged at the bottom of an ocean of the element air, which by unquestioned experiments is known to have weight' he surmised that the fluctuations were caused by changes in the downward pressure exerted by this ocean of air. In 1648, the experiment was repeated under the direction of French mathematician Blaise Pascal during an ascent of the Puy-de-Dôme Mountain in the Auvergne. Using Torricelli's newly invented 'barometer', the level of the mercury was recorded at various points in the ascent. It revealed a progressive fall at each measurement station. A second barometer, kept at the foot of the mountain, showed no change. Not only did this prove that air pressure did indeed fall with increasing height, it opened up the possibility, if the fall seen at each measurement station were found to be uniform everywhere, of using changes in air pressure to determine actual height. In England, Robert Hooke repeated the experiment atop the tower of old St Paul's Cathedral (1661), and on the Monument to the Fire of London (1668).[11] However, it was realised that even greater heights were needed. Following measurements made on the 1,831-foot Pendle Hill near Burnley in Lancashire, Snowdon came to play its part as a place 'more likely to be visited, and to have experiments repeated upon it, than the remote hills of the [far] north [of Britain]'.[12]

It was John Caswell, rather than his partner John Adams, who fulfilled Flamsteed's request and performed the Torricelli experiment atop the mountain on 17 July 1682: 'I filled a glass tube with [mercury symbol], and with a wire ... drew up the bubbles of air, then I inverted it, and found it to stand above the [mercury] in the cup 25 6/10 [of an inch, presumably].' Unfortunately, Caswell 'could not get to a place near the sea' to take another measurement on account of his surveying commitments.[13] Edmond Halley, of the Royal Society, noted this unfortunate omission and set out to rectify it on his own visit to north Wales in Whitsun week of 1697. Staying at Llanerch Hall in Denbighshire, the home of the naturalist and antiquary Robert Davies (1658–1710), whose own 'standing barometer' served as one of the control points, Halley measured the mercury in his barometer on top of Snowdon on Wednesday, 20 May 1697. Later in the day he made a measurement at Llanberis, and the next day took another, near sea level, at Caernarfon.

The difference in height of the mercury between summit and base was some '3 Inches, 8 Tenths', which, he suggested in his report to the Royal Society, may, along with the height of Snowdon as measured by Caswell using trigonometric methods, 'serve for a standard, till a better be obtained on a higher place'.[14]

Following on from Halley, other visitors attempted to measure Snowdon's height using both trigonometric and barometric methods. They included the likes of Joseph Banks, President of the Royal Society, and, in 1775, the surveyor William Roy (1726–90, FRS 1767, Copley Medal 1785). Among the Welsh contingent was Edward Lhwyd, who knew the area well from his botanical and geological excursions. In the same year as Halley took his measurements, Lhwyd declared his intention 'to try the barometer and thermometer on the top of Snowdon and Cader Idris, and to take their perpendicular height'. It is believed he estimated the height of the former at 1,300 yards (3,900 feet).[15] Lewis Morris made an estimation of the height 'from low water mark to be within a trifle of a mile [5,280 feet]'.[16] Clearly, this is a significant overestimation and may reflect a mistaken recording of his figure or, perhaps, an estimate made while he was at sea. Cajori (1929) notes how seaborne determinations are 'theoretically suspect' being 'encumbered by difficulties of such a grave nature that it may lead to extravagant error'. These difficulties included the problem of estimating the distance from a ship at sea to the base of a mountain on land, as well as difficulties caused by atmospheric refraction; Cajori suggests errors of between 121 and 149 per cent in such estimates.[17]

As the eighteenth century progressed it was increasingly realised that the mercury barometer itself might be susceptible to error through influences on it other than air pressure. These could include capillary action within the mercury tube or the effects of temperature change on the mercury, the equipment being used, and/or the surrounding atmosphere. New ideas for measuring height were therefore proposed. One came in a paper submitted in 1724 to the *Philosophical Transactions* by Daniel Gabriel Fahrenheit (1686–1736). In it he suggested using the different temperatures at which water boiled on a mountain top compared to its base as a possible method.[18]

In 1779, 'On the Variation of the Temperature of Boiling Water' appeared in the *Philosophical Transactions* by the MP, mathematician and astronomer Sir George Shuckburgh-Evelyn (1751–1804, FRS 1774, Copley Medal 1798).[19] It included a table of observations giving barometric readings and boiling point temperatures taken at various locations, including: Llanberis on 13 August 1778; the summit of Snowdon on 15 August; and the top of Carn Cwm Gafr on 16 August. It is clear from surviving correspondence (and other documentation) between Shuckburgh and his close friend John Lloyd of Wigfair, that Lloyd must have been with Shuckburgh when these readings were taken. As we shall see later, Shuckburgh, who lived at Shuckburgh Hall in Warwickshire, shared Lloyd's passion for astronomy, but between 1778 and 1793 both men indulged another passion and were often seen surveying together in north Wales.[20]

Shuckburgh left Warwickshire on 31 July 1778 and Lloyd, travelling from London, planned to meet him at Chester on 3 August. They would then travel into Snowdonia and in preparation Lloyd wrote asking his mother to have a horse ready 'to carry our instruments'.[21] Both men were recorded at the inn of John Close, at Nant Peris,[22] which is but a stone's throw from Cwm Gafr, and, on 15 August 1778, were seen surveying with a theodolite at 'Clogwen y Garedd' on Snowdon, and on 16 August at 'Cwm Gafr'.[23] Clearly, it was during these trips that the boiling-point measurements reported in Shuckburgh's paper were taken.

A year earlier (1777), William Roy (who had sponsored John Lloyd for the Royal Society in 1774),[24] published a comparison of traditional barometric and trigonometric methods in the measuring of Snowdon and nearby Mount Eilio.[25] Between 1747 and 1755 Roy had undertaken the Military Survey of Scotland – after which, in 1766, he proposed a 'General Military Survey of England'. His proposal had been rejected by Government, but in 1784 he became closely involved in another project – a triangulation survey between Britain and France aimed at precisely determining the latitude and longitude of the Paris and Greenwich Royal Observatories; an essential aid to, among other things, global navigation. Proposed by the director of the Paris Royal Observatory, César François Cassini de Thury, to Joseph Banks as President of the

Royal Society, it was Banks who delegated the running of the project in Britain to Roy.

The first priority was to establish a precisely measured baseline. From each end of this line two more lines could be projected in order to form a triangle with known angles. Their lengths could then be calculated using trigonometry. From this initial triangle all the other triangles making up a survey would be calculated; hence, the need for precision in the initial baseline measurement. In April 1784, Roy began measuring out the necessary baseline on Hounslow Heath and his work became something of a spectator sport. The King and Queen paid a visit. So too, on 31 August, near the end of the baseline measuring, did John Lloyd of Wigfair. He did so in company with Joseph Banks; Charles Bisset, the Scottish physician and military engineer; the diplomat Sir William Hamilton; and the mineralogist and founder of Milford Haven and its proposed technical college (see Chapter 6), Charles Francis Greville (who, along with William Roy, had sponsored John Lloyd for the Royal Society).[26] When Roy eventually measured the precise length of the marked-out baseline it extended 27,404.7 feet (5.190 miles), which compares to a modern GPS based measurement of it at 27,376.8 feet (5.185 miles). Given the difficulty of using the equipment of the time, Roy deserved the Royal Society's Copley medal, its highest honour, which was presented to him in 1785. Sadly, though, after a period in the mountains of north Wales to recuperate from a lung disease, Roy died in 1790 and so did not live to see the 1791 foundation of the Ordnance Survey, of which he is usually regarded as the founding father.[27]

The Ordnance Survey began with the appointment of two surveyors: the little-known Edward Williams (d.1798) as its first director, and Plymouth born William Mudge (1762–1820) as his deputy.[28] The Survey aimed at making a triangulation of the whole of Britain and it was decided that the baseline measured by Roy would be used again; but for the sake of accuracy both Williams and Mudge first re-measured the line. Having done so, a discrepancy of just 0.3845 feet was found. The Hounslow baseline now became the starting point for the initial triangulation of southern England and ultimately the whole of Britain; a task undertaken by Edward Williams, William Mudge and another recruit, Isaac Dalby.[29]

Edward Williams died in January 1798 and his obituary, in the *Sun* newspaper, is of particular interest since it not only describes him as 'well known in conducting the Trigonometrical Survey of this Kingdom', but also as 'Colonel Edward Williams, of the Royal Regiment of Artillery'.[30] It is this last fact that allows Williams to be positively identified as a Welshman. Writing on 20 April 1795 to John Lloyd at Wigfair, Joseph Banks notes that:

> We expect daily to receive from Col. Williams his continuation of Roy's measurements it is to be printed in the next part of the Phil[osophical] Trans[actions]. I hope as he is a Welshman he will do honour to his part of that interesting undertaking.[31]

By 1795, the triangulation survey of England was well underway and Banks is clearly referring in his letter to the measurements contained in the 'first Account of the Trigonometrical Survey', which was read by Williams and Mudge to the Royal Society on 25 June 1795.[32] Williams, who would be included in similar reports before his death in 1798, had also written an earlier publication on the art of Gunnery (1766; see Volume 2).[33] Though the picture of his personality and contribution to the Survey outlined by Seymour (1980) and Hewitt (2011) is not flattering, his position as the first Director of the Ordnance Survey certainly demands more investigation.[34]

Further advances in cartography and measured height within Wales came after surveyors from the Ordnance Survey advanced into the country in late 1803/early 1804. Among the survey team now was Thomas Frederick Colby (1784–1852). Born in Rochester he was, as Rachel Hewitt puts it, 'an ancestral Welshman' since he spent much of his youth with his paternal aunts at 'Rhos-y-gilwen, just north of the Preseli Mountains in Pembrokeshire'.[35] Although only recently recovered from a wound to the head and the loss of his left hand, when a gun he had been holding exploded, Colby worked on the survey with William Mudge (now promoted to Director of the Survey) using the so-called 'Great Theodolite', made by London instrument maker Jesse Ramsden (1735–1800, FRS 1786, Copley Medal 1795) who John Lloyd had helped sponsor for fellowship of the Royal Society. Originally ordered from Ramsden by Roy in 1784, the instrument would not be delivered for three years,[36] due, in part, to the kind of accidents that became a discussion point between Joseph Banks and John Lloyd. As Banks informed Lloyd on 19 August 1786:

> Roy's instrument is not yet done. Ramsden's men suffered a large brass ruler to fall on it when the divisions were nearly all cut – it shook the whole circle so severely that the divisions have been commenced again *ab initio* so that I have little hopes of much progress this year.[37]

The finished instrument eventually weighed in at some 200 pounds and had to be winched into position at each new location; during the survey of Wales this would include the summits of Yr Wyddfa (Snowdon), Arennig Fawr and Cadair Idris.

As an extension to the triangulation in the counties of southern England and the March, the survey progressed through south and central Wales from Trellech Beacon, overlooking the Wye valley. The north Wales survey had its own baseline between Rhuddlan and Abergele. Like the earlier baseline measured out by Roy for the cross-channel survey, precision was essential since the triangles based on the Rhuddlan-Abergele baseline would have to meet up with those from the south; the link-point being at Aberystwyth. When Roy had made his 1784 baseline survey, he had experienced considerable difficulty with wet weather warping and expanding the 20-foot wooden measuring rods being used. The eventual solution would be the use of glass rods.[38] While on the Rhuddlan-Abergele baseline Colby would write to John Lloyd of Wigfair discussing the problems that fluctuations in temperature had caused to the length of the metal survey chains being used for distance measuring.[39]

Colby's broad-based initial survey of Wales was followed up later by a more detailed one made at a smaller scale. During this process the survey's use (or misuse) of Welsh words and names became an issue. The botanist Hugh Davies undertook to check the surveyor's use of them, but there were still numerous complaints, not least from Lewis Weston Dillwyn over Cardiff being expressed by the survey as *Caerdiff*, and Lewis Morris who declared that Welsh place names were being 'murdered by English mapmakers'. As Rachel Hewitt remarks: 'By the 1820s Wales had become a laboratory for toponymic innovation'. The surveyors finally left Wales in 1811, and while maps of particular areas of the country appeared piecemeal thereafter, it would not be until the 'mid-1840s' that a complete OS map of Wales appeared.[40]

The Structure of the Land

As the science of cartography developed so too did a desire to know and understand the nature of the landscape being mapped, and the long eighteenth century saw its study develop into the science we recognise today as geology. The endeavour essentially began in the late seventeenth century and the first half of the eighteenth century with the cataloguing, recording, describing and illustrating, through display and publication, of the vast amounts of geological and other natural history material then being collected. For example, from his first employment as Under Keeper in the Ashmolean Museum in Oxford, Edward Lhwyd had been enhancing its collections with material gathered from the coal mining areas of Wales and England, in particular, what he would term *Lithophyta*. These were the fossil Carboniferous plants and leaves that formed a significant component of his fossil catalogue *Lithophylacii Britannici Ichnographia* of 1699. Later, on a grand tour of Britain, Ireland and Brittany, that saw him travel over 3,000 miles between 1697 and 1701 with three assistants (David Parry, William Jones and Robert Wynne), Lhwyd managed to dispatch some twenty-two chests of material to the Ashmolean by 1700.[41]

The scale of Thomas Pennant's collecting is hinted at in a letter from the mineralogist and dealer in fossils Emanuel Mendes da Costa. Writing in advance of a trip by Pennant to Ireland, da Costa expresses his hope that Pennant would 'return home safe with your Hibernian spoils for I do not doubt by your diligence & industry you will acquire many elegant curiosities & make many valuable observations'.[42] The suggested date for this letter is '*c*.1754' and, on 11 October 1754, William Morris, then working as a Customs officer at Holyhead, noted in a letter to his London-based brother Richard: 'Your friend and mine Mr. Pennant is returned from his Irish travels … His fossils, etc., are gone in a vessel from Dublin to Flintshire', and presumably then on to Downing, near Holywell, where Pennant lived. On 24 November William recorded 'some twenty-six parcels of these curiosities'.[43]

Lewis Morris too possessed a substantial and diverse collection comprising some '5 or 6 thousand articles'.[44] It may well have been to this that his brother and fellow collector, William Morris, referred in a 1755 letter to their London-based brother Richard. 'If you were to see the Museum Morrisianum at Holyhead, you would hardly trouble your head with Sir Hans's'.[45] Sir Hans Sloane had died two years earlier and his vast and varied collections, to which several Welshmen collecting overseas contributed (see Volume 2), formed part of the newly established British Museum, which opened its doors in January 1759. 'I could have staid there three days, I verily believe, without victuals or drink,' Richard Morris declared following his visit to the new museum in 1760, 'ni welodd llygad dyn erioed y fath anrhyfeddodau ['mans eye has never seen such wonders']'.[46] Perhaps it was in response to hearing from his brother of the Museum's natural history collections in their glass-covered cabinets that Lewis Morris 'laid a plan' in December 1760 'for a cabinet to put up my natural curiosities of fossils shells &c'.[47] In fact, he became something of an artisan maker of so-called 'cabinets of curiosity'. These included a substantial one for Lord Powys, among whose varied contents the 'shells, seeds and sea plants are to amuse ye Countess of Powys, the ores for ye edification of his Lordship'.[48] The interest in ores by Lord Powys seems to have rubbed off on his daughter, Lady Henrietta Clive (1758–1830). She collected minerals while living in India (see Volume 2) with her husband Lord Edward Clive, the eldest son of Robert Clive (a.k.a. Clive of India). Today her collection resides in the National Museum of Wales in Cardiff, along with two volumes of her meticulously kept catalogue of its more than 1,000 specimens.[49]

Other geological collections surviving from the long eighteenth century form important parts of the heritage of the Isles. Edward

Lhwyd's collections remain in the Ashmolean Museum. The 800 minerals collected by Thomas Pennant on a visit to Ireland are to be found in the Natural History Museum in London; and the Sedgwick Museum, in Cambridge, holds a collection of minerals made in central Wales between 1696 and 1728 by Englishman John Woodward, author of *An Essay toward a Natural History of the Earth* (1695).[50] However, such collections and their cabinets of display were not solely the province of scholars and the gentry. In 1788, for example, the naval surgeon and later apothecary of Caernarfon, Humphrey Edwards (1730–88), left to his sister Margaret his 'Cabinet of Shells and the Shells therein contained', some of these it is likely he collected between 1764 and 1766 when on board the frigate Tamar, circumnavigating the world as part of a naval expedition led by Captain John Byron, grandfather to the poet Lord Byron.[51]

For some, these collections acted as a stimulus to closer study of the material collected and, thereby, the widening of scientific horizons – both intellectual and geographical. As Lewis Morris observed when studying one of his own specimens:

> My triangular shell is a bivalve and very neat, the hinge in one of the angles very small. I never looked closely into shells till now I had business to sort them for this cabinet; and it surprizd me that our Cambrian coast produces the very same shells as are dug in ye mountains of Sicily and many of those found in Jamaica.[52]

Study led in turn to an understanding of the need to update and correct older attempts at description and classification. As a forthright Lewis Morris put it:

> I have examined into all the British fish, and have given them as many of ye Welsh names as we have from the real knowledge of ye fish, not from ignorant dictionaries. Have rectified several mistakes in [John] Ray and [Francis] Willoughby. My collection of plants, birds and fossils are better and correcter than anything I have met with.[53]

Collecting might also reveal the new. In 1779 the French mineralo-gist Antoine Grimoald Monnet in his *Nouveau Système de Minéralogie* would describe a new mineral found at Parys Mountain on Anglesey as 'vitriol de plomb'. It would later be formally named Anglesite.[54] Thomas Pennant collected some of the first specimens of the mineral pyromor-phite in Wales, where it forms as a secondary deposit associated with lead-bearing rocks.[55] Among the collections made by Edward Lhwyd on the Welsh-leg of his Grand Tour were fossil specimens that he described as a type of 'Flat fish', found in 1698 in rocks at Dynevor Castle, near Llandeilo, in south Wales: he illustrated a specimen in a letter to his friend and fellow naturalist Martin Lister (see Figure 14). Shortly there-after the illustration appeared in the *Philosophical Transactions* and is now considered the first published drawing of what later came to be called trilobites. The modern classification of Lhwyd's example, from the Ordovician rocks of Llandeilo, is *Ogygiocarella debuchii*.[56]

Acting primarily as catalogues of the natural world, collections did not immediately challenge the widely-held religious view of the earth as derived from an act of Creation and subsequently modified by the Deluge (Noah's Flood). They did, though, begin to shift the debate away from theology and towards a greater consideration of natural evidence, as Roy Porter outlines in *The Making of Geology* (1977).[57] In this respect the collections made by the likes of Edward Lhwyd, Thomas Pennant, Lewis and William Morris, Lady Clive, Martin Lister, John Woodward and John Ray form the bedrock on which later developments in the long-eighteenth-century history of geology could take place.

At the risk of oversimplifying the story of geology, we examine in the sections to follow some of its key developments, and we do so principally through the works of scientists of Wales. The nature of fos-sils and the manner of their formation are considered through the late seventeenth-century ideas of Edward Lhwyd and Henry Rowlands (see 'Formed Stones' below). The recognition that rocks comprise discrete strata, and that these can be traced in a vertical succession as well as horizontally over considerable distances, is briefly assessed through the early work of George Owen and the later work of Edward Lhwyd (see 'Layer upon Layer' below). Finally, an understanding of the earth

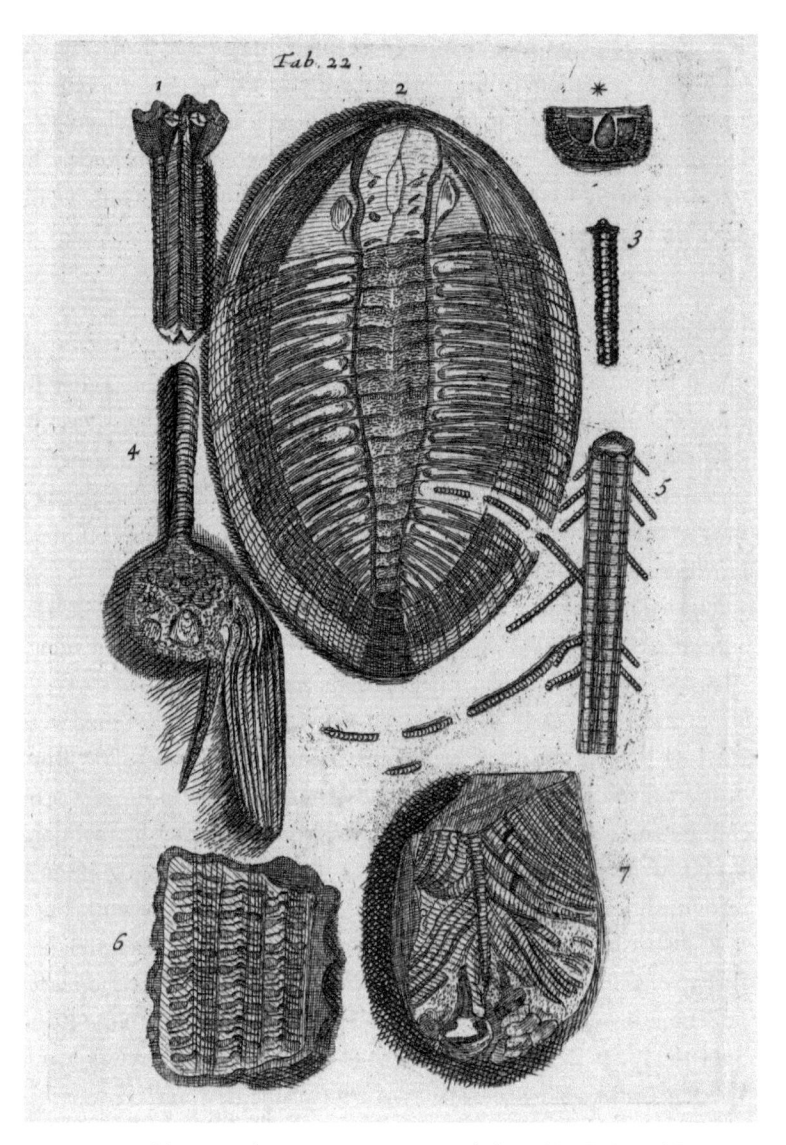

FIGURE 14 Trilobite (*Ogygiocarella debuchii*) found by Edward Lhwyd at Dinefwr and illustrated in the *Philosophical Transactions of the Royal Society* (1698) and in his *Lythophylacii Britannici Ichnographia* in 1699; the very first published illustrations of a trilobite. The image shown here is the same specimen but in a plate from a volume published in 1760. Supplied by Llyfrgell Genedlaethol Cymru/National Library of Wales.

as a dynamic place, in which physical forces operate in the short and long term, is explored through the cautious mid-eighteenth-century approach of Thomas Pennant, the Wales-born, Scotland-based geologist John Williams, and the more whole-hearted embrace of such ideas by Iolo Morganwg as the eighteenth century passes into the nineteenth (see 'The Restless Earth' below).

Formed Stones

The debate on 'formed' or 'figured stones' – as fossils were generally known in the late seventeenth century – centred on whether they represented the preserved remnants of once living creatures, or whether they were simply artefacts of nature – an origin seemingly confirmed by their appearance within a rock from which they were themselves formed. However, if they were the former, how had they formed within a rock and how did they relate to biblical Creation? In the Book of Genesis land and sea were separated on the third day, two days before living things appeared. So how could 'marine shells' appear upon mountain tops?

Among the early scientists to consider these questions were Edward Lhwyd and the Revd Henry Rowlands (1655–1723). Born in Llanedwen on Anglesey, Rowlands's interests were primarily agricultural and antiquarian.[58] However, in a surviving letter he mentions having written a work on fossils, and that this had since gone to a second edition. Until recently, little has been known of the work, but a brief discussion of it occurs in letters written in 1714 between the then Bishop of Carlisle, William Nicolson (1655–1727), and the headmaster of Newcastle-upon-Tyne Free School, and later Secretary to the Royal Society (1721–27), James Jurin (1684–1750). Nicolson refers to his 'friendly correspondence by letters' with Rowlands, whom he describes as 'my very learned friend: who amongst his other excellencies, is a great Master of Mathematics'. Nicolson then informs Jurin that 'About ten years ago [c.1704], this worthy Divine [Rowlands] publish'd an Anonymous Account of the Origin and Formation of Fossil-Shells', in which he proposed 'a new *Hypothesis* of his own'. Nicolson also notes that the work aimed to reconcile 'the different Opinions of our modern

Virtuosi.[59] Emery (1978) and Williams (2007) present convincing evidence for the work in question being *An Account of the Origin and Formation of Fossil-Shells etc. wherein is proposed a way to reconcile the two different opinions, of those who affirm them to be the Exuviæ of real animals, and those who fancy them to be Lusus Naturæ*, which was published anonymously in London in 1705.[60] Although the work has sometimes been attributed to one Charles King, Nicolson's comments regarding the work's title, its reconciling of opinions, and its possible publication date, clearly indicate Rowlands as the author. In his letter, Nicholson then went on to outline for Jurin, in some detail, the hypothesis proposed by Rowlands in his publication. We need not repeat it here for as Jurin notes in his reply to Nicolson: 'Mr Rowland's Opinion … seems to agree wth the sentiments of his friend M[r] Lhwyd.'[61]

Lhwyd had outlined his theory concerning 'The Origin of Marine Fossils and Mineral Leaves, branches etc.' in a letter to naturalist John Ray. Written at 'Rhaeadr' on 29 July 1698 it suggests the possibility that the biblical Deluge (Noah's Flood) had somehow injected spawn from marine creatures, as well as seeds and spores, deep into the earth. There they remained and grew to become the fossils found within the rock.[62] Having formulated this idea for his letter to Ray in 1698, it seems likely Lhwyd had discussed it with Rowlands when he visited him on Anglesey in August the following year.

As we have seen, when Rowlands published his version of what now appears to have been Lhwyd's theory of fossil formation he did so, according to Nicolson, with the aim of reconciling 'the different Opinions of our modern *Virtuosi*' on the subject.[63] The Lhwyd/Rowlands theory was just one among a number proposed at the time by contemporaries and correspondents of Lhwyd, including the likes of Robert Hooke, John Ray, Martin Lister and John Woodward. Although Lhwyd's theory may seem somewhat bizarre today it would not have appeared so at the time, given the then state of geological knowledge and the religious convictions he and his contemporaries held – in particular, a belief in the act of Creation and subsequent Deluge, as told in the Bible, with the Earth's creation dated to 23 October 4004 BCE by Bishop Ussher. Nor was Lhwyd's theory put forward as some sort of poorly considered suggestion. As Emery (1970) has outlined, Lhwyd, in detailing the theory for Ray, first critiqued the other theories that were circulating before presenting his own. He also described the arguments that led him to his proposal, as well as ten difficulties that he saw as arising from it. With fossils, he declared, 'what one day's observations suggest was by those of the next called in question, if not totally contradicted or overthrown'. Emery also notes that Lhwyd's overall scientific method comprised three elements: careful observation; the precise and methodical recording of material; and a 'refusal to theorize or construct hypotheses on insufficient evidence'. Nevertheless, as a true scientist Lhwyd also felt it important to make even provisional ideas known because, as he wrote in his letter from Rhayader to John Ray of 29 July 1698, 'the communicating to our friends, what carries but the least shadow of probability, does often contribute somewhat toward ye speedier discovery of the truth'.[64]

Layer upon Layer

Conybeare[65] and Phillips, in their *Outline of the Geology of England and Wales* (1822), suggest that the first person to write about how 'the same series of rock succeed each other in a regular order throughout extensive tracts of country' was George Owen (*c*.1552–1613) of Henllys in Pembrokeshire.[66] He did so, according to North (1931), in a 1595 note *On the course of the strata of coal and lime in Pembrokeshire*. This was afterwards incorporated into his *Description of Pembrokeshire* of 1603.[67] Through careful observation Owen had traced two 'veins' of limestone across south Wales, together with their associated coal 'veins'. These now represent the lower carboniferous limestone and associated coal measure outcrops that define the northern and southern boundaries of the south Wales coalfield. Owen's achievement at such an early date is clearly important in the broad history of geology, but in observing the association between the limestone and coal 'veins' he did not take the next step and formulate a true geological inference about the nature of their conjunction as beds or layers of rock. Instead, he concluded that:

Whether I shall say there are two veins of coal to be found between these two veins of limestone, or to imagine that the coal shall wreathe or turn itself in some places to one, and in other places to the other; or to think that all the land between these two veins should be stored with coals; I leave to the judgment of the skilful miners, or those which with deep knowledge, have entered into these hidden secrets.[68]

Sadly, Owen's *Description of Pembrokeshire* would not be published until its appearance in the *Cambrian Register* for 1796 (actually published in 1799), and so priority for describing the horizontal continuity and relationship between different rock strata falls to the Danish geologist, anatomist and later bishop, Nicolaus Steno (1638–86). In his *Dissertationis prodromus* of 1669 he not only established the horizontal continuity of strata but also a key law of all subsequent geology – the *law of superposition*. Simply put, all rock strata are laid down horizontally, in water according to Steno, with the oldest in any sequence of such layers at the base and the youngest at the top.[69] The law is simple but crucial because it opened up the possibility of looking back through time, from the youngest strata down through a stratigraphic section of different strata to the oldest.

It was here too that collections, the acute observations, and the scrupulous recording of material, made by the likes of Edward Lhwyd and others, came into their own. Porter (2008), for example, notes how Lhwyd observed a relationship between the basaltic formations at Giant's Causeway in Ireland and 'structures on Cader Idris'.[70] Furniss (2019) records Thomas Pennant's realisation that in Scotland 'the Giant's Causeway and [the island of] Staffa were part of a massive basaltic formation that extended all the way to Skye' (see Figure 15).[71]

It also came to be realised that individual strata, at different levels in a vertical succession of them, may contain different fossils whose recognition would allow correlation of strata across significant distances. The different fossils might even reflect different ages for the rocks. As Howells (2007) records: 'Edward Lhwyd correlated, on the basis of their faunas, the (Carboniferous) limestones at Barry with those of Caldey

FIGURE 15 Thomas Pennant and the polygonal basalt columns at Fingal's Cave, Scotland. From *Pennant's Tour in Scotland, and Voyage to the Hebrides, 1772* (Chester, 1774)

Island [offshore Tenby] and, most importantly, he recognised that they were different from the Jurassic limestones in south Glamorgan'.[72]

Such developments placed increasing value on collections, observation and records of material, but this also meant that from the middle of the eighteenth century attitudes to collecting, collections and their associated cabinets of curiosity began to change. For Lewis Morris, although his collection of shells was chiefly intended 'to show the various kinds, not to pair them, as your great collector's do, who polish and varnish them', he still, in 1757, gave his reason for collecting as: 'The study of shells, fossils, and plants is vastly natural to youth, and I suppose to people that are grown children ye second time, and that is the reason I am catchd with it now'; at the same time adding 'About forty I minded little but drinking'.[73] By 1817, though, some collectors adopted a more recognisably scientific approach. As Iolo Morganwg noted in a draft letter of that year concerning his proposed collecting of samples from the Lias (Early Jurassic) rocks of south Wales:

> I wish to be correct and as systematical as possible, proceeding on true geological principles, which is not merely to collect a number of morsels of uncommon, singular and, occasionally, beautiful fossils to be arranged orderly in a little pigmy cabinet of curiosities which may be term'd the baby houses of those who continue to be children, mere boys and girls, all the days of a long life.[74]

What is particularly interesting, too, is that Iolo sees the need for geological sections to be drawn and fossils collected in order to properly understand strata relationships:

> The gray or rag lias tract extends from Southern Down along the sea shore as far as Porthkery Bay, and thence to Penarth, the blue lias. The grey lias bassets up towards the east in an angle of about ten degrees with the horizon, with a few little anomalies, of which the causes are obvious. I have made a rough sketch of this tract in its longitudinal section … but subsequent observations and researches have convinced me that it is considerably defective. I hope to be able in the course of this coming autumn [at the age of seventy-two] to supply some of the deficiencies. Specimens are requisite.[75]

Iolo makes mention in the letter of having 'dipt into several very recent publications on British Geology', and while he makes no mention of William Smith, we might wonder whether he had seen Smith's famous geological map of England, Wales and part of Scotland published in 1815, and, perhaps, the early parts of Smith's *Strata Identified by Organized Fossils*, which appeared in four parts from 1816 to 1819.

Inevitably, as rocks were understood to be formed of layers of strata rather than being a formless mass left over from the Creation, it became clear that not all strata lay horizontally. Many were folded, deformed and fractured to varying degrees, and some were so completely over-turned that Steno's *law of superposition* was effectively reversed; with the oldest at the top and the youngest at the base. As the eighteenth century progressed a desire to understand the processes by which such deformation had taken place came increasingly to the fore. As Iolo Morganwg also wrote in his 1817 letter, it was in rocks 'wherein we find written in indelible character the history of the creation of this globe and its subsequent revolutions, the casualties that have befallen it, &c., &c.'.[76]

The Restless Earth

The earth as a place of dynamic and active forces was well known. Throughout history, earthquakes, volcanoes, landslides and tsunamis had provided often devastating confirmation. In his *A Compleat History of Europe* (1698), David Jones records the great 1692 earthquake at Port Royal in Jamaica that, through associated subsidence, consigned the grave of pirate/privateer Sir Henry Morgan to a watery eternity.[77] There were many other such events, including, in 1755, the great Lisbon earthquake that killed more than 60,000 people and destroyed much of the city.

In the absence of a rational scientific explanation for these events the inevitable recourse for many was to see them as acts of, or warnings from, God. As late as 1779, the Independent minister, Edmund Jones (1702–93), in *A Geographical, Historical and Religious Account of the Parish of Aberystruth*, in Monmouthshire, produced a valuable eighteenth-century consideration of the area's 'topography, soil, climate, demography and "natural curiosities"'.[78] At the same time, he reflects that 'the rocks, sinkings, gashes, and wounds, and scars of the earth' in and around Aberystruth can derive only from 'violent unnatural causes'. These might be 'Earth-quakes, violent shakings of the Earth, and long great Rains! But the chief cause must be the Universal Deluge'.[79] Yet, if such phenomena are related to the Deluge, why do they continue to afflict the world? Their cause, he argues, 'must be the sins of the World both Original and Actual'.[80] Thus 'the marks of God's displeasure upon the mountains of

Aberystruth are so many proofs … of the sins of the inhabitants of times past; and so many warnings not to displeasure the Lord in the time to come, to the end of the World'.[81] Jones believed '*the knowledge of Nature in every* BRANCH should be SUBSERVIENT *to Divinity*'.[82]

For other scientists of Wales in the second half of the eighteenth century it was perhaps caution with regard to theology rather than subservience that characterised their 'knowledge of Nature'. For example, Thomas Pennant, in 1750 and 1781, and John Lloyd of Wigfair, in 1783, had letters to the Royal Society describing earthquakes in north Wales published in the *Philosophical Transactions*. Their content, however, is pure reportage – from the shaking of china cups on shelves to descriptions of the geographic areas affected. There is no appeal to either a scientific or theological explanation for such phenomena. In the case of John Lloyd, aside from an apparent change in the content of his library around 1750 – away from theological texts and towards those with a more scientific content (see Chapter 8) – we know little of how his Anglicanism affected his scientific outlook. The caution of Thomas Pennant, though, is perhaps more evident. Having viewed a section of folded strata at Whitehaven in Cumbria, while *en route* to Scotland, Pennant later admitted in his *A Tour of Scotland, 1772* that such deformation might be the result of 'Operations of nature past my skill to unfold'.[83] Whether true or not, his caution at attempting to theorise might also relate to what Evans (1993) describes as Pennant's 'firm belief in the Biblical concept of the "Universal Deluge"'.[84]

The suggestion made by Furniss (2019) that Pennant's apparent 'reluctance to engage with geotheory or geohistory was characteristic of the natural history of the period'[85] may well have been true before 1780, but is, perhaps, a less certain characteristic in the period after that date. In July 1789, for example, shortly before he crossed the Channel to become embroiled in the early events of the French Revolution in Paris (see Volume 2), the one-time Dissenting minister and now teacher, George Cadogan Morgan, walked the beach in front of the White Cliffs at Dover with three travel companions, one of whom noted 'Morgan's admirable remarks on the formation of the earth, the changes which the globe has undergone, the constant fluctuation of the sea, and such other subjects'.[86] Rather than a complete lack of engagement with geotheory and geohistory, what is evident among the scientists of Wales by the 1780s is a continued attempt to unite biblical theology with geological theories and observations that, in themselves, challenged theological orthodoxy. It is a trait that we can see in the work of Montgomeryshire-born, Scotland-based mineralogist and geologist John Williams (1732–95).

Born in the parish of Kerry, in Montgomeryshire, Williams spent his early years working in the lead mines of mid-Wales, and he may have been in London in the 1740s as a cadet at the Royal Military Academy before being seconded, along with two others, to work with William Roy on the Military Survey of Scotland. Apart from spending his last years working in Italy (he died in Verona), Williams lived and worked for most of his life in Scotland.[87] In the same turbulent revolutionary year in which George Cadogan Morgan expounded his geological ideas at Dover, John Williams published *The Natural History of the Mineral Kingdom* (1789), which earned him a gold medal from the Russian empress, Catherine the Great. In his work Williams assesses two contemporary theories of rock formation. The first, Neptunism, stressed the importance of the action of water in rock formation, through 'chemical precipitation or erosion and deposition from the supposedly precipitated rocks'. The principal advocates of Neptunism were the French naturalist George-Louis Leclerc, Comte de Buffon (1707–88), who published his thirty-six-volume *Histoire Naturelle* between 1749 and 1788 and *Les Époques de la Nature* in 1778, and the German geologist Abraham Gottlob Werner (1749–1817).

The second theory, Plutonism or Vulcanism, advocated the importance to rock formation of heat; thus, rocks 'such as granite' were 'formed by the cooling and consolidation of magma deep under the earth's surface'[88] before they were forced upwards and exposed to the elements. Once exposed they were eroded with the resulting debris then transported to the ocean, deposited on the sea floor, and again formed into rock by heat from below. Heat, in relation to the formation of rocks became a topic of debate from the mid-eighteenth century, in part through an increasing interest in volcanoes. Between 1766 and

1780, for example, the British diplomat Sir William Hamilton discussed the geology of various Italian volcanoes in a number of papers in the *Philosophical Transactions*. According to Furniss (2019), Pennant too, in his *Tour of Scotland, 1772*, contributed to the concept that 'the British Isles harboured extinct volcanoes'.[89] In December 1791, Iolo Morganwg suggested the possibility of an extinct volcano at Brandon Hill in Bristol. He was mistaken in this instance, but his reasoning was based not only on antiquarian ideas – relating to the Saxon origin of the name Brandon Hill through '*Burnt Hill*' or '*Burning Hill*' – but also the evidence for 'heat' at nearby 'Hot Wells' in Bristol and, slightly further afield, at the roman baths at Bath.[90] In 1804, Joseph Banks wrote to John Lloyd of Wigfair: ' it is very curious that so many earthquakes have succeeded each other in your neighbourhood it seems as if the vast cauldron[?] under you were heating up for some future explosion which may singe the beard of some Goats of Snowdon.'[91] The leading British advocate for heat and Plutonism in the late eighteenth century is also one of the founders of modern geology, James Hutton (1726–97). His *Theory of the Earth*, which strongly advocated the Plutonist idea, was published in 1788, the year before Williams published *The Natural History of the Mineral Kingdom*.

In introductory comments to his work, John Williams references Buffon and interrogates the work of Hutton. He also makes his own position on the subject of rock formation very clear: 'I have treated of the stratification of the superficies of our globe by the agency of water.'[92] Williams presents cogent arguments for this position, and for his objection to Hutton's, based on his own ideas and field observations, but always within the constraints of the biblical Deluge and the 'prodigious height and force of the diluvial tides'. Williams was not alone in supporting the Neptunist idea in such terms. Writing under the pen-name 'Viator' in the short-lived *Cambrian Visitor* journal of 1813, the Welsh-speaking Englishman, and Swansea-based colliery owner, Charles Smith (d.1813), similarly wrote of Neptunism in relation to the Deluge, as did many others in Britain at the time.[93] Indeed, the Neptunist *versus* Plutonist debate would continue in Britain well into the nineteenth century. It only ended when it was finally realised that both water and heat are involved in the formation of sedimentary rocks (mainly by the action of water) and igneous/metamorphic rocks (through the action of heat).

Regarding another geological process – erosion by weathering – Williams 'frankly acknowledged the truth of almost all that the Count [Buffon] and the Doctor [Hutton] had advanced'; including their suggestion that the resulting 'spoils of the mountains are carried down by land-floods to the valleys and the borders of the oceans'. He did not accept, though, that these 'spoils' then passed into the very depths of the ocean. Instead, he offered an alternative, and this time purely geological scenario, whereby 'the sea purges itself by the tides of all the earthly matter carried down by the floods'. This 'earthly matter' was then 'thrown back upon the shores, in the bays and creeks, and at the mouths of great rivers, where, by degrees, it enlarges the bounds of the dry land in exact proportion to the quantity carried down by the floods'.[94] Once again both methods form part of modern geology.

The most famous and enduring part of Hutton's work is his presentation of what came to be called *uniformitarianism* by the mid-nineteenth-century. Like Steno's superposition, it is a fundamental concept for modern geology and states that geological processes at work today were also active in the past, and so the present can act as a guide to that past. It also had a profound implication for the theological dating of the Creation. Given the slow nature of most geological processes, the earth must be much older than Bishop Ussher's date of 4004 BCE; a conclusion also reached by the Comte de Buffon in his work. Here Williams responds neither through an alternative geological explanation nor through a combination of the religious and the geological. He argues solely from the constraints of his religious beliefs, since he simply cannot accept the great age of the Earth implicit in Hutton's long-term geological processes. 'That Dr Hutton aims at establishing the belief of the eternity of the world', Williams declares, 'is evident from the whole drift of his system', but:

> If once we entertain a firm persuasion that the world is eternal, and can go on of itself in the reproduction and progressive

vicissitude of things, we may then suppose that there is no use for the interposition of a governing power; and because we do not see the Supreme Being with our bodily eyes, we depose the almighty Creator and Governor of the universe from his office, and instead of divine providence, we commit the care of all things to blind chance.[95]

As with Neptunism and Plutonism, debate over the age of the earth and the significance (or otherwise) of the Deluge continued into the nineteenth century, and even beyond. The response of one scientist of Wales to the problem, as the eighteenth century came to a close, was to simply cover all the bases. Constantine (2003) has shown how the Plutonist ideas of Hutton, the Neptunist ideas of Werner and, at times, the Deluge or Noah's Flood all appear in Iolo Morganwg's 1791 manuscript notes on the supposed volcano at Brandon Hill near Bristol.[96] Nevertheless, as the new nineteenth century unfolds we also find Iolo, in the draft letter of 1817 discussed earlier, using some themes already described – collection and description of samples, close observation and study – to produce a very modern example of integrated geological thinking:

In the compact grey lias and also in the bastard lias we find the shells of the nautilus and gryphites in a petrified state. These shell fish are natives of the torrid zone [between the Tropic of Cancer and Tropic of Capricorn] and of the southern parts of the northern temperate zone [from Tropic of Cancer to Arctic Circle]; are found, I am told, in the Mediterranean along the coast of Africa, very rarely on the coasts of Italy, Spain, &c. But they are never seen in more northern latitudes. Of course, they are not natives of the British seas and, by calculating the changes of climate from the [precession] of the equinoxes, it will appear that Great Britain or its latitude could not have been within the torrid zone or in the southern part of the temperate zone much later than twelve or perhaps fifteen thousand years ago, and then it must have been under the seas where only petrifaction and stratification can possibly take place, according to what is known of the laws of nature at present.[97]

Also present in the passage is his acceptance of the new; it is evidenced by his adoption of an age for the earth well beyond that of biblical orthodoxy. Jenkins (2012) has suggested that 'By at least the early 1780's ... Iolo harboured grave reservations about the validity and the mission of the established church'.[98] Perhaps, therefore, it is his subsequent 'affinity with the doctrines and intellectual tradition of dissent' that allowed him to accept some of these new tenets of geology. Equally uncertain is the source of his ideas on the precession of the equinoxes and their impact on climate as a way of explaining the various fossil types that he is collecting in the south Wales Lias.

Precession of the equinoxes relates to the fact that the earth's axis of rotation (a line though the poles) wobbles, in the manner of a spinning top, due to the gravitational pull of the Sun and Moon on the Earth's equatorial bulge. As a consequence, the axis of rotation also traces out a circle on the sky, making a complete cycle once in $c.26,000$ years. 'The climatic effect of precession', as David Smith explains:

> is down to the fact that it moves the seasons around the Earth's elliptical orbit, so that, for example, the northern hemisphere summer (in the sense of longer days) may take place when the Earth is a bit nearer to the Sun, or a bit further away – hence summer can be a bit warmer or a bit cooler; the main period over which this happens being c. 26,000 years. Iolo clearly knows all this, as he has somehow worked out that the climate could have been warmer up to 12,000 to 15,000 years ago, so he knows that's when Wales would have been a bit nearer to the sun than it is now, enough to shift the climate belts north.[99]

Although the mathematics of precession was of interest to the likes of Isaac Newton and French mathematician Jean le Rond d'Alembert (1717–83), they made no detailed connection to climate. Indeed, it is not until references to geology and climate appear with Charles Lyell in 1830, John Herschel in 1832, and especially James Croll in 1864 that climate properly enters the picture. Yet, as David Smith also observes, Iolo refers to precession and its links to climate in a way that suggests common knowledge. So, beyond the possibility that the idea actually originated with Iolo, what were his possible sources for the link between precession and climate? He certainly could have learned about precession from Abraham Rees's monumental *Cyclopædia* (see Chapter 16), which was published in parts between 1802 and 1819. Part 55, published in 1814, contained a detailed account of precession but it made no comment on its links to climate. The only other association of precession with climate before Iolo's time seems to be in the 'Lectures and Discourse on Earthquakes' delivered by Robert Hooke to the Royal Society in 1668 and 1687.[100] Whatever might be the source of Iolo's ideas on precession they represent an early, and possibly the earliest, modern reference to its influence on climate. When added to his acceptance of a new age for the Earth, they also represent an example of integrated thinking and a 'widening of scientific horizons – both intellectual and geographical' – as the nineteenth century dawned.

Notes

1. George Cadogan Morgan, *Lectures on Electricity*, 2 vols (Norwich, 1794), I, pp. xli–xlii.
2. *ODNB*; Oliver M. Ashford, *Prophet or Professor: The Life and Work of Lewis Fry Richardson* (Bristol, 1985), p. 39, 244.
3. For all details, see Adrian Robinson, 'Introduction', in Geoffrey F. Budenberg (ed.), *Lewis Morris Plans in St. George's Channel: 1748* (Beaumaris, 1987). Also, Hugh Owen, *Life and Works of Lewis Morris* (Anglesey, 1951), p. xvii.
4. Rachel Hewitt, *Map of a Nation: A Biography of the Ordnance Survey* (2010; London, 2011), p. 311.
5. See *https://blog.library.wales/lovemaps-mary-ann-constantine-2* (accessed 7 August 2021); also, Derek Williams, 'John Evans' Map of North Wales, 1797', *Ystrad Alun*, 11 (Nadolig/Christmas 2010). Among the subscribers to the Evans map were Joseph Banks, and John Lloyd of Wigfair who ordered 100 copies.
6. As note 5, but see 'lovemaps', *https://blog.library.wales/lovemaps-mary-ann-constantine-4* (accessed 7 August 2021);
7. *ODNB*; Adams began his surveying career 'about 1672 plotting Welsh market towns within 100 miles of Aberdovey to help a Mr. Lloyd of Llanvorda, 2.5km west of Oswestry, plan his fish marketing': Blake Tyson, 'John Adam's Cartographic Correspondence to Sir Daniel Fleming of Rydal Hall, Cumbria, 1676–1687', *The Geographical Journal*, 151/1 (March 1985), 21–39 (at 21). This is possibly a reference to the father of Edward Lhwyd. The survey of England and Wales was never completed.

8. RSC LBO/11A/19 letter from John Caswell 'Giving an Account of the Torricellian Experiment, try'd on the Mountains Snowdon Cader Idris etc. with the height of these mountains taken by the said John Caswell then employed by Mr John Adams in the Survey of England and Wales', 3 August 1696.

9. RSC LBO/11A/19, letter from John Caswell 'Giving an Account of the Torricellian Experiment', in which he writes: 'Mr Flamsteed Astronomer Royal desired Mr Adams to have the torricellian experiment try'd here and accordingly sent all us all things required.'

10. Vincenzo Viviani (1622–1703), he too acted as an assistant to Galileo and edited the first edition of Galileo's complete works.

11. Lisa Jardine, 'Monuments and Microscopes: Scientific Thinking on a Grand Scale in the Early Royal Society', *Notes and Records: The Royal Society Journal of the History of Science*, 55/2 (May 2001), 289–308 (at 289–90); Robert Hooke, diary entry for Thursday 16 May 1678 'at Fish Street piller [*sic*] try'd experiment it [the mercury] descended at the top about 1/3rd of an inch'.

12. William Roy, 'Experiments and Observations made in Britain, in order to obtain a Rule for Measuring Heights with the Barometer', *Philosophical Transactions of the Royal Society*, 67 (1777), 653–787 (at 723).

13. William Roy, 'Experiments and Observations', see note 12.

14. Edmond Halley, 'A Letter from Mr Halley of June the 7th 97 concerning the Torricellian Experiment tryed on the top of Snowdon-hill and the success of it', *Philosophical Transactions of the Royal Society*, 19 (1697), 582–4.

15. Gunther, pp. 308–10 (at p. 310).

16. NLW MS 821C [William Williams], *A Survey of the Ancient and Present State of the County of Caernarvon by a Landsurveyor* (1806), p. 3.

17. Florian Cajori, 'History of Determinations of the Heights of Mountains', *Isis*, 12 (1929), 482–514.

18. Daniel Gabriel Fahrenheit, 'Barometri Novi Descriptio', *Philosophical Transactions of the Royal Society*, 33 (1724–5), 179–80.

19. George Shuckburgh-Evelyn, 'On the Variation of the Temperature of Boiling Water', *Philosophical Transactions of the Royal Society*, 69 (1779), 362–75.

20. See various entries at *https://sublimewales.wordpress.com/attractions/snowdon/ snowdon-scientists/snowdon-measuring-the-height* (accessed 23 May 2024).

21. NLW MS 12418D, 31 July 1778; NLW MS 12423, 24 July 1778.

22. See *https://sublimewales.wordpress.com/attractions/snowdon/snowdon-scientists/ snowdon-measuring-the-height* (accessed 23 May 2024): see entry '1778, Anon [J. M. John Matthews], 'A Surveyor of Wrexham, Chester and later Aberystwyth, "Tour through N. Wales in the Year 1778"', Central Library, Cardiff, MS 1.549, pp. 26–8, 47–9.

23. NLW MS 12461A.

24. Roy also knew James Lind and Charles Blagden, both of whom also sponsored Lloyd for the Royal Society. Lloyd, along with Roy and Blagden, was a member of the Royal Society Club, a select dining club established by Banks in 1775. It ran until 1784. Lloyd also attended Banks's famous scientific breakfasts when he was in London.

25. William Roy, 'Experiments and Observations made in Britain, in order to obtain a Rule for Measuring Heights with the Barometer', *Philosophical Transactions of the Royal Society*, 67 (1777), 653–787.

26. Greville was said, at this time, to be the lover of Emma Hamilton, Sir William Hamilton's wife, and later Nelson's mistress. Emma herself had started life in some poverty at Hawarden in Flintshire. See Royal Museums Greenwich, *www.rmg.co.uk/stories/blog/emmas-mother-unseen-power* (accessed 23 November 2022).

27. For full details of the Paris-Greenwich project, see Rachel Hewitt, *Map of a Nation*, at Chapter 3.

28. Hewitt, *Map of a Nation*, p. 114.

29. The ends of the baseline are marked by ships cannon set in the ground on the northern perimeter road at Heathrow and in Roy Close in Hampton, near Twickenham.

30. Hewitt, *Map of a Nation*, p. 152.

31. NLW MS 12415C, f. 23.

32. Edward Williams, William Mudge, Isaac Dalby, 'An Account of the Trigonometrical Survey carried on in the Years 1791, 1792, 1793, and 1794, by Order of His Grace the Duke of Richmond, late Master-General of the Ordnance', *Philosophical Transactions of the Royal Society*, 85 (1795), 414–591.

33. Edward Williams, *The Theory and Practice of Gunnery, treated in a new and easy manner. With the construction and use of an Instrument for readily solving the several cases, also rules for calculating the charges of mines, with remarks on Mr. Belidor's method. And various problems, of use to the practical gunner. To which are prefixed the elements of vulgar and decimal arithmetic, &c* (London, 1766).

34. See W. A. Seymour (ed.), *A History of the Ordnance Survey* (Folkestone, 1980), p. 23, available online; Hewitt, *Map of a Nation*, pp. 114–5. The genealogical site of Charles Olivier Blanc at *https://gw.geneanet.org* includes useful basic information detailing Williams' marriage on 30 October 1766 at Largo in Fife to Arabella Mallet (1745–1816), his Welsh heritage and elements of his career in America.

35. Hewitt, *Map of a Nation*, pp. 177–9, 187. According to Seymour (1980) he 'was the eldest son of a military family with an estate near Newcastle Emlyn in South Wales'.

36. Anita McConnell, *Jesse Ramsden (1735–1800): London's Leading Scientific Instrument Maker* (2007; Abingdon, 2016), pp. 194, 200.

37. NLW MS 12415C, f. 12.

38. See Hewitt, *Map of a Nation*, pp. 74–7.

39. NLW MS 12417C, f. 19.

40. See Hewitt, *Map of a Nation*, pp. 182–9 for details of the survey in Wales; pp. 189–91 for the internal survey; and pp. 191–5 for the toponymic questions and final appearance of the Welsh maps.

41. F. V. Emery, '"The Best Naturalist now in Europe": Edward Lhuyd, FRS (1660–1709)', *THSC* (1970), 59.

42. R. Paul Evans, '"A Round Jump from Ornithology to Antiquity": The Development of Thomas Pennant's Tours', in Mary-Ann Constantine and Nigel Leask (eds),

Enlightenment Travel and British Identities: Thomas Pennant's Tours in Scotland and Wales (2017; London, 2019), p. 34 n. 8.

43. *ML*-I, pp. 311, 318.

44. AML-II, p. 499.

45. *ML*-I, p. 336.

46. *ML*-II, p. 220. As required by the Museum rules, Richard Morris had booked his visit in advance, and he did so 'with a set of the Tower gentry'. This is likely to be a reference to the family of Joseph Harris. He was an Assay-Master at the Royal Mint, which was within the Tower at this time. Joseph died and was buried at the Tower in 1764.

47. AML-II, p. 499.

48. *ML*-I, p. 428.

49. See Tom Cotterell, *The Fabulous Mineral Collection of Lady Henrietta Antonia Clive, Countess of Powis*, at *https://museum.wales/articles/1104/The-fabulous-mineral-collection-of-Lady-Henrietta-Antonia-Clive-Countess-of-Powis* (accessed 23 May 2024).

50. For Pennant, see Evans, 'A Round Jump from Ornithology to Antiquity', 17; for Woodward, see David E. Bick, *The Old Metal Mines of Wales: Part 4, West Montgomeryshire* (Newent, 1977), Appendix II, p. 62.

51. G. T. Roberts, 'Humphrey Edwards (1730–1788)', *Transactions Caernarvonshire Historical Society*, 25 (1964), 13–22.

52. *ML*-I, p. 433.

53. *ML*-I, p. 111. John Ray (1627–1705, FRS 1667) completed the work on *Historia Piscium* ('History of Fishes') begun by Francis Willoughby (1635–72, FRS 1663) who died in 1672. It was published under the auspices of the Royal Society in 1686. Ray also completed Willoughby's *Ornithologiae libri tres* (1676; English edition, 1678). These are probably the two works referenced by Lewis. Ray and Willoughby had undertaken to reform the study of natural history, with Ray on plants and Willoughby birds, fishes and insects. See Sachiko Kusukawa, '"Historia Piscium" (1686)', *Notes and Records: The Royal Society Journal of the History of Science*, 54/2 (2000), 179–97. Lewis Morris's personal annotated copy of *Historia Piscium* can be seen at *library.wales/discover-learn/digital-exhibitions/manuscripts/early-modern-period* (accessed 23 May 2024) (NLW MS 24052E).

54. See National Museum Wales Mineralogical Database, *https://museum.wales/mineralogy-of-wales/database/?mineral=31&name=Anglesite* (accessed 23 May 2024).

55. See National Museum Wales Mineralogical Database, *https://museum.wales/mineralogy-of-wales/database/?mineral=402&name=Pyromorphite* (accessed 23 May 2024).

56. See Richard Fortey, *Trilobite: Eyewitness to Evolution* (2000; London, 2001), pp. 24, 43–6. Also, Edward Lhwyd, 'Part of a Letter from Mr. Edw. Lhwyd to Dr. Martin Lister, Fell. of the Coll. of Phys. and R. S. Concerning several regularly figured stones lately found by him', *Philosophical Transactions of the Royal Society*, 20 (1698), 279–80.

57. Roy Porter, *The Making of Geology: Earth Science in Britain 1660–1815* (1977; Cambridge, 2008).

58. In 1704 he wrote a description of Anglesey agriculture in his *Idea Agriculturae,* though this would not be published until 1764, and his antiquarian interests were reflected in *Antiquitates Parochiatus* (1710) and *Mona Antiqua Restaurata* (1723).

59. Andrea Rusnock, *Correspondence of James Jurin (1684–1750)* (Amsterdam and Atlanta GA, 1996), p. 67.

60. Frank Emery, 'Edward Lhuyd and A Natural History of Wales', *Studia Celtica,* 12–13 (1977–8), 247–58; T. P. T. Williams, 'The Lost Geology Book of the Reverend Henry Rowlands of Llandinam', *Transactions of the Anglesey Antiquarian Society & Field Club* (2007), 11–24.

61. See Rusnock, *Correspondence of James Jurin,* pp. 65–9 (at p. 67 and p. 69).

62. Emery 'Best Naturalist', 65–7.

63. Rusnock, *Correspondence of James Jurin,* p. 67.

64. See Emery, 'Best Naturalist', 60, 65–7; Gunther, p. 382.

65. William Daniel Conybeare (1787–1857, FRS 1832) was born in London. A geologist and Anglican minister he undertook much work on the geology of south Wales; including laying the foundations of our understanding of the south Wales coalfield. He was appointed Dean of Llandaff Cathedral (1845–57); see *DWB* entry.

66. Cited in B. G. Charles, *George Owen of Henllys: A Welsh Elizabethan* (Aberystwyth, 1973), p. 147 and n. 4.

67. F. J. North, *Coal, and the Coalfields in Wales* (Cardiff, 1931), p. 162.

68. See 'A History of Pembrokeshire, by George Owen, Esq. of Henllys, Lord of Kemes; with additions and observations by John Lewis, Esq. of Manarnawan … and now first published from the original by his great grandson, Richard Fenton, Esq.', in *CR*-II, 53–231 (at 88).

69. See David Oldroyd, *Thinking About the Earth: A History of Ideas in Geology* (London, 1996), pp. 63–7.

70. Porter, The *Making of Geology,* p. 58.

71. Tom Furniss, '"As if Created by Fusion of Matter After Some Intense Heat": Pioneering Geological Observations in Thomas Pennant's Tours of Scotland', in Mary-Ann Constantine and Nigel Leask (eds), *Enlightenment Travel and British Identities: Thomas Pennant's Tours in Scotland and Wales* (2017; London, 2019), p. 169.

72. M. F. Howells, 'British Regional Geology: Wales' (Keyworth, Nottingham, British Geological Survey, 2007); taken from *earthwise.bgs.ac.uk/index.php/History_of_the_geological_research_of_Wales* (accessed 5 May 2021).

73. *ML*-I, p. 458.

74. *CIM*-III, p. 438.

75. *CIM*-III, p. 438.

76. *CIM*-III, p. 439.

77. [David Jones], *A Complete History of Europe* (London, 1699), pp. 475–6.

78. *ODNB* entry for Edmund Jones.

79. Edmund Jones, *A Geographical, Historical and Religious Account of the Parish of Aberystruth* (Trevecka, 1779), p. 29.

80. Jones, *A Geographical, Historical and Religious Account,* p. 31.

81. Jones, *A Geographical, Historical and Religious Account,* pp. 33–4.

82. Jones, *A Geographical, Historical and Religious Account*, p. 41.

83. Furniss, '"As if Created of Matter After Some Intense Heat"', pp. 165–6.

84. See Ronald Paul Evans, 'The Life and Work of Thomas Pennant (1726–1798)' (unpublished PhD thesis, University College of Wales, Swansea, 1993), 7; Evans also notes that Pennant titled one of his notebooks *Reliquiae Diluvianae or a catalogue of such Bodies which were deposited in the Earth by the Deluge* (p. 130).

85. Furniss, 'As if Created of Matter After Some Intense Heat', pp. 165–6.

86. See Mary-Ann Constantine and Paul Frame (eds), *Travels in Revolutionary France & A Journey Across America* (Cardiff, 2012), p. 78. Having been living in Norfolk, Morgan may well have been developing ideas presented in 1746 to the Royal Society by 'the Norfolk gentleman naturalist William Anderon'. The strata of the Norfolk cliffs, Anderon suggests, were 'frequently the same kind many times repeated; as if at one time dry Land had been the Surface, then the Sea; after, morassy Ground; then the Sea, and so on, till these Cliffs were raised to the Height we now find them'; Porter, *Making of Geology*, p. 122 citing BM Add. MS 27966, f. 62–3.

87. For biographical and work details see *DWB*, *ODNB*, and especially Hugh Torrens, '"Mineral Engineer" John Williams of Kerry (1732–95)', *Montgomeryshire Collections*, 84 (1996), 67–102.

88. Definition quotations taken from Oldroyd, *Thinking about the Earth*, pp. xxii, xxiv.

89. Furniss 'As if Created of Matter After Some Intense Heat', p. 177.

90. NLW MS 13089E, pp. 174–6, *Cursory Observations on Brandon Hill, Clifton Hill etc near Bristol – Decr 19th 1791. By E. Williams*. See transcription and discussion in Mary-Ann Constantine, *'Combustible Matter': Iolo Morganwg and the Bristol Volcano* (Aberystwyth, 2003).

91. NLW MS 12415C f. 42; Joseph Banks to John Lloyd, 22 January 1804.

92. John Williams, *The Natural History of the Mineral Kingdom in Three Parts*, 2 vols (Edinburgh, 1789), I, p. xviii.

93. *Cambrian Visitor* for 27 February and May to August 1813, available at 'Welsh Journals Online'; for a brief obituary of Charles 'Viator' Smith, see issue for May 1813, pp. 290–1.

94. Williams, *The Natural History of the Mineral Kingdom*, I, p. xxvi.

95. Williams, *The Natural History of the Mineral Kingdom*, I, p. lix.

96. Constantine, *'Combustible Matter'*, pp. 8–9.

97. *CIM*-III, p. 436. In the published letter the word precession has been rendered as 'precision'. A re-examination of the original indicates 'precession' as the more correct reading of Iolo's handwriting.

98. Geraint H. Jenkins, *Bard of Liberty: The Political Radicalism of Iolo Morganwg* (Cardiff, 2012), pp. 66, 69.

99. David Smith (personal Communication, 14 July 2021). I am very grateful to Dr Smith for his help in understanding Iolo's writing on this subject.

100. For a discussion of the papers by Hooke, and the relation of earthquakes to precession and possible climate effects, see Oldroyd, *Thinking About the Earth*, pp. 61–2.

10

THE SCIENCE OF NATURE

At a time, when the study of natural history seems to revive
in *Europe*; and the pens of several illustrious foreigners
have been employed in enumerating the productions
of their respective countries, we are unwilling that our
own island should remain insensible to its particular
advantages; we are desirous of diverting the astonishment
of our countrymen at the gifts of nature bestowed on
other kingdoms, to a contemplation of those with which
(at least with equal bounty) she has enriched our own.[1]

Thomas Pennant, *British Zoology* (London, 1761–77)

In the previous chapter we surveyed the contributions made by scientists of Wales to the study of the physical world. In this chapter, we focus on their contributions to the study of nature. These contributions were principally made in the sciences of botany (flora) and zoology (fauna), together with agriculture, horticulture and forestry. Their study in Wales, as elsewhere, followed a similar path to the physical sciences – from intense collecting of specimens, through their display and classification, to identification of the new. It is this path to the expansion of knowledge that is reflected in the opening quotation to this chapter, with its language of enumerating the productions and gifts of nature. It is also the path that underpins the structure of each of the sections below, and we begin with the study of soil – an often-ignored link between the physical and natural sciences.

On Soil

Composed of both organic and inorganic materials, soils of various chemical and physical properties cover most land surfaces and the *Encyclopaedia of Wales* (2008) provides us with a succinct description of their importance and mode of formation:

> They are the foundation of all terrestrial ecosystems, including agriculture and forestry, providing a vital link between physical environment and living world. Formed at the surface of the earth, soils are natural entities resulting from the interaction of climate and vegetation with weathered geological materials.[2]

When outlining key historical concepts in the development of soil studies, Bockheim et al. (2005) suggest that the pre-1880 concept of soil can be summarised as being 'a medium for plant growth and as a weathered rock layer'.[3] Yet some scientists of Wales in the long eighteenth century did attempt to go further by delineating different soil types, albeit at a basic level. Among the topics Edward Lhwyd included in the 1695 'design' for his proposed, but never realised, *Natural History of Wales*, was 'A General Description of the Country, in respect of its situation, and quality of the soyl'. Consequently, his widely distributed 'parochial queries', which were sent out in 1696 with the aim of creating a database for the *Natural History*, requested respondents to provide not only details of the colour and fertility of their soils, but also whether they could be classed as 'Mountainous or Champion Ground? Woody, Heathy, Rocky, Clay-Ground, Sandy, Gravelly, &c?'[4]

Neither Lhwyd's questionnaire nor his approach to basic soil classification was unique. Something similar had appeared in the very first volume of the *Philosophical Transactions of the Royal Society* in 1665; here, though, the questions were linked to agriculture rather than to a broad natural history. Published as *Enquiries concerning Agriculture* the questionnaire contained a section 'For Arable'. In it, respondents were asked to state, in terms very similar to those in Lhwyd's questionnaire, the soil type most common in 'your country'.[5] These *Enquiries* were intended to be part of a survey of English agriculture made under the auspices of the Royal Society's *Georgical Committee* (from the Greek *geōrgika* meaning agricultural things, also *Georgics*, a poem by Latin poet Virgil on agriculture). Established in March 1664, an opportunity arose in late September 1665 to widen the survey when Sir Robert Moray, a founding Fellow of the Society who we first met in Chapter 2 as patron to the Welsh alchemist Thomas Vaughan, proposed making a personal 'incursion into Wales'. Shortly before making his incursion, Moray received a letter from Henry Oldenburg (1618–77, FRS 1660), a founding Fellow and the first Secretary of the recently formed Royal Society, and the foundational editor, in 1665, of its *Philosophical Transactions*. In the letter Oldenburg is clearly keen to obtain information to widen the survey:

> I am assured your journey into Wales will be also of advantage to our society, of which you will, I know, remember yourself to be a fellow, wherever you are. I wish you might in that country [Wales] meet with an acquaintance, yet could and would impart to you ye state and practice of their husbandry, and answer our queries; of which if you have no copy with you, either per [Robert] Boyle, or Dr [John] Wallis, may furnish you with a copy of theirs.[6]

Within Wales, soil and its condition had long been considered a part of general husbandry. As early as 1599, George Owen of Henllys wrote a 'Treatise on Marle' and summarised it in his later *Description of Pembrokeshire* (1602/3). The treatise discussed the value of marling as an aid to growing 'very good and deep hay and the finest grass that may be'; Owen even invented a 'marl slice', as an alternative to the more commonly used hatchet, for digging in marl pits.[7] Marl is essentially claystone mixed with calcareous material such as shell fragments. It is the decay of these fragments that acts to counter the acidic nature of the soils found in many parts of Wales. Increasingly important too was the more concentrated application of calcareous material in the form of lime slaked with water (calcium hydroxide); the lime being derived from burning limestone in kilns scattered across the country. As Jenkins (1989) has noted, 'from the 1680s onwards the livelier spirits within the farming community were becoming more responsive to new techniques in husbandry'; including the more extensive use of 'lime, marl, sand, dung and muck as manures', with seaweed, shells and sand used nearer the coasts. Along with agricultural theories of the day and accounts of local experiments, the cleric Henry Rowlands (1655–1723), in his *Idea Agriculturae: the principles of vegetation asserted and defended* (written in 1704 but not published until 1764, in Dublin), presented 'cogent arguments' in favour of wider use of these new farming techniques, including the use of marl and lime.[8]

Liming became a particularly favoured method of soil treatment in the mid to late eighteenth century as the desire for land 'improvement' became a priority.[9] 'Improvement' is defined by David Ceri Jones (2005) as 'the management and cultivation of land to make it more profitable' and, from the mid-eighteenth century onwards, it was being driven by a number of social and political developments. These included an increasing population and occasional crop failures; wars restricting international trade and the importation of goods; and the growth in tenant farming resulting from the enclosure of once common land by landed gentry to create estates whose owners, unlike many of their tenants, had the time, money and sometimes the expertise to experiment with new techniques.[10]

One 1750s' response to the desire for 'improvement' came with the establishment of agricultural societies; an early formation in Britain, and the first in Wales, being the Brecknockshire Agricultural Society of 1755. Others soon followed and by 1818 every Welsh county had one. Their early membership comprised mainly gentry landowners, and it is probably fair to say that when 'meeting monthly for bucolic evenings around (or sometimes under!) the dining table'[11] agricultural and soil science did not figure prominently, if at all. Consequently, even though the societies adopted the need for improvement, the response through the second half of the eighteenth century largely remained one of offering prizes for the most 'improved' crops, vegetables and well-bred livestock. Improvements achieved through quasi-scientific trial and error, rather than a true scientific understanding of the chemical and biological processes involved in their production. Nor were the rewards of improvement always forthcoming, as Lewis Morris notes in a 1765 letter to his London-based brother Richard:

> You think me happy because I have no hard labour, and I think you happy because you have it, for that you have the consequences of it which is profit, and I have nothing but a dog's life, except the unprofitable business which I cut out for myself of gardening and farming, by which I lose yearly, though I improve the land, etc., for my successor in it.[12]

Nevertheless, as the nineteenth century dawned, some of the older generation of landowners did begin to take greater note of recent scientific advances. For example, in 1796 the Merthyr ironmaster and landowner, Richard Crawshay (1739–1810), wrote to gentleman farmer John Franklen (1734–1824), of Llanmihangel, recommending *A Treatise, shewing the intimate connection that subsists between agriculture and chemistry* (1795). Franklen had been one of the principal movers in the foundation of the Glamorgan Agricultural Society in 1772.[13]

In the main, though, Lewis Morris was right. It would be a younger generation of landowners who would benefit when they, and some tenant farmers, began to promote and adopt a more scientific approach to 'improvement'. Scotsman Sir John Sinclair (1754–1835), another correspondent of Richard Crawshay,[14] was eleven when Lewis Morris wrote his letter of 1765 to his brother Richard. Later, as an advocate of land enclosure and a prime mover in the establishment of the London government's Board of Agriculture, Sinclair commissioned the early nineteenth-century agricultural surveys undertaken in Wales by Gwallter Mechain (Walter Davies, 1761–1849). Aged four at the time of the Lewis letter, Mechain was born in Montgomeryshire and educated at Trinity College, Cambridge. Although he took holy orders in 1795, and became a poet and antiquarian, he also kept some 300 field notebooks during his later travels around Wales. These, together with material contributed by Iolo Morganwg, became the basis for Mechain's Board of Agriculture reports, of which there were three: one covering the counties of north Wales, published in 1810, with two on south

Wales following in 1815. Although essentially surveys of existing agricultural practice, they had a basic scientific rationale and included sections on climate, meteorology and soil types. They also concerned themselves with obstacles to improvement, as well as identifying those areas where improvements could be implemented.[15]

Another of the younger generation of landowners, Thomas Johnes of Hafod, was seventeen when Lewis Morris wrote his 1765 letter. In 1800 Johnes published his *Cardiganshire Landlord's Advice to his Tenants*, along with a translation of it into Welsh; the latter a not insignificant fact given the limited material then available on agricultural and scientific topics in the indigenous language. Comprising a series of extracts on modern farming techniques taken from publications of the day, the work included practical advice as well as details of Johnes's own crop and animal experiments. He would also voice his opinion on the dangers of the indiscriminate use of liming, with confirmatory evidence coming from his own observation and experimentation. The published letters of Johnes also reveal his interest in the most recent developments in agricultural science.[16]

Eighteen at the time of the 1765 Lewis Morris letter, Iolo Morganwg had become a stonemason and knowledgeable proto-geologist by 1817 when he expressed, in recognisably scientific terms, his understanding of how soil formed, and the importance of such knowledge for agricultural husbandry:

It has been observed and tolerably well established that the soils, or superstrata, that lie on different kinds of rock are only those rocks decomposed by the operations of the atmosphere, and that but few of those soils contain all the ingredients or substances that constitute a soil of the highest degree of fertility. Some of those substances are very defective in some places; in others they are redundant. These things well understood would enable gentlemen to form such judgments of various soils on their estates that would enable them to discover what substance was deficient and what redundant in them, and clearly point out the most effectual means of permanently improving their various soils.[17]

Consequently, Iolo declared: 'I am fully convinced that a knowledge of the substrata, and of their properties is very necessary ... Thus we should collect data whereas to ground subsequent agricultural studies.'[18] One of the remedies he proposed in 1817 for countering the soil husbandry problems in his own county of Glamorgan was the formation of a Geological Society, the London-based Geological Society having been founded not long before (1807). Iolo, too, voiced concern over the indiscriminate use of liming, while at the same time observing and describing its consequences: 'Soils that are too strongly calcareous are always very productive of coltsfoot, a baneful plant, extremely hurtful to every kind of crop.'[19]

As an acute observer and recorder of fauna and flora, as well as geology, Iolo had noticed how certain types of flora seemed to prefer a particular subsurface geology:

Some botanical characteristics are almost peculiar to the grey lias. The elm, viburnum guelder rose, maple leaved service or azarole, wild gooseberry, wild apricot, acanthus or bearsfoot, &c., &c. are indigenous and some of them amongst the rarest vegetable production of our island ... On the other hand, the mountain ash, beech, hornbeam, linden, birch, alder, birdcherry, broom, heath, foxglove, &c., &c. are no where to be found, of indigenous growth, on the grey lias.[20]

Beyond confirming his encyclopaedic knowledge of the geology and natural history of his county of Glamorgan, the quote again places Iolo within the scientific developments of his time. For example in France, Jean Étienne Guettard (1715–86), dubbed 'the father of all the national Geological Surveys',[21] noted the relationship between plants, soils and subsoil, when working with the French chemist Antoine Lavoisier (1743–94), to produce his 1780 mineralogical map of France.[22] In the early nineteenth century, as Iolo was writing, the German traveller, explorer and naturalist Alexander von Humboldt (1769–1859) was concentrating on much larger, continent-wide plant distributions in relation to geology.[23] However, the origins of Iolo's more localised perspective

– linking plant types to specific local geology – may actually have come from his knowledge of the work of the Swedish botanist Carl Linnaeus (1707–78), who discussed 'the habitats of plants, with reference to the physical conditions by which they appeared to be determined'.[24]

Iolo certainly knew of Linnaeus because he briefly mentions him in a letter written in 1811 to the Caerphilly-born, London-based philosopher David Williams.[25] One possible source of this knowledge is the translation of Linnaeus's Latin work into English that was published in Swansea between 1800 and 1806 (see below). Furthermore, John Morris-Jones, in his study of traditional Welsh poetry, *Cerdd Dafod* (1925), suggests the classification scheme for the natural world established by Linnaeus in *Systema Naturae* (1758) is also reflected in a revision of the traditional metres of Welsh poetry proposed by Iolo.[26] The poetic link is entirely plausible. Another name mentioned in Iolo's letter to David Williams is the botanist and natural philosopher Erasmus Darwin (1731–1802), who declared the work of Linnaeus to be 'unexplored poetic ground, and an happy subject for the muse'.[27] Darwin even used the Linnaean scheme in a famous 1789 poem, which Iolo also mentions in his letter to David Williams, called 'The Loves of the Plants'.[28]

Flora and Fauna

The collecting of flora and fauna by scientists of Wales in the long eighteenth century was carried out with as much intensity as that recorded in the previous chapter for minerals and fossils. However, botanical and zoological collections are more prone to decay and the depredations of time than rocks and minerals, and, for that reason, more likely to be lost or dispersed on the death of their owner. Nevertheless, some idea of their nature can be gained from what remains; and we begin here with the botanical and zoological collecting of two pre-eminent Welsh naturalists of the long eighteenth century who we also met in the previous chapter: Edward Lhwyd and Thomas Pennant.

One of the principal aims of Edward Lhwyd during his time at the Ashmolean Museum in Oxford was to expand its botanical and zoological collections, as well as to contribute to the botanical collection of the Oxford Physic Garden, which had been founded in 1621 for the growing of plants for medicinal purposes. One of Lhwyd's earliest foraging expeditions had been to Snowdonia in 1682. In 1686, he presented to the Oxford Philosophical Society 'a paper containing an account of some plants, which grow in North Wales, & are omitted in Mr Ray's catalogue'. John Ray (1627–1705), the foremost botanist of the period, had collected in Wales in 1658 and 1682.[29] Another source of specimens for Lhwyd was his 'Great Tour'. Undertaken between 1697 and 1701, this again took him through Wales, and to the south-west of England, Scotland, Ireland and Brittany. Comprehensive details of a journey that covered some 3,000 miles can be found elsewhere.[30] Suffice to say here, it was a mammoth undertaking that involved viewing and/or collecting geological, botanical and archaeological material, together with items of antiquarian interest, including ancient manuscripts and inscriptions. The scale of some of Lhwyd's collecting intentions are well illustrated by a pre-departure order he placed for some '10 Grosses of round boxes [=1440]' and 'half a grosse of square [= 72]' to hold expected specimens.[31] As with his fossil and mineral collections, Lhwyd used his zoological and botanical specimens for comparative purposes, and as an aid to developing rational explanations for otherwise mysterious or unexplained events.[32]

In collecting specimens, Lhwyd wanted to do more than follow established tourist routes. When giving instructions to a north Wales correspondent – David Lloyd (1660–1703) of Blaenyddôl, Corwen – on how and where to collect, Lhwyd recommends using a mountain guide 'who must not direct you in the easiest way of goeing up; but must bring you to all the steep and craggie cliffs, yt are, (tho but difficulty) accessible'.[33] Difficult to reach areas, away from well-scavenged tourist trails, were clearly the places for a scientist to find rare or new specimens. Taking his own advice, it was on 'the highest rocks of Snowdon' that Lhwyd discovered the Snowdon Lily, subsequently named *Lloydia serotina* and today, *Gagea serotina*, which is a species of spiderwort and now known from some of the highest mountains in various parts of the globe. The same advice saw him climbing 'the bare quartzite peak of Nephin' (2,646 feet) in County Mayo, Ireland. His finds in such difficult country helped render his collections from Ireland 'a landmark in the history of Irish botany'.[34] In all, Lhwyd would provide 'the first British record of seventeen plants, as well as the first Welsh record of very many mountain plants'.[35]

Lhwyd's fields of interest were primarily geology and botany, but one of his first jobs as Underkeeper at the Ashmolean Museum had been to catalogue its zoological collections for the Principal of Brasenose College.[36] That his zoological interest continued is evident in his surviving letters. For example, in the 1690s we find him transcribing for the naturalist Martin Lister references in existing literature to cuttlefish, squid and jellyfish.[37] At the same time, he was always on the lookout for the new, and, in his short *Philosophical Transactions* paper – 'An account of some uncommon Plants growing around Pensans [*sic*] & St. Ives in Cornwall' – he records how he had 'met with no birds or fish since our coming hither, that I suspect for undescribed: Only two or three *Stellae* ['starfish'] and some other *Exanguia marina* ['aquatic crustacea'?] have occur'd which I have not seen before on our British Coasts'.[38] Lhwyd would lecture on starfish in a series of natural history lectures that he gave at Oxford.[39]

When making his own travels in the mid-to-late eighteenth century the naturalist and antiquary Thomas Pennant did not intend being seen as just another 'Topographer' perusing landscapes in the romanticised manner then becoming common. He wanted to be viewed as 'a curious traveller willing to collect all that a traveller may be supposed to do in his voyage', and with an accompanying desire for 'accuracy'.[40] Pennant made excursions into England, Ireland and the continent, but it was his travels through Wales and Scotland that resulted in two of his most famous works – *A Tour in Scotland* (1769 and 1772–6) and *A Tour in Wales* (1778–83).[41] Both contain antiquarian details interlaced with perceptive observations on natural history and geology. Some of his other publications, though, have more rigorous zoological titles. They include *British Zoology* (1761–6, published in parts by the Cymmrodorion Society, with eleven plates of mammals and 121 of birds); *Indian Zoology* (1769); *The Synopsis of Quadrupeds* (1771); *The History of Quadrupeds* (1781); *Genera of Birds* (1781), which was originally written in 1772 for

Robert Ramsay, Professor of Natural History in Edinburgh, to use as a teaching aid; and *Arctic Zoology* (1784–5). Pennant also published a number of papers in the *Philosophical Transactions* of the Royal Society,[42] to which he was elected a Fellow in 1767. From his home at Downing, in Flintshire, he became associated with the Cymmrodorion in London, and was honoured with membership of the Royal Swedish Society of Sciences in Uppsala (on the recommendation of Linnaeus), the Royal Swedish Academy of Sciences in Stockholm, the Royal Physiographic Society in Lund, the Society of Antiquaries of Scotland (he was also given the Freedom of Edinburgh), and the American Philosophical Society. Numerous zoological taxa were first described by Pennant and many others named after him, and, like Edward Lhwyd before him, he used his collections as a valuable scientific resource. When Joseph Banks sent him the skin of a penguin collected in the Falklands to add to his collection, Pennant went on to write a paper for the *Philosophical Transactions* describing all the various species of 'Pinguins'.[43]

In writing *Indian Zoology* (1769), *Arctic Zoology* (1784–5) and a twenty-two-volume unpublished manuscript titled 'Outlines of the Globe' (today in the National Maritime Museum at Greenwich),[44] Pennant did not actually visit the places they mentioned. However, these works were not products of his imagination. He travelled to the places mentioned intellectually, with a mind fuelled by reading, by maps and by an enormous correspondence and acquaintance with other natural historians and collectors, some of whom provided him with relevant scientific information, access to their own collections, and actual specimens from the places he was writing about.[45] For *Arctic Zoology*, Joseph Banks provided ornithological data from a visit he had made to Iceland in 1772. Banks had also visited Newfoundland and Labrador in 1766. Thomas Hutchins (d.1790), a surgeon in the Hudson Bay Company, provided manuscript observations 'in a large folio volume: in every page of which his extensive knowledge appears'.[46] Pennant originally intended what became *Arctic Zoology* to have been a work on the *Zoology of North America*, but the latter title was changed as a result of his dislike of the 1776 American Revolution: 'the most deplorable event in the annals of Great Britain'.[47] Consequently, *Arctic Zoology* contains material from well south of the Arctic Circle, a considerable part of which came from Pennant's access to the personal collection/museum created by the largely self-taught naturalist and correspondent of Linnaeus, Anna Blackburne (1726–93) of Orford Hall, in Lancashire.[48] This included some 100+ bird specimens from New York (Blackburne received material from her brother in America), as well as a number of fish, insects and a mammal. For *Indian Zoology* Pennant received help from the collections made by John Gideon Loten (Johannes Gideon Loten, 1710–89, FRS 1760) during service in the Dutch East India Company. Loten's collections were subsequently donated to the British Museum.

So, what has become of the collections of Edward Lhwyd and Thomas Pennant today? The remains of Lhwyd's botanical specimens are to be found in a number of different collections currently held in Oxford and London. They include those of Robert Morison (1620–83; a Professor of Botany at Oxford), William Sherard (1659–1728), James Petiver (*c*.1663–1718), and the herbarium of Sir Hans Sloane held in the Natural History Museum in London.[49] Sadly, and despite his early cataloguing of zoological collections in Oxford, Lhwyd's own zoological material appears to have suffered considerable loss. By contrast, the collections of Thomas Pennant, which now reside in the Natural History Museum in London, remained mostly untouched at Downing, his home in Flintshire, from the time of his death in 1798 until 1912 when their then owner, the Earl of Denbigh, presented them to the trustees of the British Museum. A 1913 report in the journal *Science* notes that apart from 'several volumes of a manuscript catalogue', there were *c*.860 minerals, more than 1,000 fossils, 'a few mammals, fishes and crustaceans' and some '140 birds'. Among the last were 'the only two known specimens of the extinct British race of capercailzie [a member of the grouse family], as well as the originals of many birds figured in the "British Zoology" (1766)'. The collection also contained many recent shells, including '16 type-specimens and 70 figured specimens, all described in the "British Zoology"'.[50]

Shell collecting ('conchology') proved popular throughout the long eighteenth century and another of the early acts of Edward Lhwyd at the Ashmolean Museum was to make a complete catalogue of the museum's shell collection (1684). With little previous interest in shells this involved him in extensive reading of available works, with a particular reliance placed on Martin Lister's *Historiæ Animalium Angliæ* (1678). Lhwyd later described his cataloguing work in a letter to Lister:

> I must confesse yt when I began this Catalogue, I was alltogether ignorant in ye Historry & Method of Shells; but having a good Collection at hand I first disposed them in such Method as seemd to me most agreeable to their Nature making all Shells congenerous, wch I thought to agree in figure; & then consulted all Authors yt had treated on yt Subject for every Species wch are about 800. But I finde yt none besides yr self ... have hitherto distinghd shells secundum genera et Species [according to genus and species], much lesse reduced ym under method, as Plants & ye other animals are done.[51]

The catalogues were essentially 'inventories with little analysis or discussion' beyond Lhwyd's own occasional comments. He would ultimately catalogue 'all the natural history collections of the museum'.[52]

When issuing directions to his would be collector friend in Wales, David Lloyd of Corwen, Lhwyd had urged him to 'Put up all Sorts of Snayl Shells you meet with; all sorts of River Muscles, or any other Sweet Water Shell'.[53] In the mid-eighteenth century, Lewis Morris, writing to his brother William, described his own eventful 1753 weekend collecting shells in 'Dovey'; the letter also illustrating how the pursuit of natural history is sometimes made against a backdrop of now forgotten social difficulty and economic possibility:

> Just returned from [Aber]Dovey, endeavouring to settle things there. Things are there in better order than the rest, only they riot a little now and then, break our windows and threaten our officers, etc. [Lewis was collector of customs at Aberdovey between 1746 and 1756]. Staid there from Friday to Monday gathering shells for Lady Lincoln and some money for self etc. If you have idle people about you, gather as many shells as you can of all sorts, and sort

them in strong paper bags, and fill boxes with them, and I will tell you how to make the shells your friends by recommending you to great folks … Gather also all the sea spoils, bron alarch, sponges, white honey comb, skate and dog spawn bags, sea mosses etc., to throw among the shells for variety.[54]

As one of the great folk, 'Lady Lincoln' was Catherine Pelham (1727–60), daughter of the Prime Minister, Henry Pelham (PM 1743–54), and wife to Henry Fiennes Pelham-Clinton (1720–94), the ninth earl of Lincoln. Lewis Morris had dealings with both Henry Pelham and the Earl at this time; including being appointed the Earl's steward for Shropshire, Montgomeryshire and Denbighshire on 23 October 1750.[55]

In the later eighteenth century, Huntingdonshire-born William Lyons (1766–1849), who lived in Tenby from at least 1796 until his death in 1849, amassed a substantial collection of local Pembrokeshire shells between 1808 and 1831; together with a few additions from Ireland and the coast of England. Some 216 species are represented, and the collection is today held in Tenby Museum. It is believed to constitute the oldest surviving natural history collection still in Wales. Also present in the museum's archive is a manuscript by Lyons entitled: 'A List of shells found on the seashore at Tenby in Pembrokeshire'. Although unpublished, the manuscript is of great value since, in our age of environmental change, it allows a comparison to be made between the taxonomic diversity of the area 200 years ago and that of the present-day.[56]

Another Pembrokeshire shell collector, John Adams (1769–98, FLS 1795), came from an established Welsh gentry family in the area. Educated at Christ's College, Cambridge, where he studied mathematics, Adams became a Collector of Customs and seems to have been self-taught with regard to natural history, via a useful library and a correspondence with acknowledged specialists. Although his shell collection has not been found, his descriptions of some fifty-three new species appeared in four papers published in the *Transactions of the Linnean Society* between 1797 and 1800. These included his first: 'The specific characters of some minute shells discovered on the coast of Pembrokeshire, with an account of a new marine animal' (1797); and the last, which was published posthumously: 'Descriptions of some marine animals found on the coast of Wales' (1800). Besides shells, Adams also studied foraminifera and made descriptions of 'sea-anemones, starfish, isopods, sea-spiders, hydroids and bryozoans'. As an aid to collecting he is said to have undertaken dredging operations from a boat in Milford Haven. It would be at Pennar Gut, where the River Pembroke enters Milford Haven, that Adams drowned in 1798, either by falling from the shore into the sea or, as reported elsewhere, when his boat 'bearing a heavy stress of sail ... unfortunately upset'.[57]

Like many collectors of the time, John Adams did not limit his interest to shells and his other passion was for botany. Although he did not publish on the subject, he did correspond with a number of eminent botanists of the time, including James Edward Smith, the Linnean Society founder, to whom Adams also sent specimens. Adams was elected to the Linnean Society in 1795 with the eminent botanist James Sowerby (1757–1822) as one of his sponsors.[58]

Two other botanist friends of James Edward Smith were Thomas Johnes of Hafod and his young daughter Mariamne (1784–1811). Johnes would be elected a Fellow of the Linnean Society in 1794 and his extensive correspondence with Smith, together with numerous letters from Mariamne, are preserved in the Society archives. Preserved there too are botanical specimens collected by Thomas Johnes, and a number of entomological (insect) specimens labelled as being collected by 'Johnes'.[59] Whether the latter came from Mariamne, or her father, remains unclear. However, we know that Mariamne's interest in natural history extended to both entomology and botany from two pieces of evidence.

First, James Edward Smith, who acted as both a doctor and mentor to Mariamne during her short illness-prone life, notes in one of his letters that 'Miss Johnes, though not above ten years of age, has taken a wonderful turn for botany and entomology'. In 1796 Mariamne thanked Smith for a cabinet full of insects that he had sent her. A year later, in 1797, Smith dedicated to Mariamne his co-authored and lavishly illustrated work, *The Natural History of the Rarer Lepidopterous Insects of Georgia* (1797) – co-authored with John Abbot: 'Miss Johnes, of

Hafod ... When you look over this book, it will remind you of many hours we have passed together, in the practical investigation of similar objects to those which it illustrates.' Smith had edited *Insects of Georgia* from the drawings and text of his co-author, the English traveller and naturalist John Abbot (1751–*c*.1840), who 'devoted his adult life to documenting the flora and fauna of an untamed southeastern North America'. Furthermore, the publisher of *Insects of Georgia*, James Edwards (1720–1816), was a friend to Mariamne's father. Probably as a gift for her fifteenth birthday in 1799, Edwards presented Mariamne with a bound volume of the *original* 106 drawings made in America by John Abbot. Endorsed 'To Miss Johnes in testimony of sincere regard from James Edwards', her father noted it as 'the most magnificent & beautiful present I have ever seen'. The volume survived the 1807 fire at Hafod and it now resides in the John Work Garrett Library, a part of Johns Hopkins University in America.[60]

The second evidence for Mariamne's entomological interest is indicated in a letter her father sent to Smith in which he asks for Smith's opinion of the natural history collection of Charles Alexandre de Calonne (1734–1802). Dismissed as finance minister of France in the years running up to the French Revolution of 1789, Calonne had fled to Britain. There he later sold his large natural history and art collections. Dubbed the *Museum Colonnianum*, the auction catalogue announcing 'beautiful and rare subjects in Entomology, Conchology, Ornithology, Mineralogy, &c. ... All of which are now exhibiting at Saville House ... Leicester Square, previous to the sale thereof' appeared with a publication date of 1 May 1797. No date for the actual auction was given and on 26 May 1797 Thomas Johnes wrote asking Smith for his opinion of the collection. 'I have had the offer of it', he informed Smith. 'It would be a fine acquisition for my dear girl [Mariamne]. If I could compass[?] it easily as [to] the time and terms of payment.'[61] Curiously, the auction of some 3,037 lots did not actually take place until May and June 1801, when a new and less detailed catalogue was also issued.[62] Could this long delay have been the result of negotiations with Johnes as he attempted to raise finance? Since *individual* Calonne specimens did enter other collections[63] it is clear Johnes could not have purchased everything. However, the whereabouts of the vast bulk of the Calonne collection remains unknown. Might it be that Johnes negotiated and purchased selected parts of the collection, those that were of particular interest to Mariamne? We do not currently know but, if he did, it is likely the collections would have been destroyed in the devastating fire that swept through Johnes's home at Hafod on 13 March 1807.[64]

Beyond widespread availability and their appeal as objects of beauty, a major attraction of shells and botanical specimens must have been their ease of preservation and display. This was in contrast to larger zoological specimens, some of which required the services of a taxidermist. For botanical specimens, both the seeds and their host plant could be successfully preserved, the latter through drying and pressing into collections known as a *hortus siccus*. In this manner the Welsh-speaking and possibly north Wales born horticulturalist, and one-time 'Herbarist to the Physick Garden of Westminster', Edward Morgan (*c.*1619–*c.*1689), amassed between 1672 and 1682 some 2,000 specimens in a *hortus siccus*. The collection was ultimately donated to the Ashmolean Museum in Oxford.[65]

Of course, botanical and zoological collecting in person, or on organised field trips, was not the only means of obtaining specimens. The following quote from a 1753 letter of William Morris in Anglesey to his London-based brother Richard, gives a flavour of the interaction between scientists of Wales and others in the mid eighteenth century:

> I wrote to you per last post to which I refer, but I forgot therein to desire of you to wait of *Mr. John Ellis*, merchant, in Lawrence Lane, Cheapside, with my compliments and to let him know that I have had Mr Meyrick's commands concerning sea plants, but as I don't pretend to any extra knowledge in that article, it would be ridiculous in me to send him what would be of no account, therefore it would be necessary he should send me some specimens as he proposed to brother Lewis, and a small glass to view them with. I have engaged our oyster drudgers to procure me all the sea plants they can meet with in deep water, but they have brought me nothing hitherto worth notice.

John Ellis (*c.*1710–76, FRS 1754, Copley Medal 1767, APS 1774) would publish *An Essay towards a Natural History of the Corallines* in 1755. He also worked on zoophytes, which Linneaus would define in 1758 as 'a composite small organism, with both animal and plant characteristics'. *A Natural History of Many Uncommon and Curious Zoophytes, Collected by the late John Ellis* would be published posthumously by the Swedish-born naturalist, Daniel Solander, in 1776.

William's letter to his brother continues:

> I have borrowed Dilenius's History of Mosses to enable me to make some little progress in that study. If Mr. Ellis thinks proper to send me specimens, there will be a good opportunity to have them in the box with the garden roots, etc., and if he has any curious seeds or roots that may be managed without a greenhouse, should be glad of a few. When I can pick up any sea plants, [I] shall observe the directions given in Mr. Meyrick's letter.[66]

The archives of collectors such as Edward Lhwyd, William and Lewis Morris, and Iolo Morganwg, often record the receiving or giving of specimens as gifts, in the manner noted by William. They also purchased seeds from suppliers at home and abroad, and from the sale of other collectors' collections. In one of his letters to Hans Sloane, Edward Lhwyd even notes receiving 'a Glasse of Indian Lizards' out of 'a Dutch ship wracked in Glamorganshire'.[67] In 1757, Lewis Morris records the washing up of a whale that had lived in Cardigan Bay, where it 'often frightened the small craft of Aberystwyth who used to meet with him asleep, and sometimes ranging about, chiefly near Bardsey and in Pwllheli Bay'. When the whale eventually stranded, near Aberdovey, the naturalist Lewis managed to obtain 'one of the joints of his backbone near the tail, ten or twelve inches diameter'; some of the other large bones being used less scientifically 'as stools in Merionethshire'.[68] In 1761, William Morris informed his brother Richard, who conveniently worked in the Navy Office in London: 'Some hundreds of foreign shells I have, gwae fi na wyddwn pa rai sydd gennwch ['woe is me I do not know which ones you have']; many, very many of these came through your hands from London'.[69] In 1761, William told Lewis Morris of the shooting of a 'churn owl [nightjar]' in north Wales, following which he 'took the dimensions, weight, etc., and sent the account, with yours of its hatching, etc., this day to Mr Pennant'. All suggesting William and Lewis had been studying the bird with an ornithological interest for some time before it was shot.[70] In 1802, the Revd Hugh Davies, better known for his botanical studies (see below), sent Thomas Pennant the wings of a 'Tawny Bunting', later renamed by Pennant a 'Snow Bunting'; the wings were found within a folded envelope placed within Pennant's copy of his 1802 edition of *British Zoology*.[71]

A more unusual, dead, and significantly larger specimen came the way of Flintshire-born (Hawarden) surgeon and anatomist Honoratus Leigh Thomas (1769–1846, FRS 1806, member of the Imperial Academy of Sciences in St Petersburg). In 1801, he made the first full dissection of a male rhinoceros.[72] The animal, a single-horned Indian rhino, had been brought to London as part of Gilbert Pidcock's menagerie of chained and unchained animals to be exhibited on the upper floor of the Exeter 'Change, a building in the Strand in London (how did they get the animal up there?). Having been sold on by Piddock to the German Emperor Francis II, the animal died in a stable yard in Drury Lane before it could be shipped abroad.[73] Thomas undertook an autopsy and presented his results in a paper read to the Royal Society on 29 January 1801. It then appeared in the *Philosophical Transactions* as the 'Anatomical Description of a Male Rhinoceros', complete with a plate of figures illustrating 'parts of the eye' and 'a section of the jejunum (small intestine)'.[74]

Living zoological specimens were also studied at first hand. For example, John Keys, of the still surviving Bee Hall, near Pembroke, published *The Practical Bee-master* in 1780. As its lengthy subtitle suggests, it was a manual of bee 'management', together with an outline of 'improvements to the hives, Boxes, and other Instruments to facilitate the operations' all 'without injuring the bees'. The last was a significant innovation since at this time the bee colonies were usually killed-off when honey was collected. That John Keys had scientific as well as apiarist aspirations is evident in the introduction to his later publication: *Keys on Bees: The Ancient Bee-master's Farewell* (1796). Encouraged by the favourable reception of his first publication, and now being 'in the vale of life', his new illustrated treatise was submitted to the public:

> as a result of all my researches; drawn from a much longer and more assiduous experience, and from cooler judgment, ripened by numberless experiments, which have led me to new observations and improvements, and to differ also not more from myself than from ALL OTHERS.[75]

As with collections of fossils and minerals, the display of botanical, shell and small zoological material usually took place in cabinets of curiosity. The diverse nature of such collections is well illustrated by the contents of the drawers in a cabinet made by Lewis Morris for the Earl of Powys in 1756 (see Table 2).[76] However, as more

TABLE 2 Cabinet of curiosities made for Lord Powys by Lewis Morris

---

Cabinet of curiosities made for Lord Powys by Lewis Morris (1756)

'The content of this cabinet is as follows ... The drawers are lettered from ye top: –'

 a. Aberdovey sea plants, with glasses over ye card boats [individual card containers made by Lewis to hold the specimens in a drawer];

 b. copper ores from different parts of ye world;

 c. sea shells from Aberdovey;

 d. seeds of trees, plants etc., from Jamaica;

 f,g,h,i, shells from Jamaica and Portmahon;

 e. iron ores, amber, ambergrease;

 k. empty;

 l. petrifactions and fossil shells from Sicily, Portmahon, Norway, Wales, England;

 m. several native sulphurs, productions of Mount Etna, and coals of different countrys;

 n. earths, boles stones, sands, from different parts of ye world;

 o. lapis calam. or ores of zink, a great variety;

 p. lead ores: blue, of all ye various kinds; green; white, a great variety and transparent;

 q. various spars, several with ores mixt, of lead, copper, tin; chrystals, selenites, belemnites, talc, etc.;

 r. a few large fossil shells from Mahone, etc.

---

scientific approaches to the classification and display of specimens developed in the mid-eighteenth century such cabinets began to fall out of favour.

Until the 1750s, the naming of fauna and flora took account of the basic appearance of specimens. Consequently, names could become long-winded affairs; as in the case of the humble tomato: *Solanum caule inermi herbaceo, foliis pinnatis incises* ('Solanum with the smooth stem, which is herbaceous and has incised pinnate leaves').[77] Writing to Martin Lister, in 1694, Edward Lhwyd felt that if all plants were named by this same method it would 'discourage many from becoming botanists'. Instead, and so that 'men might more easily converse about natural things', he proposed, with a considerable degree of prescience, the giving of 'short titles, leaving the rest of their properties to their figures, descriptions, &c. or else (if ye contrary seem necessary) to give them two titles: ye one nominal shorter … ye other descriptory, longer, when requisite'. A policy that Lhwyd adopted when naming an African locust that had appeared in, and been sent to him from, Wales. Terming it the 'Pilgrim-Locust' he gives it an initial binomial – *Locusta erratica* – followed by a lengthy physical description: *alis ichthyocollæ adinstar pellucidis, reticulates maculis conspersis.*[78]

A number of different classificatory methods and schemes would be established before taxonomic salvation finally arrived with the binomial system of genus and species introduced by Carl Linnaeus in the mid-eighteenth century. First described for botanical specimens in his *Species Plantarum* of 1753, it was expanded to include zoological taxa in the tenth edition of his *Systema Naturae* of 1758. Also in 1758, we find *Systema Naturae* being mentioned in a letter of William Morris in Anglesey. A couple of years later, in March 1761, his brother Lewis would write 'to enquire after Linnaeus [in] English published in Ireland, that I must have, though I wanted victuals'.[79] The use of the Linnaean scheme would take time to be adopted and Thomas Pennant would use the classification scheme of the naturalist John Ray for his *British Zoology*, though he does mention recent innovations by Linnaeus in his Preface. To whom Lewis Morris wrote in Ireland with his request for an English translation of Linnaeus is not known, but in the summer of that same year, 1761, his brother William made a collecting trip to Ireland. There William learnt first-hand how the Linnaean innovation affected the work of at least one botanist.

William Morris had a day job as a customs officer in Anglesey, but he was also an enthusiastic botanist whose correspondence with his brother Lewis is littered with botanical discussion and listings of plants. Having arrived in Dublin in the summer of 1761, William took up residence in the Phoenix Park home of Dr John Nichols, the Surgeon-General of Ireland. Clearly impressed with his circumstances, William made 'excursions for 3, 4 or 5 days' with Nichols and his family in 'a chariot and four-wheeled chase', complete with saddle-horses and servants. Nichols also introduced him to a number of fellow naturalists, whose capabilities William described, after his return home, in a letter to his brother Lewis with perhaps tongue-in-cheek Morrisian forthrightness.

Introduced to the Wiltshire-born, Dublin-based Quaker, naturalist and physician, John Rutty (1697–1775), William declared that he 'proved no extraordinary botanist or fossilist', even though Rutty was then engaged in 'making preparations for publishing a Natural History of the County of Dublin [it appeared in 1772]'. The Waterford-born topographer, Dr Charles Smith (1715–62), who had 'published the Natural History of Cork, Waterford and Kerry', proved 'not supernatural and no botanist to signify, but a person of extraordinary memory and address, but in a languishing way, therefore leaves the Natural History of the County of Dublin to be done by his competitor, my friend Dr Rutty'. So, William continues:

> It was necessary upon trials to find out a proper person to encounter the Cambro-British botanist [William] (as to fossils, shells, etc., they gave them up), and who do you think was the person pitched upon for so great an undertaking (you see by my style what country I've been at), and capable of I know not what? Guess if you can. Who but Dr Jenkins, whom you have seen here [in Wales] about 26 or 28 years ago, who was ready to swear that I was you and that you was I![80]

In another letter William hints at Jenkins's Welsh ancestry when he describes him as 'ŵyr i Gymro' (a 'grandson to a Welshman').[81] It was this Jenkins too, who, having 'studied botany for 30 years at home and abroad', was now working toward 'publishing an Irish Herbal on the Tournefortian system' – a flawed but then still popular classification scheme outlined by French botanist Joseph Pitton de Tournefort (1658–1708) in his *Eléments de botanique, ou Méthode pour reconnoitre les Plantes* (1694). Alas, poor Jenkins had now 'adopted the Linnaean system' and been forced 'to alter' his Tournefortian efforts and re-prepare all 'his labours for the press'.[82]

Despite Lewis Morris having believed an English-language version of Linnaeus might be available in Ireland, it would not be until the Lichfield Botanical Society (comprising its founder, Erasmus Darwin, and two friends) published *A System of Vegetables* between 1783 and 1785, and *The Families of Plants* in 1787 that part of Linnaeus's *Systema Naturae* appeared in English. The actual translation of the complete work did not occur until 1800, when Swansea-based conchologist, William Turton (1762–1835), published the first volume of an English translation that he had made from the thirteenth Latin edition of *Systema Naturae* edited and published by Johann Friedrich Gmelin between 1788 and 1793.

Born in Olveston, Gloucestershire, Turton practised as a physician in Swansea, probably from the 1790s until at least 1807 (based on the dated prefaces to his published works).[83] His English translation of Linnaeus – *A General System of Nature, through the three grand Kingdoms of Animals, Vegetables, and Minerals* – appeared in seven volumes between 1800 and 1806, with the printer of the first volume in 1800 given as 'Voss & Morris, Castle Street [Swansea]. For Lackington, Allen and Co., Temple of the Muses, Finsbury-Square, London' (see Figure 16).[84]

By the start of the nineteenth century, zoological and botanical collectors in Wales had firmly adopted the Linnaean system. Both John Adams and William Lyons used it for their shell collections, and Lyons also moved away from cabinets of curiosity to adopt more modern methods of display; as the presence of shells attached to slats of wood for presentation in glass-topped cabinets in his Tenby collection bear

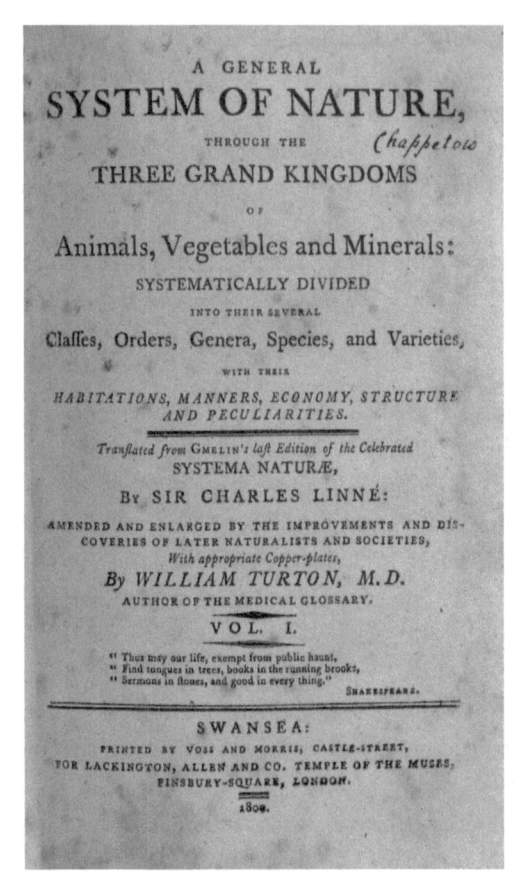

FIGURE 16 Title
page of Volume 1
of the first English
translation of the
complete *Systema
Naturae* by Carl
Linnaeus (1735).
Translated from
the original Latin
by William Turton
and published in
Swansea between
1800 and 1806.
Courtesy of the Bath
Royal Literary and
Scientific Institution.

witness. In 1807, the printer 'J. Evans, of Wind Street, Swansea', would publish William Turton's *British Fauna: Containing a Compendium of the Zoology of the British Islands: arranged according to the Linnaean system.*

Another to use the binomial nomenclature was Lewis Weston Dillwyn (1778–1855, FLS 1800, FRS 1804). Though born in Walthamstow, he was descended from a Breconshire family who had emigrated in 1699 from Breconshire to the Welsh Tract in Pennsylvania.[85] His father, William Dillwyn, having returned to Britain during the American Revolution, bought the Cambrian Pottery in Swansea and put his son Lewis in charge. At the same time, Lewis became an enthusiastic naturalist, with particular interests in botanical and molluscan studies. In 1801, he published a paper in Tilloch's *Philosophical Magazine* on 'the effects of oxymuriatic acid on the growth of plants' (a subject also of interest to Thomas Johnes),[86] and his 'Catalogue of rare plants found in the environs of Dover, with occasional remarks' appeared in the 1802 *Linnaean Society Transactions*. Jointly authored with banker, naturalist and antiquarian Dawson Turner (1775–1858), *The Botanists Guide through England and Wales* was published in 1805, followed by Dillwyn's use of binomials in his important work on the *British Confervae* ('freshwater algae'). Begun in 1802 and published in 1809, it contained numerous species new to science;[87] however, he notes that further work needed to be done on their taxonomy because Linnaeus had been 'too busily engaged in the immense field he had entered on, to spare the time necessary for an investigation of the submerged algae'. In the last years of our period of interest, Dillwyn published a *Descriptive Catalogue of Recent Shells, arranged according to the Linnaean method* (1817). His collections are today spread through numerous institutions from Kew Gardens, the Botanical Museum in Copenhagen and the Natural History Museum in London to the National Museum in Cardiff, and the Royal Institution of South Wales in Swansea.[88]

Also in the last years of our period came publication by the Revd Hugh Davies (1739?–1821, FLS 1790) of his *Welsh Botanology giving a Systematic Catalogue of the Native Plants of the Isle of Anglesey in Latin, English, and Welsh* (1813). A landmark in Welsh botanical studies, it was the first detailed flora of any Welsh county (see Figure 17). As Thomas Pennant put it, 'By his labours a Flora of the Island is rendered as complete as is possible to be effected by a single person in one season of the year. The number of plants he observed amounted to 500';[89] and these, as Davies notes in his Preface to *Welsh Botanology*, were 'classed according to the sexual system of Linnaeus'.[90] Davies, like other Welsh botanists of the period, made contributions to the work of others. An engraving of a shark taken from one of his drawings 'in material relating to Thomas Pennant' suggests broad natural history interests, and he also contributed to the botanical work of Dillwyn and Turner, to William Hudson's *Flora Anglica*, and to James Edward Smith and

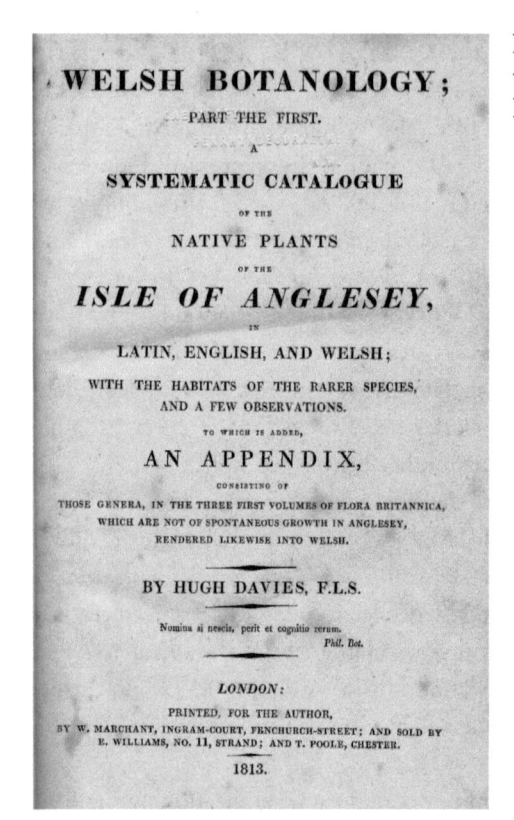

FIGURE 17 Title page to *Welsh Botanology* by Hugh Davies (1813)

James Sowerby's *English Botany*. Known to his fellow British botanists as 'the Welsh Linnaeus', the herbarium of Hugh Davies is today in the Natural History Museum in London and a further 192 specimens in the National Museum of Wales in Cardiff.[91] Another largely forgotten botanist, John Wynne Griffith (1763–1834),[92] sent specimens to the likes of James Sowerby, wrote on the periwinkle to Joseph Banks, and aided the botanist, geologist, chemist and physician, William Withering (1741–99, FRS 1785), with his *An Arrangement of British Plants*.

Thus far, most of the collecting recorded in this chapter happened 'in the wild'. Yet, for both botanists and zoologists, the landscape of the domestic garden was also available for scientific study. Not only were gardens living cabinets of curiosity, they also acted as handmaidens to another scientific endeavour – horticultural studies.

Horticulture

Many in the long eighteenth century saw gardens as no more than 'a meditative retreat' or a place of recreation and floral display. Some, though, saw them as philosophical spaces in which the scientific study of the garden and its plants – horticulture – only added to its other joys.[93] One such person was the horticulturalist Sir Thomas Hanmer (1612–78). As a royalist, who had been appointed a page and later a cupbearer to Charles I, Hanmer was exiled to Paris in 1644 as the Civil War developed. In 1646, however, he revealed to the Cromwellian parliament that the king had been in negotiations with the French and with the Scots. In light of this 'signal service to the Commonwealth', Hanmer regained his lands in 1650 after payment of a fine. He then retired to his ancestral family home at Bettisfield, Flintshire (today in the Wrexham County Borough, and close to the border with England) to concentrate on his garden.[94] There, in 1653, he completed the manuscript of his *Garden Book*.[95]

A reference to 'the Reader' in his manuscript suggests that Hanmer intended the work for publication, though this would not actually occur until 1933.[96] Essentially an instructional horticultural handbook – it provides comprehensive details on planting, methods of propagation, importance of the correct placing of plants, watering, etc. – the work also has a basic scientific rationale. This is evident on the first page where, in a discussion 'Of the Distinctions of Flowers', Hanmer notes how those authors who (in this age before Linnaean classification) 'distinguish FLOWERS by their ROOTES make but THREE SORTS of them, BULBOUS, TUBEROUS, and FIBROUS'. To these Hanmer adds another classification, one 'which I thinke as considerable as any of the other, and that is GLANDULOUS, and thus shall they bee here treated of, and not according to their Seasons of Flowering, though they shall also bee particularly mentioned'.[97] He is also keen to dispel old wives' tales and gardening folklore:

> By good cultivation and defence only shall you arrive at the great-
> est perfection flowers are capable of; for there is no known Art to
> give them such colours as wee desire by steeping their rootes or

seedes in colour'd waters, or by putting any ingredients into them that shall worke that effect, nor any way by observing the Moone or heavens to make flowers larger or double, or to worke such wonders as are both sayd and written to amuse and deceave the unexperienced and credulous.

A section on viticulture introduces 'My owne observations of Vines', and his writing on this subject pre-dates works such as *The Complete Vineyard* (1665) by William Hughes. Special attention is also given to tulips, a speciality of Hanmer. They include the variety 'Agate Hanmer', which was described by English horticulturalist John Rea (d.1681), in his *Flora Ceres and Pomona*, as having 'three good colours, pale gredeline, deep scarlet and pure white'. Rea also notes that 'This gallant Tulip hath its name from that ingenious lover of these rarities, *Sir Thomas Hanmer*, who first brought it into *England* [and Wales], from whose free community, my self and others partake the delight of this noble flower'.[98] As Rea suggests, Hanmer was generous in giving plants to others.[99]

Also included in the *Garden Book* is a calendar of 'Remembrances of what is to bee done in a GARDEN every Moneth of the Yeare'. Prior to the 1933 publication of Hanmer's 1653 manuscript, the earliest known example of a gardening calendar had been thought to be John Evelyn's *Kalendarium Hortense* of 1664. A founding Fellow of the Royal Society, Evelyn's *Kalendarium* actually formed part of a much larger manuscript that remained unpublished until 2000. The *Elysium Britannicum* is a monumental instructional and scientific compilation on gardening. It runs to some 1,000 manuscript pages and Thomas Hanmer is believed to have contributed the sections on tulips and daffodils.[100]

Hanmer describes in the *Garden Book* the use of a winter house or room as an early form of conservatory and is regarded as 'the first English [*sic*] author to mention the cedar of Lebanon, predicting its future importance'.[101] He also acted as an older mentor to John Evelyn during the establishment of the younger man's garden at Sayes Court in Deptford, London. In 1668, Evelyn enquired of Hanmer as to 'what gardens there are in Wales'. In replying, Hanmer said he did not know of 'any noble ones', and nobody had ventured upon the then increasingly popular 'large spacious ones with costly fountains … or great parterres'; the kind seen in France and Italy by many royalist exiles during the Civil War. In this Hanmer seems to have been ignorant of the spectacular fountain and parterre garden established by Mutton Davies at Llannerch Hall, in the Vale of Clwyd, Denbighshire, some thirty miles north-west of Bettisfield as the crow flies. The garden is pictured in the very first bird's-eye-view painting of a British garden. Dated to *c*.1667 the picture is today in the Yale Centre for British Art.[102] Created by Mutton Davies (1634–84), the garden, in all its formal glory, would have been seen by Edmond Halley at Whitsun 1697 when he stayed at Llanerch on his way to measure the height of Snowdon (see Chapter 9).

By the middle of the eighteenth century, the fashion for grand fountains and parterres had waned and less formal gardens were the passion, sometimes with an associated passion for silviculture. John Evelyn makes only a brief mention of Wales in *Sylva, or a Discourse on Trees*, whose subtitle revealed its aim – the 'Propagation of Timber in his Majesties Dominions'. The book, which was the very first publication of the Royal Society, appeared in 1664. It was followed by many others on the subject of trees and forests, but the first comprehensive work on forest management from within Wales appears to be *A Treatise on Forest-Trees* (1753) by the Cambridge-educated Breconshire curate William Watkins (*fl.*1750–62). Watkins, like John Evelyn, wrote in response to 'the continual Devastation of wood in this Kingdom' and its likely consequences for the country. Along with encouraging landowners to plant trees, his work is a manual of tree management, tree planting, propagation and overall cultivation based on what Watkins had 'gathered from undeniable Experience myself, and from Authors, as I am fully satisfied, may be depended on'.[103]

One landowner who took to tree planting on a vast scale was Thomas Johnes of Hafod and his efforts won him five gold medals from the Royal Society of Arts between 1800 and 1810. The more scientific aspect of Johnes planting of trees lies in what Linnard (1970) describes as his use of 'what were, in his day, very progressive techniques, in particular his use of small planting stock, and the care taken to protect the roots before planting. Indeed, some of the techniques have scarcely been improved upon'. This assertion has recently been modified by Peters (2015) who notes that such techniques were noted in works such as Evelyn's *Silva* from the seventeenth century.[104] They are evident too in the *Treatise* by Watkins, who describes the care of tree roots from his own experience.

Horticultural matters were also an area of science where women scientists of Wales were involved. Jane Johnes, the wife of Thomas Johnes, and her daughter Mariamne, each had gardens designed for them by the agriculturalist and inventor of the Scottish plough, Dr James Anderson.[105] In Gower, south Wales, Mary Lucy Fox-Strangways (1776–1855), the new wife of Thomas Mansel Talbot (1747–1813), came to her home at Penrice having already become an acknowledged expert in horticultural matters in her own family in England, an expertise reflected in the content of the £300-worth of books Talbot bought for her on the eve of their wedding in February 1794 (see Table 3).[106]

In his *Garden Book* of 1659, Thomas Hanmer had written of the value of the warm-room as a place for raising tender plants. It must have been

TABLE 3 Natural history books bought by Thomas Mansel Talbot for his soon to be wife Mary Lucy Fox-Strangways

Natural history books bought by Thomas Mansel Talbot for his soon-to-be wife Mary Lucy Fox-Strangways; December 1793/January 1794

Flora Londinenses, 67 nos (£18)
Botanical Magazine, 84 nos (£4 11s.)
William Lewin's *English Birds*, (£31 10s.)
Oeuvres de Buffon, 38 vols (£42)
James Bolton's *Fungi*, 4 vols (£8 8s.)
James Bolton's *Ferns*, 2 vols (£2 2s.)
Da Costa's *English Shells* (£2 10s.)
Thomas Pennant's *Tour in Wales*, 2 vols (£3 3s.)
Mark Bloch's *Histoire de Poissons*, 6 vols (£22 14s.)
James Maddock's *Florists' Directory* (10s.)
Sir William Hamilton's *Campi Flegri* [on volcanoes], 3 vols in one (£20)
Revd William Hanbury's *Gardening*, 2 vols (£3 3s.)
William Bailey's *Machines*, 2 vols in one (£4)
Benjamin Wilkes's *English Moths* (£9 9s.)
Moses Harris's *English Insects* (£5 5s.)
Cramer's *Papillons Exotiques*, 4 vols (£26 5s.)
Cramer's *Papillons d'Europe* (£25)

just such a place at Wynnstay, in Denbighshire, the home of Sir Watkin Williams-Wynn that led to the making of a horticulturalist first in Britain. According to the Scottish botanist, garden designer and writer John Claudius Loudon (1783–1843), in his *Encyclopaedia of Gardening* (1822), 'The horticultural and floricultural establishments [at Wynnstay] are very complete; and here the banana was fruited and its fruit used at the dessert, for the first time in [Wales and] England'.[107] *The Church of England Magazine* adding the interesting detail from Loudon that:

> Specimens were sent to the Horticultural Society in 1819, which were between four and five inches long, and possessed an agreeable, luscious, and acid flavour; and the produce from a single plant is so abundant as to entitle the banana to be considered as useful for the table. The plant at Wynnstay was planted in a stove in 1811.[108]

Although they lacked the warm-room facilities and designer gardens of the grand houses, the small gardens of Wales whose owners had botanical and horticultural interests also made, or intended to make, botanical and horticultural contributions. For example, like his predecessor Thomas Hanmer, William Morris in Anglesey created a Gardening Calendar and, by 1740, well before publication of Hugh Davies's *Welsh Botanology* in 1813, he began to make a catalogue 'in English, Welsh, and Latin of the plants, etc., growing in and around Holyhead'. He also made 'a kind of a dry garden, or specimen of each plant' (a *hortus siccus*, see above), and had:

> lately taken in hand and finish'd (with a design of adding the same to Dr Davies's Botanologium*) a catalogue of all the plants (in Latin, Welsh, and English) out of Mr Ray's *Synopsis Stirpium Britanniciarum* – gwaith pwysfawr oedd hwn ['a weighty work']. I intend to add the Exotics os caf ennyd ['if I can spare a moment']. And all for the good of the publick.[109]

[*The *Botanologium* of Dr John Davies of Mallwyd (*c.*1567–1644) included in his *Antiquae Linguae Britannicae Dictionorum* of 1632][110]

William also catalogued the contents of a garden he shared in Holyhead with the curate of the area, the Revd Thomas Ellis. Among the 153 entries was an exotic, tobacco; a plant that seems to have done well in north Wales given a letter to Sir Thomas Hanmer of 27 July 1660 'touching the growing of tobacco in co. Flint contrary to law'.[111] Of tulips in William's garden there was 'no great stock'. Of course, the same could not be said of Thomas Hanmer's garden at Bettisfield, but one absence that did concern Hanmer related to the possible use of his plants for medicinal purposes. In a preface presumably intended for the *Garden Book* he notes how 'the ensuing Catalogue of choice plants ... is exhibited to the public with short directions for their preservation and increase not meddling with their medicinal qualities whereof so many volumes have been written'.[112]

Notes

1. Thomas Pennant, *British Zoology* (London, 1768), I, p. i.
2. John Davies, Nigel Jenkins, Menna Baines and Peredur I. Lynch, *The Welsh Academy Encyclopaedia of Wales* (Cardiff, 2008), p. 823.
3. J. G. Bockheim, A. N. Gennadiyer, R.D. Hammer and J. P. Tandarich, 'Historical Development of Key Concepts in Pedology', *Geoderma*, 124 (2005), 23–36.
4. Dewi W. Evans and Brynley F. Roberts (eds), *Edward Lhwyd Archæologia Britannica: Texts and Translations* (Aberystwyth, 2009), pp. 37, 44.
5. Anon., 'Enquiries touching agriculture for arable and meadows', *Philosophical Transactions of the Royal Society*, 1 (1665), 91–4.
6. RSC EL/OB/36a. Oldenburg, Robert Boyle and John Wallis were members of the Society's *Georgical Committee*.
7. B. G. Charles, *George Owen of Henllys: A Welsh Elizabethan* (Aberystwyth, 1973), pp. 149–50.
8. Geraint H. Jenkins, *Foundations of Modern Wales 1642–1780* (Oxford and Cardiff, 1987), pp. 114–5.
9. See R. J. Moore-Colyer, 'Of Lime and Men: Aspects of the Coastal Trade in Lime in South-west Wales in the Eighteenth and Nineteenth Centuries', *WHR*, 14/1 (1988), 54–77.
10. David Ceri Jones, 'Iolo Morganwg and the Welsh Rural Landscape', in Geraint H. Jenkins (ed.), *The Rattleskull Genius: The Many Faces of Iolo Morganwg* (Cardiff, 2005), pp. 227–50 (at p. 228).
11. R. J. Colyer, 'Early Agricultural Societies in South Wales', *WHR*, 12/4 (1985), 567–81 (at 568–9).
12. *ML*-II, p. 324.
13. Chris Evans, *The Letterbook of Richard Crawshay 1788–1797* (Cardiff, 1990), pp. 154–5. In 1775 Franklen received a silver medal from the Glamorgan

Agricultural Society for 14 acres of turnips and was commended in 1776 for his acres of cabbage, one of which weighed in at 18¾ pounds; see *Cowbridge and District Local History Society Newsletter*, 54 (April 2004), available online at 'Peoples Collection Wales'.

14. See, for example, Evans, *Letterbook of Richard Crawshay*, p. 153.
15. See *library.wales/discover/digital-gallery/printed-material/gwallter-mechains-reports-for-the-board-of-agriculture* (accessed 23 May 2024) for details and open access to the digitised reports.
16. See, for example, Richard J. Moore-Colyer, *A Land of Pure Delight: Selections from the Letters of Thomas Johnes of Hafod 1748–1816* (Llandysul, 1992), pp. 138–9 (on muriatic acid), pp. 260–1 (on Fuller's Earth).
17. *CIM*-III, pp. 434–5.
18. Jones, 'Iolo Morganwg and the Welsh Rural Landscape', pp. 242–3, citing NLW MS 13115B, pp. 60–1.
19. *CIM*-III, p. 435.
20. *CIM*-III, p. 438.
21. *Nature*, 6 January 1921, p. 617.
22. *Atlas et description minéralogiques de la France* (1780).
23. *Essai sur la Géographie des Plantes* (1805).
24. V. M. Spalding, 'The Distribution of Plants', *The American Naturalist*, 24/285 (1890), 819–31 (at 819).
25. *CIM*-III, pp. 37–46 (at 40–1).
26. See discussion and references in Ceri W. Lewis, 'Iolo Morganwg and Strict-Metre Welsh Poetry' in Jenkins (ed.), *A Rattleskull Genius*, p. 86 and n. 63.
27. See Janet Browne, 'Botany for Gentlemen: Erasmus Darwin and "The Loves of Plants"', *Isis*, 80/4 (1989), 593–621, (at 601), which cites Darwin's quote as being taken from Anna Seward, *Memoirs of the Life of Dr Darwin, Chiefly during his residence at Lichfield; with Anecdotes of His Friends and Criticisms on His Writings* (London, 1804), pp. 130–1. Available online.
28. See Lewis, 'Iolo Morganwg and Strict-Metre Welsh Poetry', p. 86, citing John Morris-Jones, *Cerdd Dafod* (Rhydychen, 1925), pp. 372–9.
29. *Lhwyd*, p. 54.
30. See *Lhwyd*, pp. 145–70.
31. *Lhwyd*, p. 159.
32. For an example of Lhwyd's comparative use of his collections, see Gunther pp. 223–4, and 'Part of a letter from Mr. Edward Floyd, Cim. Ashm. Oxon. To Dr M. Lister, giving an account of locusts lately observed in Wales', *Philosophical Transactions of the Royal Society*, 18 (1694), 45–7. For a discussion of the locusts and their relation to 'the mysterious Merionethshire fires of 1693–4', see *Lhwyd*, pp. 125–6, and Gunther, pp. 211, 213–14, 219–24.
33. Gunther, p. 69, *Lhwyd*, p. 39.
34. F. V. Emery, '"The Best Naturalist Now in Europe": Edward Lhuyd, F.R.S. (1660–1709)', *THSC*, I, (1969/70), 62–3.
35. *Lhwyd*, p. 235.

36. Available at *http://drc.usask.ca/projects/ark/public/catalog.php?catalog=1695Bra* (accessed 14 April 2022).

37. Gunther, pp. 163–6, 250–3.

38. Gunther, pp. 433–4.

39. See Gunther, pp. 446–7.

40. Quoted in Mary Ann Constantine and Nigel Leask, 'Introduction: Thomas Pennant, Curious Traveller', in Constantine and Leask (eds), *Enlightenment Travel and British Identities: Thomas Pennant's Tours in Scotland and Wales* (2017; London, 2019), p. 1.

41. Publishing dates for Pennant's works are complex. For those used here see 'A Short Bibliography of Thomas Pennant's Tours in Scotland and Wales', in Constantine and Leask (eds), *Enlightenment Travel*, pp. 245–8, and, for other works, the index to the same volume under 'Pennant'.

42. For example, Thomas Pennant, 'An Account of two new tortoises; in a letter to Matthew Maty, M.D. Sec. R.S.' (1771), *Philosophical Transactions of the Royal Society*, 61 (1771), 266–73; also 'An account of the Turkey', *Philosophical Transactions of the Royal Society*, 71 (1781), 67–81.

43. Thomas Pennant, 'Account of the different species of the birds, called Pinguins', *Philosophical Transactions of the Royal Society*, 58 (1768), 91–9.

44. Seemingly intended for his personal use, the manuscript of 'Outlines of the Globe' was sold, along with his library, in 1938. Today the work resides in the National Maritime Museum, Greenwich. See Ronald Paul Evans, 'The Life and Work of Thomas Pennant (1726–1798)' (unpublished PhD thesis, University College of Wales, Swansea, 1993).

45. See Evans, 'The Life and Work of Thomas Pennant', p. 618.

46. W. L. McAtee, 'The North American Birds of Thomas Pennant', *Journal of the Society for the Bibliography of Natural History*, 4/2 (1963), 101.

47. See McAtee, 'North American Birds of Thomas Pennant', 100–24 (at 100).

48. See V. P. Wystrach, 'Anna Blackburne (1726–1793) – a neglected patroness of natural history', *Journal of the Society for the Bibliography of Natural History*, 8/2 (1977), 148–68 (at 156–8).

49. For details, see Mark Seaward, 'Fielding-Druce Lichen Collections', *https://herbaria. plants.ox.ac.uk/bol/Content/Projects/oxford/resources/Fielding_Lichens.pdf* (accessed 24 May 2024); Charles E. Jarvis, '"The Most Common Grass, Rush, Moss, Fern, Thistles, Thorns or Vilest Weeds You Can find": James Petiver's Plants', *Notes and Records: The Royal Society Journal of the History of Science*, 74 (2020), 303–28; J. E. Dandy, *The Sloane Herbarium* (London, 1958).

50. See 'The Thomas Pennant Collection', *Science*, 37/950 (14 March 1913), 404–5.

51. Gunther, p. 89, quoted in *Lhwyd*, p. 57.

52. For all comments and quotes see *Lhwyd*, p. 57.

53. Gunther, p. 69.

54. *ML*-I, pp. 255–6.

55. See E. D. Evans, 'Lewis Morris's Aristocratic Connections', *NLWJ*, 31 (1999–2000), 121–4.

56. All details from P. Graham Oliver, K. Talbot, B. Fredriksson, V. Tomlinson, M. Lewis, D. Fraser, 'William Lyons of Tenby (1776–1849) and his conchology collection in the Tenby Museum & Art Gallery with recognition of type material', *Colligo*, 3/1 (2020).

57. All details from P. Graham Oliver, 'John Adams FLS of Pembroke (1769–1798): a forgotten Welsh naturalist and conchologist', *Archives of Natural History*, 46/2 (2019), 183–202.

58. Oliver, 'John Adams FLS'.

59. For example, LINN639 *Carabus inquisitor* and LINN3524 *Silpha vespillo*. See *https://linnean-online.org/view/collector/insects* (accessed 28 May 2024).

60. All details from John V. Calhoun, 'A Glimpse into a "Flora et Entomologia": *The Natural History of the Rarer Lepidopterous Insects of Georgia* by J. E. Smith and J. Abbot (1797)', *Journal of the Lepidopterist's Society*, 60/1 (2006), 1–37.

61. Digital Archive of Linnean Society, James Edward Smith Correspondence: item GB-110/JES/COR/16/38.

62. See *https://mineralogicalrecord.com/new_bibliography/calonne-cahrles-alexandre-de/* (accessed 16 April 2022).

63. For example, James Parkinson (see his *Organic Remains of a Former World*, 3, p. 344) and the British Museum (see *List of the Specimens of British Animals in the Collection of the British Museum*, Part VII, Mollusca, Acephala and Brachiopoda (London, 1851), pp. 17–8, 46, 62).

64. For the 1797 catalogue, see *https://archive.org/details/gri_33125010895072/mode/2up* (accessed 16 April 2022). Also, *The Gentleman's Magazine*, 77 (1795), 286–8, suggests the collection, 'a labour of thirty-five years; and at an expense of above sixty thousand guineas' had previously been mortgaged to assist the overthrown 'Princes and French Nobility'.

65. All details on Morgan from R. H. Jeffers, FLS, 'Edward Morgan and the Westminster Physic Garden', *Proceedings of the Linnean Society of London*, 164/2 (1953), 102–33.

66. *ML*-I, p. 260.

67. BL Add MS Sloane 4039 f. 59.

68. *ML*-II, p. 21.

69. *ML*-II, p. 312.

70. *ML*-II, p. 377. For more ornithological interest, see *ML*-II, p. 360.

71. Hellen Pethers, 'Spitting Feathers: Causing a Stir Amongst the Pages', *https://naturalhistorymuseum.blog/2017/06/22/spitting-feathers-causing-a-stir-amongst-the-pages* (accessed 22 April 2022).

72. A two-horned rhino had previously been studied by James Parsons in 1743, but as Thomas notes in his paper, Parsons was only concerned with external features, all of which Thomas notes as being correct.

73. Kees Rookmaaker, John Gannon and Jim Monson, 'The lives of three rhinoceroses exhibited in London 1790–1814', *Archives of Natural History*, 42/2 (2015), 279–300; available online from the 'Rhino Resource Center'.

74. H. L. Thomas, 'An anatomical description of a male rhinoceros', *Philosophical Transactions of the Royal Society*, 91 (1801), 145–52.

75. All details from '*Book of the Month – Keys on Bees*', from Tenby Museum (accessed 16 February 2018), and the original volumes.

76. *ML*-I, p. 428.

77. See *https://scientiaandveritas.wordpress.com/2013/04/15/carl-linnaeus-the-man-of-many-names* (accessed 23 May 2024).

78. Gunther, pp. 248, 224.

79. *ML*-II, pp. 94, 324.

80. *ML*-II, p. 372.

81. *ML*-II, p. 375.

82. *ML*-II, pp. 370–3, 375.

83. See *ODNB* entry.

84. The various editions of this work and the variety of publishers involved are complex and have not been definitively determined here. However, given Turton's residence in Swansea it seems likely the Swansea editions appeared before the London publication later the same year.

85. See Kirsti Bohata, 'Pioneers and Radicals: The Dillwyn Family's Transatlantic Tradition of Dissent and Innovation', *THSC*, 24 (2018), 47–63.

86. This was also of interest to Thomas Johnes of Hafod. He discusses it in a letter of 1798 to James Edward Smith, see Moore-Colyer, *A Land of Pure Delight*, pp. 138–9 and notes.

87. For a volume of algal specimens (more than 280) collected by Dillwyn, see Katherine Slade, 'Dillwyn's Book of Algae. A Glimpse into the scientific life of a 19th century philanthropist in Wales', *https://museum.wales/blog/2216/Dillwyns-Book-of-Algae-A-glimpse-into-the-scientific-life-of-a-19th-century-philanthropist-in-Wales* (accessed 19 April 2022).

88. See Peter S. Dixon, 'Notes on Important algal herbaria, IV, The Herbarium of Lewis Weston Dillwyn (1778–1855)', *British Phycological Bulletin*, 3/1 (1966), 19–22.

89. Quoted in Dewi Jones, *The Botanists and Guides of Snowdonia* (Llanrwst, 1996), pp. 48–55 (at p. 48).

90. Hugh Davies FLS, *Welsh Botanology* (London, 1813), p. xii.

91. See Jones, *Botanists and Guides of Snowdonia*, pp. 48–55; Mark Lawley, *Hugh Davies (1739–1821)*, *www.britishbryologicalsociety.org.uk/wp-content/uploads/2020/12/FB105_Bygone-Bryologists-Hugh-Davies.pdf* (accessed 28 May 2024); T. J. Owen, 'Hugh Davies: The Anglesey Botanist', *Transactions of Anglesey Antiquarian and Field Club* (1961), 39–52.

92. See Jones, *Botanists and Guides of Snowdonia*, pp. 42–7.

93. For a discussion of the garden as a philosophical and experimental space see Vera Keller, 'A "Wild Swing to Phantsy": The Philosophical Gardener and Emergent Experimental Philosophy in the Seventeenth-Century', *Isis*, 112/3 (2021), 507–30.

94. See *ODNB* entry

95. See Eleanour Sinclair Rhode (ed.), *The Garden Book of Sir Thomas Hanmer* (London, 1933).

96. Rhode, 'Introduction', in Rhode (ed.), *The Garden Book*, p. xxxi

97. Rhode, 'The Garden Book of Sir Thomas Hanmer', in Rhode (ed.), *The Garden Book*, p. 3.

98. See Rhode, 'Introduction', in Rhode (ed.), *The Garden Book*, p. xxi.

99. Among those to whom Hanmer gave plants in June 1656 were: 'Lord Lambert', 'Mrs Thurloe' and 'Mr Hygens' ('Introduction', in *The Garden Book*, p. xx). A parliamentarian general, politician and keen gardener, Lambert (1619–84), had been one of the negotiators of the Treaty of Westminster (1654), which ended the first Anglo-Dutch war. 'Mrs Thurloe' is conceivably the wife of John Thurloe (1616–68), politician and spy-master to Cromwell. 'Mr Hygens' is not identified but given that Hanmer is not known for consistent spelling – 'Hanmer rarely spelt the same word twice alike' – could 'Mr Hygens' be a miss-spelling of the renowned Dutch family of 'Huygens'? If so, there are two main possibilities: the diplomat Lodewijck Huygens (1631–99), who between 1651 and 1652 travelled in England and south Wales (staying at Rhiwperra Castle); or his brother, the diplomat and statesman Constantijn Huygens Jr (1628–97), who also worked on scientific matters with his more famous scientist brother, Christiaan Huygens (1629–95, FRS 1665). Between 1649 and 1650 Constantijn Jr is said to have been in Britain, along with the diplomat Adriaan Pauw (1583–1653) who had been almost bankrupted during the tulip craze in Holland in the 1630s. Pauw is said to have displayed blooms in his garden using mirrors to further reflect their beauty.

100. See Margaret Willes, *The Curious World of Samuel Pepys & John Evelyn* (2017; New Haven CT and London, 2018), pp. 167–8, 265 n. 21.

101. See *ODNB* entry.

102. See *https://collections.britishart.yale.edu/catalog/tms:337* (accessed 28 May 2024); also, Bettina Harden, *The Most Glorious Prospect: Garden Visiting in Wales 1639–1900* (Llanelli, 2017), p. 79.

103. All details from W. Linnard, 'The First Treatise on Forest Trees in Wales', *Journal of the Welsh Bibliographical Society*, 11 (1975–6), 247–50.

104. Andrew J. Peters, *An Investigation into the Treescape of Hafod Uchtryd Using Targeted Dendrochronology* (unpublished MA thesis, University of Wales Trinity St David, 2015), pp. 22–3, available online.

105. Virginia van der Lande FLS, 'The Linnean Society of London's Smith Correspondence and Dr James Anderson LLD (1738–1808): Some Observations', *The Linnean*, 33/1 (April 2017), 17–19. Also, E. Inglis-Jones, *Peacocks in Paradise* (Llandysul, 1990).

106. From Joanna Martin (ed.), *The Penrice Letters 1768–1795* (Cardiff, 2013), pp. 43, 47 n. 57.

107. John Claudius Loudon, *An Encyclopaedia of Gardening* (London, 1822), p. 265.

108. *The Church of England Magazine*, 16 (1844), 405.

109. *ML*-I, pp. 37–8.

110. See Caryl Davies, 'The Dictionarium Duplex (1632)', in Ceri Davies (ed.), *Dr John Davies of Mallwyd: Welsh Renaissance Scholar* (Cardiff, 2004), pp. 146–70 (at pp. 154, 162–3).

111. NLW MS Elwes Papers, Item 1249.

112. Rhode, 'Introduction', in Rhode (ed.), *The Garden Book*, p. xviii.

11

LOOKING UP

Now the *Astronomers* pretend to a strange *familiaritie*
with the *starrs*, the *Natural Philosophers* talk as much: and
truly an Ignorant man might well think they had been
in heaven, and conversed … with *Jove* himself.[1]

Thomas Vaughan, *Magia Adamica* (1650)

Just as geology and minerals are important to the generation of soil
and the vast range of fauna and flora that it supports, so too are the
weather and the distant heavens. In this chapter we consider those
scientists of Wales who looked up to scrutinise not only the weather –
that perpetual Welsh preoccupation – but the starry realm beyond it.

Weather Watchers

In September 1663, Sir Kenelm Digby, founding Fellow of the Royal
Society and the alchemist friend of Bassett Jones, told John Aubrey
that the Tudor Magus John Dee (1527–1609) had studied the weather
for seven years and 'developed such skill in predicting [it] that he was
accounted a witch'.[2] This could be a potentially dangerous accusation in
England between 1560 and 1706 when some 2,000 were tried for witchcraft and 300 executed. In Wales, thirty-seven prosecutions are recorded
between 1598 and 1698, with eight found guilty and five death sentences handed down.[3] The Witchcraft Act would not be repealed until
1736, but happily most weather-watchers in the long eighteenth century
could employ less dangerous methods of weather prediction. Some utilised scriptural guidance, as in Matthew 16.2–3: 'When it is evening, ye

say, *It will be* fair weather: for the sky is red. / And in the morning, *It will be* foul weather to day: for the sky is red and lowring.' Others relied on astrological prediction of the sort published in the almanacs readily available throughout Wales (see Figure 7). There were also local 'prognosticators' who, like the almanac compilers, linked weather to astrology and astronomy in the service of farmers, villagers and townsfolk. One of them, William Augustus ('Wil Awst'),[4] published in Carmarthen in 1794 his *Erra Pater, neu Ddarogynydd yr amserau ...* with an English language version appearing as *The Husbandman's Perpetual Prognostication for the Weather.* In it he sought to foretell rain, wind, fair weather and foul together with cold, frosts, snow and times of plenty and scarcity.[5]

In his local area of Cil-y-cwm, near Llandovery, Wil Awst is said to have been able to forecast to 'within the hour' the onset of 'rain, frost, gales or thunderstorms'.[6] His published work contains sections on 'the birth of children with respect to the age of the moon', and 'infallible signs of rain and drizzling weather, taken by observations of the planets, and other stars, elements, creatures &c'. With our advantage of hindsight, it is easy to dismiss such abilities as wholly unscientific. Yet their astrological nature implies, at the very least, a basic understanding of the astronomical data on which astrology and its predictions are based. Wil Awst certainly includes 'A Table of the Planetary Hours for every day of the week', and a degree of acute observation is evident when he asks his readers to 'Mark the sun rising, and if it look broader than usual, then many moist vapours are gathering from the sea'. Mathematics plays a part too, particularly with reference to so-called Golden Numbers and Sunday Letters. The Golden Number is 'the date of the full moon on or after the spring equinox of March 21, according to a nineteen year cycle' and the Sunday Letter is used to identify 'the days of the year when Sundays occur'. Having both numbers for any particular year allows the date of Easter to be found in a calendar.[7]

How far Wil Awst relied on other texts for his ideas is uncertain, but the title page of *The Husbandman's Perpetual Prognostication* certainly advertises his knowledge of the teachings of 'Albert, Alkind, Haly and Ptolemy'. 'Albert' probably refers to the Italian architect and all-round Renaissance man, Leon Battista Alberti (1404–72), who is credited with making the first anemometer for measuring the strength of the wind. In all probability 'Alkind' is the polymath scientist Al-Kindi (801–73) – to use the Latinised version of his Arabic name. He wrote a number of letters on the weather and applied to it 'arithmetical astronomy, which enables the expert in meteorology to make mathematical calculations to predict the weather'.[8] 'Haly' is likely to be Edmond Halley (1656–1742) who produced the first worldwide weather map illustrating the monsoon and Trade Winds. Finally, among the works of Claudius Ptolemy (second century CE) we find *Phases of the Fixed Stars and Collection of Weather Signs* and *Tetrabiblos*, a second century book 'on astrology that had paramount authority among Greek, Arabic and Latin readers'.[9]

Beyond astrological prediction, some weather events simply went unexplained or became the subject of local dispute, with a scientific explanation occasionally posited as an alternative to one based on religion or folklore – as in this diary entry for the period 25 to 29 June 1783 by William Thomas of Michaelston-super-Ely in Glamorgan: 'All these days very misty weather. Morning and evening the sun to be seen red as blood, which makes the Vulgar to think the world at an end, others that the Sun is eclipsed.'[10] In fact, the misty weather and the blood-red sun resulted from an event taking place in faraway Iceland. The Lakagígar fissure on the island erupted on 8 June and continued for eight months, 'killing at least half the country's livestock and one-fifth of its people'. Moved by prevailing winds in a spiral pattern across Europe, the path taken by its gaseous emissions has now been mapped using diary entries akin to those of weather-watcher William Thomas. For example, the naturalist Gilbert White, in Selborne, Hampshire, observed its effects between 23 June and 20 July in terms similar to those of Thomas's: 'Sun looks all day like the moon, and shed a rusty red light' so that alarmed 'country people began to look with a superstitious awe at the red, louring aspect.'[11]

Descriptions of normal and unusual weather events occurred throughout the long eighteenth century. In a surviving unpublished manuscript Henry Rowlands describes 'minutely and scientifically' a strange hailstorm in Anglesey and Caernarfonshire during May 1697.[12] Others described their experiences in letters addressed to the Royal

Society, with some of them finding their way into the *Philosophical Transactions*. A letter from Edward Lhwyd to his friend Tancred Robinson described a storm similar to the one related by Rowlands, but this time in Pontypool, a month later, when eight-inch hailstones 'ruined as much glass at Major *Hanbury's* house, as cost four pounds in repairing'.[13] In 1706, an anonymous letter to the then Secretary of the Royal Society, Hans Sloane (Secretary 1693–1713), described a great deluge of rain in Denbigh.[14] In March 1726, Dr Perrot Williams recounts, in an unpublished letter to a later Secretary, James Jurin (Secretary 1722–7), how floods at Haverfordwest had swept away the town bridge and cost four lives. Interestingly, Williams went on to suggest this particular deluge might be the result of a change in wind direction from 'SW to NE whereby the air became at once several degrees colder than twas before'. Perhaps, he argues, this contracted the air, so squeezing 'the spongy surface of the earth, before saturated to a high degree with water, and by that means force it out in so stupendous a manner'. He was happy though to 'entirely submit' his thoughts to Jurin's 'more knowing judgment'.[15]

One Evan Davies, of Pencarreg, in Carmarthenshire, writing to John Eames, a mathematician friend of Newton, described the consequences of a severe lightning strike on a house in his area in December 1729. Three children were hit and lost their senses; there was a cracked hearth, singed hair, and one woman lost an eye, her teeth and the tip of her tongue. She was rendered speechless for a week. What conclusion might be drawn from this, Davies was unsure. Whether caused by a thunderbolt or ball lightning he felt 'the learned are best able to distinguish'. Some in the area had been quick to see it as God's judgement on them. Davies, though, believed it unwise to 'dive too far into these insoluble Arcana's, but only do pray to God to preserve us evermore, and to avert such heavenly judgments our sins may just derive by such or like terrible visitations'.[16]

At the same time as these general weather impressions were being recorded, there was a growing awareness of the need for more accurate information, preferably gathered over a lengthy period. In anticipation of writing his *Natural History of Wales*, alas never completed, Edward

Lhwyd had entertained including 'An account of Meteors [i.e., meteorology] with comparative Tables of the weather in general places'. In a questionnaire he sent out as part of a data-gathering exercise in 1696 one of his queries aimed at 'A Register of the Weather for the space of one year at least, kept by one or two in each county'. This he felt 'would be of considerable use' if it showed 'the figures of snow and hail: the time it generally begins to snow on our highest mountains, and when it desists'. Any other curious remarks about meteorology would also be welcomed. As Emery (1985) has shown, some of Lhwyd's correspondents kept such accounts. They included Thomas Evans, vicar of Llanberis, in his unpublished 'Scheme of the Wind and Weather at Llanberis' for the period 1 March 1697 to 28 February 1698. David Lewis of Llanboidy in Carmarthenshire kept a similar record 'for ten months in 1698'.[17] Aside from unwittingly noting the consequences of faraway events in Iceland, the Vale of Glamorgan diarist William Thomas also recorded general weather comments from 1762 to 1795.[18] Wales-born, London-based Richard Price made occasional weather remarks too, and on 21 December 1788 noted in his private journal a period of ten days of severe cold and 'no rain of any consequence since *Sept*[embe]*r*'. Though Price generally loved frosty weather, the winter of 1788/9 would be particularly severe, with his thermometer falling to 'a little below 10 degrees [−12° Celsius if Price is using the Fahrenheit scale, see below]'.[19] It was this hard winter, by compounding the effects of an earlier drought, a poor harvest and a destructive hailstorm, that would contribute to the start of the revolution in France in July 1789, an event whose opening events Price controversially welcomed in his later published *A Discourse on the Love of Our Country*.[20] Other weather watchers included observations in their travel journals, poems and songs: such as those produced by Walter Davies (Gwallter Mechain) from 1797 to 1846, Thomas Pennant between 1773 and 1781, and Dafydd William of Llandeilo in a ballad of 1785. Despite their value today for helping us understand past weather events and broader climate changes, these were mainly general impressions of the weather, rather than accurate day-to-day records. As such, they did not usually include the long-term and regular measurement of temperature, pressure, wind strength or rainfall essential to the development of accurate weather prediction.[21]

A particular Welsh scientist who did record such data was Perrot Williams of Haverfordwest. In 1723 he answered a call for weather recorders sent out by the then Secretary of the Royal Society, James Jurin. Born in London and a physician and writer of some fifty papers on various science topics, Jurin made his meteorological request in *Invitatio ad Observationes Meteorologicas communi consilio* (1723).[22] The *Invitatio* was written in Latin to attract as wide a European audience as possible. Published in the *Philosophical Transactions* and also distributed separately, it included instructions about the desired method of collecting and recording data. It proved eminently successful. Weather watchers were enrolled from as far afield as North America, Ireland, Sweden, the Netherlands, Germany and Russia, as well as Wales and the rest of Britain. The meteorological measurements made by Perrot Williams, in Haverfordwest, were forwarded to Jurin each year between 1724 and 1728. They are now preserved in the archives of the Royal Society (the measurements for the year 1724 are lost).[23] Neatly arranged in columns, Williams records the date and time of his twice daily readings (8 a.m. and 4 p.m.), the barometric pressure, the temperature, the wind direction and a general comment such as *Pluit* ('raining'), *colum nebulosum* ('foggy or cloudy sky') and *nubibus obductum* ('covered by clouds').

When trying to collate the large amounts of meteorological data he gathered, Jurin faced a number of problems. These mainly resulted from the lack of appropriate standardised measurements and instruments in the eighteenth century. One particular issue concerned the great variety of thermometers and thermometer scales then in use.[24] To modern eyes, an odd feature of Perrot Williams's temperature records is the fact that his winter temperatures are consistently higher than those of summer. Jurin noted this, and in May 1725 wrote to enquire where Williams had purchased his thermometer and to request a description of its scale.[25] Williams replied in August. He had bought the thermometer from Richard Glynne, a mathematical instrument maker of Fleet Street in London, and he enclosed a sketch of the scale.[26] It is unusual in having 'cold' at 55, 'temperate' at 45 and 'warm' at 35. Neither

Williams nor Jurin mention how these figures relate to actual degrees of temperature. The scale is certainly not the one invented by Gabriel Daniel Fahrenheit (1686–1736) in 1714, or that of his new and accurate thermometer, which appeared in 1724 just as Williams was making his measurements. Unusual scales, though, were not unique at the time. For example, Lansberg (1964) records thirty-five different scales being proposed and used between 1641 and 1780 and 'whoever constructed a new type of thermometer threw a scale in for good measure'.[27] Even Anders Celsius (1701–44), when he first developed his now almost universal thermometer scale in 1742, had 0 degrees as the boiling point of water and 100 degrees as its freezing point. Carl Linnaeus, on receiving in 1744 a Celsius thermometer having this original scale, reversed it to the one we know today, with 0 degrees as the freezing point.

By the late eighteenth century, scales and measuring equipment had improved considerably and in 1771 another example of scientific weather recording occurred in Wales. It concerned 'Experiments made in North Wales, to ascertain the different Quantities of Rain which fell in the same time, at different heights'. For this purpose, rain gauges were positioned on top of 'Rennig [Arennig Fawr or Arennig Fach], which is about four miles west of Bala', and another lower down at a house called 'Bochyraidr', some half a mile from Arennig. The experiment was sponsored by English lawyer, antiquary and naturalist Daines Barrington (1727–1800, FRS 1767, FSA 1767), who made a number of notable contributions to antiquarian and natural history studies in Wales. However, the setting up of the gauges and actual recording of data were undertaken by Mr Meredith Hughes of Bala: 'a very ingenious land-surveyor' who 'from his philosophical turn, would be very pleased with executing the commission, though a very troublesome one'. One of the troubles for Hughes was to measure the height of Arennig using the barometric method and 'Dr Halley's method of computation'. Barrington published the resulting rainfall and height measurements in the *Philosophical Transactions* with the conclusion drawn that the quantity of rain depends 'scarcely at all upon the comparative heights of places'.[28]

Science also began to provide practical ideas aimed at guarding against the worst effects of weather. For example, since at least 1753 the American polymath Benjamin Franklin had advocated the use of conducting rods with pointed ends as a way of protecting buildings from lightning strikes. In the light of Franklin's advocacy, and a recommendation by a 1772 committee of the Royal Society, pointed rods had been installed on a number of government buildings. These included the Ordnance Board gunpowder store at Purfleet, on the banks of the Thames near Tilbury. Yet despite Franklin's advocacy and the Royal Society advice, lightning still struck the store in May 1777. This created considerable consternation in the Ordnance Board given the importance of the stores to a country by then at war with its rebellious American colonies.

By the summer of 1777, one of the prime opponents of Franklin's pointed conductors, Benjamin Wilson (1721–1809, FRS 1751, Copley Medal 1760), had started giving electrical displays at the Pantheon in London to illustrate the advantages of his preferred blunt conductor rods. At this point the subject of pointed *versus* blunt conductors became a subject of scientific gossip between John Lloyd of Wigfair and his friend from the days of the Roy baseline measurement, Charles Blagden. At the time (1778), Blagden was serving as a surgeon with the British forces in America. From New York he wrote to Lloyd expressing his view of the Purfleet strike: 'Why the building at Purfleet was at all damaged, I have not yet been able to comprehend to my perfect satisfaction.'[29]

Blagden returned to private life in England in 1780 and in 1782 became more directly involved with the lightning-conductor controversy, and with another Welsh scientist, George Cadogan Morgan (1754–98, see p. 101). Born in Bridgend and nephew to Richard Price of Llangeinor and Newington Green, Morgan went from Cowbridge Grammar school to Jesus College, Oxford but left without taking his degree. He then entered the Dissenting ministry. Eventually he would leave this too in order to become a freelance teacher, but in 1781/2 he was living in Norwich as minister to the congregation of the city's famous Octagon Chapel. On 12 June 1781 the local Heckingham House of Industry, a recently built workhouse for the poor situated a few miles from Norwich, was damaged by a lightning strike and consequent fire. This occurred despite its being defended by 'eight pointed conductors each above half an inch in diameter', as Morgan's uncle, Richard Price, informed Benjamin Franklin in January 1782. Franklin by this time was resident in Paris as part of the American team trying to negotiate a peace treaty with Britain. His position there – as an American rebel and advocate of pointed conductors that had again failed to prevent lightning damage – almost certainly influenced some in Britain, including the King, to turn towards Benjamin Wilson and his blunt conductors. Furthermore, as Price also informed Franklin, members of the Board of Ordnance were again apprehensive 'of danger to the magazine at Purfleet from the pointed conductors there'. As a result, they had once more 'apply'd to the Royal Society for advice; and a Committee is just appointed by the Society to examine the facts'.[30]

The Society had already received reports on the events at Heckingham from several Norwich gentlemen. They included George Cadogan Morgan, who had visited the establishment and quizzed some of its inmates and staff, though Morgan little trusted their testimony. As he later wrote:

> The contradictory absurdities which they asserted and maintained, are scarcely conceivable. Indeed, the human fancy at all times is abundantly prolific of deceits, but never equally so, as when it is embarrassed, or stupefied by terror; it then conjures up spectres it never saw, and is insensible to the impressions it might have retained.[31]

Such truisms might, of course, be found among gentleman scientists too, as subsequent events at Heckingham would reveal.

The Royal Society and its President, Joseph Banks, sent two representatives to Norfolk: Edward Nairne (1726–1806, FRS 1776), a strong supporter of Franklin; and Charles Blagden, who recorded his Norfolk visit in a diary entry for 26 January 1782. Not only did he see 'Mr Wilson's [blunt] conductors to the cathedral terminating in the weathercock', he also viewed 'some houses which had been struck by lightning in the town, particularly some at a distance from one another, which were damaged by the same explosion'. He then dined with Samuel Cooper, the governor of Heckingham and, in the evening, went 'to Mr Brookes [a seller of books and electrical and optical equipment in Norwich] and Mr Morgan's' where he viewed electrical experiments demonstrating the degree of conductivity of various materials.[32]

On the basis of his earlier comment to John Lloyd concerning pointed *versus* blunt conductors, we might expect Blagden to have maintained an open mind on the subject. Yet, as Shaffer (2010) has detailed,[33] the reputation of the Royal Society and the value of its scientific advice were now at stake. The Society's delegates, therefore, 'needed to show the Heckingham lightning rods were badly set up' and towards this end they gathered their evidence. As a result, the report issued to the Ordnance Board in February 1782 was 'managed by Nairne, Blagden and Banks' to show 'the imperfections of the Norfolk lightning rods and strengths of the received theory of their behaviour'.[34] Morgan and Brooks, however, followed their own course in relation to the official report. Despite his passionate support for Franklin's 'politics and experiments' Morgan did not agree with his theory on pointed lightning conductors. As he later concluded in his *Lectures on Electricity*: 'It appears to me, that the best means of protection can be known only in consequence of a thorough search into the various combinations of perilous circumstances which can occur'. He even suggested that 'by guarding your house, you make it of all objects, that which is most likely to become the circuit of a cloud'.[35]

Modern science suggests that whether they are pointed or blunt makes very little difference to the performance of conducting rods, a possibility that Richard Price had already posited to his revolutionary friend, and pointed-rod advocate, Benjamin Franklin. Although events such as those at Purfleet and Heckingham 'have a tendency to discredit conductors', Price wrote, 'Mr Wilson's triumph seems improper, because there is no reason for believing the same or worse would not have happened had the conductors been blunt'.[36]

Star Gazers

'Despite the Clouds',[37] and a consequent preoccupation with weather, Welsh interest in astronomy has a long history. It stretches from the orientation of standing stones and medieval Cistercian churches,[38] to the writing of the very first introduction to astronomy in English – *The Castle of Knowledge* (1556) by Tenby-born, London-based mathematician Robert Recorde, in which a very early record of the Copernican model of the solar system is made. On 29 March 1742, Anglesey-born, London-based mathematician William Jones Sr connected to that past when he purchased at auction, for the princely sum of eleven shillings, a volume of astronomical tables compiled, expanded, and improved by his medieval countryman, Lewis of Caerleon (*fl*.1490s).[39] The volume entered the library of Jones Sr, one of the finest mathematical libraries in Britain at the time, and subsequently came to George Parker, second Earl of Macclesfield, who Jones had tutored and to whom he bequeathed his books. In 2020 the British Library bought the Lewis of Caerleon volume (for £300,000) and it is now available online.[40]

Some twenty-two years after Jones made his purchase, the Vale of Glamorgan diarist William Thomas, whose weather records have already been mentioned, made his own astronomical observation: 'the sun was eclipsed with us about 10 digits,' he wrote on 1 April 1764, 'the greatest darkness about 35 minutes after ten. Then the air looked of a lead colour, which taught me if but lost for a while the heat of the sun this world would soon turn to its former chaos'.[41] In ensuing years, Thomas would also record the passing of three comets, though he did not always see them in person.[42]

Beyond the nationality of their originators, what unites the medieval tables of Lewis of Caerleon with the eighteenth-century observations of William Thomas is their reliance on naked-eye stargazing. Thomas, for example, notes how the good folk of Michaelston-super-Ely in Glamorgan viewed the 1764 eclipse 'by a common window glass turned black in the smoke of a candle'.[43] Additional aids to astronomical observation then available were the cross-staff, the astrolabe and the octant; with the last morphing into the modern sextant in the course of the eighteenth century. However, instrumental observation truly begins in Wales with the arrival of the telescope. Invented in the Netherlands by Hans Lippershey in 1608, a so-called 'Dutch-trunk' entered the country in 1610 (see Chapter 1), but between its arrival and the recorded use of a reflecting telescope in Breconshire in 1761 (see 'Transits of Venus' below), the extent of astronomical observation in Wales is little known and under-researched. While we may presume the use of telescopes by amateur astronomers among the gentry and by military personnel, surveyors and seafarers, the main evidence for continued telescopic awareness between 1610 and 1761 comes from literary sources.

According to the *OED*, the first recorded use of *telescope* in English is attributed to Welsh alchemist Thomas Vaughan, when he writes in *Magia Adamica* (1650) of 'spots and darkness' on the sun 'as it hath been discovered by telescope'.[44] The use of the word 'telescope' in a 1651 civil war tract by Alexander Griffith has already been noted (Chapter 4). Published in the same year, the engraved title page of Thomas Powell's *Elementa Opticæ* bears the figure of 'Optica' peering heavenward through a telescope (see Figure 4). By 1652, the word *ysbien-ddrych* (lit. 'spy-glass' or 'telescope') had appeared in Welsh, with *telesgob* appearing in 1721. An appreciation of the results of telescopic observation is evident too in the many almanacs being published in Wales, principally because of the importance of such observations to astrological prediction (see Figure 7). In the early eighteenth century more substantial works, such as *Golwg ar y Byd* ('A View of the World', 1725) by the Revd Dafydd Lewys, introduced details of planetary arrangements in the Ptolemaic, Copernican and newer Newtonian 'System Newydd' ('New System'). Matters such as the size of the heavenly bodies, the distance separating them, and their motion, were also dealt with, though usually in very general terms and within an overarching religious context.

There were, of course, writers who 'failed to absorb the essence of Newtonianism'[45] and who rejected the heliocentric planetary system. The complex attitudes of alchemist Thomas Vaughan to this issue, in the 1650s, have already been explored in Chapter 3. In the Welsh-language classic *Gwledigaetheu y Bardd Cwsc* ('The Visions of the Sleeping Bard', 1703) its author, Ellis Wynne (1670/1–1734), has his narrator resting in the empyrean on a ledge of white cloud so that 'When the sun waxed strong, I beheld … our great, encircled earth as a tiny ball in the distance below'; this despite the narrator later being provided with an *ysbien-ddrych* ('telescope'). Here, of course, we can invoke the defence of artistic licence, since the work is one of fiction. Less easily defended is the view held by almanacker Thomas Jones (1648?–1713) that 'The world stands in the middle of the heavens like the yoke in the middle of an egg and the sun and moon and stars in the heavens turn constantly around it'.[46] Yet we should not see this opinion as *necessarily* implying parochialism, or of Wales as a late seventeenth, early eighteenth-century scientific backwater. Jones, after all, had lived for a number of years among the newspaper and pamphleteering hacks of Grub Street in London, where some knowledge of the scientific doings in the city might be expected, even if only to be satirised. Nor was his opposition to heliocentrism unique. In a study of Jewish reaction to the heliocentric system, Jeremy Brown (2013) notes how responses varied among the London-based rabbinic community of the eighteenth century. Some rejected it out-right; others modified their position over time, while others accepted it from the beginning.[47] As late as 1753, English clergyman Samuel Pike (1717?–73), who received his training from Newton's friend John Eames, wrote in *Philosophia Sacra* (1753) that:

> many common Christians to this day firmly believe that the earth really stands still, and that the sun moves around the earth once a day: neither can they be easily persuaded out of this opinion, because they look upon themselves bound to believe what scripture asserts.[48]

How far Thomas Jones's view resulted from a religious conviction is impossible to say, but religious constraints notwithstanding, by the eighteenth century most scientists of Wales had accepted heliocentrism, as is reflected in an early eighteenth-century poem written by Jane Brereton (née Hughes, 1685–1740). She was born near Mold, in north Wales, and when the following lines occur in her poem *Merlin* (1733) its protagonist is perusing the heavens from the summit of Pumlumon (Plynlimon):

> The Summit gain'd, I sought with naked Eye
> To penetrate the Wonders of the Sky.

No telescopic Glass known in that Age
T'assist the Optics of the curious Sage.
Tho' lov'd *Astronomy* oft charm'd my Mind,
I now erroneous all my Notions find.
I thought bright *Sol* around our *Globe* had run,
Nor knew Earth's Motion, nor the central *Sun*.
And had I known; could I Belief have gain'd,
When Ignorance, and Superstition reign'd?[49]

Of Navigators

Astronomers of the long eighteenth century were also concerned with another aspect of astronomical observation increasingly important to the times – the science and art of navigation. In the early years of the eighteenth century the two principal Welsh scientists to contribute to its study were probably inspired to do so by their early career experiences of life at sea. Having arrived in London from Anglesey, William Jones Sr, who later purchased the astronomical tables of Lewis of Caerleon, began work in a commercial enterprise and travelled to the West Indies. Afterwards, in his mid-twenties, he began teaching mathematics and navigation aboard naval vessels. As a result, on 12 October 1702, he found himself taking part in the British naval assault on Vigo, in north-west Spain.[50] Leaving the navy after this action Jones returned to a life in London. There he immediately put his mathematical and navigational knowledge to good use by writing *A New Compendium of the Whole Art of Practical Navigation* (1702).

According to Jones, the *Compendium* collected together 'many very useful things, not to say improvements, which before lay scattered about in several volumes'. Designed 'purely for practice' it centres on applying mathematical solutions to the navigational problems it poses. Given the speed of its production (presumably between late October and December 1702) it seems likely that Jones had worked on the book while at sea. Yet, when dedicating the book to his English friend and fellow mathematician, the Revd John Harris (*c*.1666–1719, FRS 1696), Jones relates that it was composed under Harris's roof, 'attempted by your encouragement, and supervised and corrected by your friendly care'. Between 1702 and 1704, Harris gave a series of mathematical lectures at the Marine Coffee House in Birchin Lane in London. He also advertised himself as a mathematical tutor working at Amen Corner. It is perhaps not too fanciful to see Jones returning to London from his adventures in Vigo, attending one of Harris's lectures or tutorials in 1702, and being offered a place to stay and to compile his treatise on navigation. John Harris had a particular interest in mathematics, astronomy, navigation and travel, for all of which Jones's own maritime adventures could have provided useful knowledge and background colour. Harris would later write a work on globes and orreries, and a compendium of travels. His most famous work, the *Lexicon Technicum* (1704 and 1710), was the first alphabetical encyclopaedia in English (see Chapter 16). Harris issued a three-page proposal for the *Lexicon* in 1702 and later declared the material within it to be 'from the best Original Authors I could procure'. It has been suggested that Jones Sr 'probably contributed the navigational articles in the volume'.[51]

Our second Welsh mariner, Joseph Harris, the Breconshire born brother of preacher Howell Harris, made two overseas journeys, each having astronomical observation as one of their purposes. One was to Vera Cruz in Mexico from 1725 to 1727, and another, between 1730 and 1732, to Jamaica (discussed in Volume 2). On returning to London from Verra Cruz, Harris wrote and published two works: *The Description and Use of the Globes and the Orrery* (1729; see Figure 18) and *A Treatise of Navigation* (1730). The *Treatise*, which he notes as having been 'examined by Dr. Halley, who was pleased to approve of it', includes 'A succinct Treatise of Plain Trigonometry' and presents similar content to William Jones's earlier *Compendium*. Given his belief that 'Middle-Latitude sailing is erroneous; and that the common method of reckonings in Meridional Distance, is grossly false', Harris suggests various improvements to these aspects of navigation. He also introduces 'a new method for finding the bearings of places on charts, without the confusion of rhumb-lines [a line crossing all meridians at the same angle]', and describes improvements to a number of maritime instruments.

FIGURE 18
Title page to
*The Description and
Use of the Globes and
Orrery* by Joseph
Harris (1729;
London, 1763)

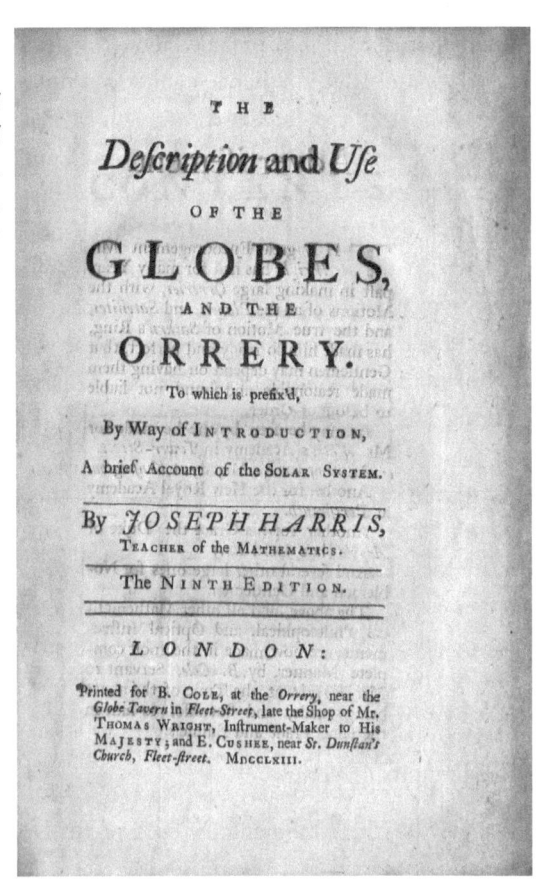

THE

Defcription and Ufe

OF THE

GLOBES,

AND THE

ORRERY.

To which is prefix'd,

By Way of INTRODUCTION,

A brief Account of the SOLAR SYSTEM.

By *JOSEPH HARRIS*,

TEACHER of the MATHEMATICS.

The NINTH EDITION.

L O N D O N:

Printed for B. COLE, at the *Orrery*, near the
Globe Tavern in *Fleet-Street*, late the Shop of Mr,
THOMAS WRIGHT, Inftrument-Maker to His
MAJESTY; and E. CUSHEE, near St. *Dunftan's*
Church, Fleet-ftreet. MDCCLXIII.

Both Jones Sr and Harris include sections on astronomy in their
works. Jones does so in a section on 'the most useful and necessary
problems in Astronomy', to which were appended tables of the sun's
declination (its angular distance north or south of the celestial equator)
provided by the Astronomer Royal, John Flamsteed. In *Globes and the
Orrery,* Joseph Harris has a forty-one page 'Introduction, Containing A
Brief Account of the Solar System, and of the Fixed Stars'. In a section
on 'Astronomical Problems' in his *Treatise on Navigation,* he proposes
improvements to the 'common sea-Quadrant', as well as describing 'a
new Fore-Staff, which is much more accurate and commodious than the
common one'; both instruments were used at the time for astronomical

observation aboard ship. Included too was 'A Table of the most eminent fixed stars, contrived in such a manner, as to shew by Inspection, which is the properest Star for Observation, in all Places, and at all Times of the Year'. Together with the engraver John Senex (1678–1740), Harris later went on to produce two undated star maps illustrating the position of fixed stars for the northern hemisphere (*Stellarum Fixarum Hemisphaerium Boreale*) and southern hemisphere (*Stellarum Fixarum Hemisphaerium Australis*). Based on star positions recorded by Edmond Halley, who Harris had met when he first arrived in London, the maps are referenced by William Herschel in his diary as 'Harris's star charts'. Herschel would be using them on 13 March 1781 when he discovered the planet Uranus.[52]

Fixed stars were crucial to navigation at this time, and their angular relation to the movement of the moon across the night sky would form one of the methods tried during the great eighteenth-century endeavour to measure longitude at sea with accuracy. This method, and the various others being proposed, will be discussed in Chapter 15, together with participation in the endeavour by scientists of Wales.

The Transits of Venus

Besides Herschel's momentous discovery of Uranus in 1781, the first planet to be discovered beyond those known since antiquity, the other great astronomical event of the second half of the eighteenth century was the transit of Venus across the face of the sun in 1761 and 1769.[53] Edmond Halley, in 1691, had urged the Royal Society to send observers to view the transits at different locations in the world as a means to obtain measurements of parallax. These measurements could then be used, via geometry and trigonometry, to determine the distance from the earth to the sun; a distance known today as the astronomical unit, the fundamental unit of distance in astronomy.

In 1761, a major international project aimed at studying the first Venus transit got underway. It involved astronomers in places as far flung as America, Europe, Russia, India and Indonesia, as well as England. Determined to be part of this great scientific enterprise,

Joseph Harris wrote on 25 April 1761 from London to his brother Howell at their family home of Trefeca in Breconshire. Joseph planned to make his own observation of the Transit from Wales. In his letter he related to Howell his dispatch of three boxes of mathematical instruments by the Hereford carrier, and he arrived himself at Trefeca on 28 April. Among the mathematical instruments he sent were a number of clocks for accurate timekeeping and a five-foot reflecting telescope, in all probability the one that remains on display at Trefeca today (see Figure 19).

On 6 June 1761, Harris observed through his five-foot reflector the latter stages of the transit of Venus across the disc of the sun. As the planet prepared to emerge from the sun's disc, having almost completed its transit, he recorded the time that the planets leading edge first touched the rim of the sun. Then, as Venus completed its transit and emerged from the sun's disc, he recorded the time of the last contact of the planets trailing edge with the sun's rim; the difference in time between these two positions being eighteen minutes and fifteen seconds. Harris then wrote up his observations, which were completed by 19 November.[54] On 25 November he submitted them to George Parker,

FIGURE 19 Reflector telescope at Trefeca, Breconshire, believed to have been used by Joseph Harris to view the Transit of Venus from Trefeca in 1761. Courtesy of Martin Griffiths.

the Earl of Macclesfield, President of the Royal Society and one time student of William Jones Sr:

> If the annexed account meets with your Lordship's approbation, I should be glad to have it laid before the Royal Society. I should have communicated it much sooner, but that I waited till I could find leisure to compute the Parallaxes, without doing which my observations could have been but little use.

For reasons unknown, Parker failed to pass on the observations for publication in the *Philosophical Transactions*. Following his observation of the Venus transit, Joseph Harris returned to his house and work at the Royal Mint in the Tower of London. He died on 26 September 1764 and is buried within the Tower, in the chapel of St Peter ad Vincula; the resting place of Anne Boleyn, Thomas More, Lady Jane Grey and others.[55]

The second transit of Venus took place on 3 June 1769. No observer in Wales is recorded this time, but in 1770 Richard Price published a paper in the *Philosophical Transactions* on 'the Effect of Aberration of Light on the Time of a Transit of Venus over the Sun'.[56] Aberration in astronomy relates to an apparent motion of celestial objects because of the motion of the observer, in this case the astronomer on the moving earth. The displacement of the celestial object occurs in the direction of travel by the observer. It would not occur if the earth was stationary, or light travelled instantaneously. Price's paper on the subject was actually a printing of a letter that he had sent to Benjamin Franklin, who had previously shown Price a letter that he had received from the American astronomer and mathematician John Winthrop (1714–79, FRS 1766). Winthrop had viewed the first transit of Venus from Newfoundland and had suggested in his letter to Franklin that the transit time of Venus would be retarded by the aberration of Venus. Price, in writing to Franklin, agreed with Winthrop: 'The aberration of Venus must, I think, affect the phases of a transit by retarding them, and not by accelerating them.' But that was not all. Price also believed there was 'a retardation arising from the aberration of the Sun, as well as that from Venus', and that 'these observations have not been attended to by astronomers'.[57] The letters by Winthrop and Price were both published in the same volume of the *Philosophical Transactions*. Franklin, though, had already forwarded an early draft of Price's letter to Winthrop who commented to Franklin that it gave him 'pleasure to find himself supported by so judicious a person' as Richard Price.[58]

At some point in his career Richard Price came to own a telescope, since it was one of the articles that he willed to his nephew George Cadogan Morgan. It was made by Dolland, a fine London maker, but whether Price actually used it to observe the second transit from his home in Newington Green, then a village just outside the City of London, is an open question, although the detailed measurements of transit times discussed in his paper on the Venus transit certainly suggest the possibility.

The Observatories

Price is not recorded as having a purpose-built observatory at his home in Newington Green, nor after his later move to Hackney, but astronomical observatories were certainly established by several other scientists of Wales in the long eighteenth century.

Whether Joseph Harris had an observatory is also unknown. However, the sale of his library in London by Samuel Baker of York Street in Covent Garden on 11 February 1765, and on 'eight following evenings at 6pm', certainly gives evidence of his observational astronomical interests. Along with a mass of general instruments such as compasses, levels, theodolites, and a pair of globes by Senex, we also find a range of telescopes. They include a series of four sold together, including a 'reflecting telescope by Short', which could be made into 'either a Newtonian or Gregorian, or a Cassegraine' form. The four were sold as one lot and possibly to Henry Cavendish (1731–1810, FRS 1765, Copley Medal 1766) who discovered hydrogen and measured the gravity of the earth. He also reviewed the meteorological instruments of the Royal Society and those of the Greenwich Observatory. Another instrument 'A brass apparatus carrying an 18″ Transit Instrument and a 12″ Quadrant, with a micrometer for measuring to four seconds, by Bird' was sold for £27 7s. to 'Canton', probably the electrical experimenter John Canton.[59]

When George Shuckburgh-Evelyn paid a visit to John Lloyd in 1778 he found his friend much taken up with the 'construction of the astronomical edifice at Hafodunnos', the second north Wales estate owned by Lloyd in addition to that at Wigfair.[60] This 'edifice' may have been to house the 'EQUATORIAL & ASTRONOMICAL CIRCLE, 20 inch radii, in brass, with adjusting levels, microscopes, lamp bracket, &c by Troughton' advertised in the 1816 sale catalogue of Lloyd's library and scientific equipment. The structure itself was advertised as 'A CONICAL OBSERVATORY, OAK ROOF, for the EQUATORIAL, with double Slit on double friction Rollers, and Points of the Compass letter, by Smeaton'.[61] No details of any other edifice are currently known.

When the sale catalogue seeks to 'impress upon the mind of the public' the 'value, magnitude, and superiority' of the astronomical instruments in Lloyd's collection, it is no salesman's overreach. The collection included instruments by the finest makers, both of 'departed excellence' – 'Smeaton, Ramsden, Troughton, Wollaston, Whitehurst of Derby, Bird, etc' – and 'those of the present day' – 'Herschel, Dolland, Holmes, Adams, Jones etc.'. Among the eleven telescope-related entries in the catalogue are 'Ten brass tubes for telescopes, of different diameters' (2s. 6d); 'a very fine brass micrometer in shagreen case, new'; and 'A very fine 10 inch Slide Telescope, 4 inch aperture, oak tube and frame with double rack-work, block-screws, mahogany supporter, with screw, &c by Dolland' (£84). Among the ancillary equipment listed is 'A sextant, with Reflectors, 4½ inch radii, with two telescopes, in mahogany case, by Ramsden' (£4); a quadrant with three 16-inch telescopes; a rotary planisphere; a pair of 23-inch globes; a fine micrometer and a transit instrument made by Smeaton, and bought from 'Mr Smeaton's observatory'; and a whole range of clocks, and barometric and surveying instruments.[62] There was also a 'very fine SEVEN FEET REFLECTOR [telescope], Speculum 7 inch diameter, with finder, double-vertical rack-work, horizontal motions, mahogany tube and frame, by [William]

HERSCHEL'.[63] Sold for £99 15s., it was made to order for Lloyd and is briefly discussed in two letters to him from Herschel in 1791 and 1796. In the first, their acquaintance and friendship is evident (Lloyd sponsored Herschel for the Royal Society in 1781). 'It will always give me great pleasure to hear from you', Herschel writes, 'still more to see you here at Slough that I may have an opportunity to thank you personally for the good Christmas Pie, which is arrived very safely tho' it had been above three weeks on the road'.[64] In his second letter Herschel apologises for a delay in sending the completed telescope. The delay was 'occasioned by a little accident which happened to the object mirror in packing it up some time ago, which required my wiping it, and as I chanced to make some considerable scratches I even repolished it again',[65] a repair whose undertaking is seriously understated by the phrase 'a little accident'. Herschel polished his own mirrors and this could take many hours to complete. It is said that his sister Caroline, an astronomer herself, would feed him during the polishing so that he did not have to stop and lose his feel of the progress of the operation.

Between 1777 and 1781, another observatory operated in Wales at Frampton House near Llantwit Major in the Vale of Glamorgan. Established on one of their estates by English brothers Nathaniel and Edward Pigott, it contained clocks, mural quadrants as well as a '6ft Dolland achromatic, a 2.5 inch achromatic by [Francis] Watkins [see Chapter 14], and a 2.5ft reflector by Heath and Wing with a 5 inch aperture'. While in Wales they established the longitude of Frampton as 1 degree 29 minutes and 11 seconds west of Greenwich[66] and made a number of significant astronomical discoveries. Edward discovered a nebula in the constellation of Coma Bernices, and Nathaniel found three previously unrecorded double stars, details of their discoveries being presented in two papers in the same volume of the *Philosophical Transactions*.[67] In 1781 the brothers moved away from Frampton to York, where they established another observatory.[68]

As we saw in Chapter 9, when John Lloyd of Wigfair visited the baseline measurement on Hounslow Heath being undertaken by William Roy in 1784, he did so in company with, among others, Charles Greville. In 1805, Greville began construction of the technical college he wished to see established in his new town at Milford Haven. The central part of the college building formed the Hakin Observatory. Built from local stone and comprising a number of adjoining small rooms it also had a house for the superintendent. This was to be John Firminger who, until 1807, had been an assistant at the Greenwich Observatory to the then astronomer Royal, Nevil Maskelyne. He had also been a correspondent of John Lloyd. By June 1808, the observatory equipment, made by the very best makers, had arrived, but it did not remain there long. With the death of Greville in 1809 the college project stalled, and the astronomical equipment was removed. At least one instrument eventually made its way into the possession of the Revd Lewis Evans (1755–1827).

Born into a family from Bassaleg in Monmouthshire, Lewis Evans took up the vicarage at Froxfield in Wiltshire in 1788. An accomplished mathematician and astronomer, he is believed to have established an observatory there.[69] Evans remained at Froxfield until his 1799 appointment as mathematical master to the Royal Military Academy in Woolwich. There he established another observatory, though, contrary to many reports, it did not form part of the Academy. As Evans pointed out in a letter to the *Monthly Magazine* in 1816, it was a private observatory attached to his house on Woolwich common and, he notes, 'built by myself'. In his manuscript archive in the Oxford Museum of Science (founded by his great grandson and namesake, Lewis Evans 1853–1930) we find his detailed instructions on how to build just such an observatory.[70] Also in his letter to the *Monthly Magazine*, Evans gives a detailed description of the contents of his Woolwich observatory. It included 'one of Mr Troughton's best transit circles', which, he notes, formed 'One of the returned instruments from Milford [Haven]'. In the wake of Greville's death the instrument had lain at Milford 'in packing cases unopened for about twelve months' before being 'sent to Mr Troughton's, in Fleet Street, for sale'.[71]

One of the first observatories to be built in Wales in the early nineteenth century belonged to Robert Roberts, who we first met as the author of *Daearyddiaeth* ('Geography'), a volume whose 548 pages opens with a lengthy chapter on astronomy and the solar system together with an 'Enwau Seryddawl'; an astronomical dictionary covering everything from 'Andromeda' and 'perihelion' to 'Ymbellhad' ('Distance'). His observatory is said to have been at his home in a stone tower on Hill Street in Caergybi ('Holyhead').[72] All trace of it has now disappeared, but evidence of Roberts's importance as an astronomer has not. In 1828 at Dolgellau, and later Carmarthen, he gave the first known lectures on astronomy to be delivered in Cymraeg ('Welsh'). These were 'illustrated by a beautiful Orrery, conducted by his son, a promising youth', all of which led to Roberts becoming locally known as 'the grandfather of the ether'.[73]

In this, and the previous two chapters, we have concentrated on the practical contribution made by scientists of Wales to the science of the earth, of nature, and of the heavens. In the next two chapters we reflect on the relationship between science and ourselves through science's contributions to medicine and social well-being, and the uses of nature and the development of economics.

Notes

1. *WTV(MA)*, p. 225.
2. Ruth Scurr (ed.), *John Aubrey: My Own Life* (2015; London, 2016), p. 140.
3. Lisa Tallis (ed.), *Cas Gan Gythraul: Demonology, Witchcraft and Popular Magic in Eighteenth-Century Wales* (Newport, 2015), p. 5.
4. I am grateful to Ken Brassil for drawing my attention to William.
5. Available online at NLW website.
6. See *DWB* entry.
7. See *Book of Common Prayer Online*, in 'Tables and Rules for Finding the Date of Easter' (accessed 7 October 2021).
8. See Gerrit Bos and Charles Burnett (eds and trans.), *Scientific Weather Forecasting in the Middle Ages: The Writings of Al-Kindi* (2000; London and New York, 2016), p. 15.
9. Bos and Burnett (eds and trans.), *Scientific Weather Forecasting in the Middle Ages*, p. 5.
10. R. T. W. Denning (ed.), *The Diary of William Thomas (1762–1795)* (Cardiff, 1995), p. 322.
11. See Alexandra Witze and Jeff Kanipe, *Island of Fire: The Extraordinary Story of Laki that Turned Eighteenth-Century Europe Dark* (2014; London, 2017), Chapter 5 and p. 108 (map).
12. See F. V. Emery, 'Edward Lhuyd and Snowdonia', *Nature in Wales*, new series, 4 (1985/86), 6.

13. Edward Lhuyd, 'A note concerning an extraordinary hail in Monmouthshire', *Philosophical Transactions of the Royal Society*, 19 (1697), 579–80.

14. Anon., 'An Account of a storm of rain that fell at Denbigh in Wales: communicated to Dr Hans Sloane, R. S. Secr.', *Philosophical Transactions of the Royal Society*, 25 (1706), 2342–4.

15. RSC LBO/18/125, Perrot Williams to James Jurin, 21 March 1726.

16. RSC EL/D2/52, Evan Davies to John Eames, 6 December 1729; Evan Davies, 'An Account of what happened from Thunder in Carmarthenshire … communicated to the Royal Society, By John Eames, FRS as he received it in a Letter from Mr Evan Davies', *Philosophical Transactions of the Royal Society*, 36 (1729–30), 444–8.

17. See Emery, 'Lhuyd and Snowdonia', 3–11 (at 5–7).

18. Aside from the most interesting entries, the weather observations are omitted from Denning (ed.), *Diary of William Thomas* but are noted on p. 29 as having been transcribed in full and made available 'as part of the microfiche edition' of the diary.

19. See Beryl Thomas and D. O. Thomas, 'Richard Price's Journal for the Period 25 March 1787 to 6 February 1791. Deciphered by Beryl Thomas with an Introduction by D. O. Thomas', *NLWJ*, 21/4 (1980), 366–413; at entries for 21 December 1788 (p. 388) and 4 January 1789 (p. 389).

20. For details see Paul Frame, *Liberty's Apostle, Richard Price his Life and Times* (Cardiff, 2015).

21. For all details see Neil Macdonald, Cerys A. Jones, Sarah J. Davies and Catheryn Charnell-White, 'Historical weather accounts from Wales: and assessment of their potential for reconstructing climate', *Weather*, 65/3 (2010), 72–7 (at 72, Table 1).

22. James Jurin, 'Invitatio ad Observationes Meteorologicas communi consilio instituendas', *Philosophical Transactions of the Royal Society*, 32 (1723), 422–7.

23. RSC MA/100.

24. See Andrea Rusnock, *Correspondence of James Jurin* (Amsterdam and Atlanta GA, 1996), pp. 29–30.

25. Rusnock, *Correspondence of James Jurin*, p. 298, and Wellcome Collection MS. 6146.

26. Rusnock, *Correspondence of James Jurin*, pp. 313–4, and RSC LBO/18/95.

27. H. E. Landsberg, 'A Note on the History of Thermometer Scales', *Weather*, 19/1 (1964), 2–6 (at 2).

28. See 'Letter from Hon. Daines Barrington FRS to William Heberden, MD, FRS, giving an Account of some experiments made in North Wales, to ascertain the different quantities of rain, which fell in the same time, at different heights', *Philosophical Transactions of the Royal Society*, 35 (1771), 294–7. The barometer change, as noted by Hughes, was 'one inch and six tenths'. Using Halley's computation method this gave a height of 450 yards (1,350 feet). This, however, is well below the modern recorded height of Arennig Fawr (2,801 feet) and Arennig Fach (2,259 feet).

29. See NLW MS 12416 D, f. 53 and full transcript in Paul Frame, 'Charles Blagden in Revolutionary America: Two Unpublished Letters to John Lloyd', *Notes and Records: The Royal Society Journal of the History of Science*, 71 (2017), 361–9.

30. W. Bernard Peach and D. O. Thomas, *The Correspondence of Richard Price*, 3 vols (Durham NC and Cardiff, 1983–94), II (1991), p. 114.

31. Morgan, *Lectures on Electricity*, 2 vols (Norwich, 1794), II, p. 234.

32. MSS Osborn fc16, Box 2 f.18; Diaries/1782, Jan–Nov in Sir Charles Blagden, Diaries. James Marshall and Marie-Louise Osborn Collection, Beinecke Rare Books and Manuscripts Library, Yale University.

33. For comprehensive details of the Heckingham and Purfleet episodes, see Simon Schaffer, 'Charged Atmospheres: Promethean Science and the Royal Society', in Bill Bryson (ed.), *Seeing Further: The Story of the Royal Society and Science* (2010; London, 2011), pp. 132–54.

34. Schaffer, 'Charged Atmospheres', in Bryson (ed.), *Seeing Further*, p. 150.

35. Morgan, *Lectures on Electricity*, II, p. 290 and 298.

36. Peach and Thomas (eds), *The Correspondence of Richard Price*, II (1991), p. 114.

37. Bryn Jones, 'Despite the Clouds: A History of Wales and Astronomy', *The Antiquarian Astronomer*, 8 (April 2014), 66–96.

38. Bernadette Brady, Darrelyn Gunzburg, Fabio Silva, 'The Orientation of Cistercian Churches in Wales: A Cultural Astronomy Case Study', *Cîteaux – Commentarii cistercienses*, t. 67, fasc. 3–4 (2016), 275–302.

39. See 'The Macclesfield Copy of the works of Lewis of Caerleon' at the blog of Peter Kidd (of the Bodleian Library and later curator of illuminated manuscripts at the British Library), *https://mssprovenance.blogspot.com* (accessed 28 May 2024).

40. See *blogs.bl.uk/digitalmanuscripts/2020/11/lewis-of-caerleon-manuscript-saved-for-the-nation.html* (accessed 15 September 2021).

41. Denning, *Diary of William Thomas*, p. 101.

42. Denning, *Diary of William Thomas*: 8–10 April 1766 (p. 160); 6 September 1769 (p. 229); 18 August 1783 (p. 324).

43. Denning, *Diary of William Thomas*, p. 101.

44. *WTV(MA)*. Sunspots were discovered by Christoph Scheiner and Galileo in 1611 with priority for the discovery unclear.

45. Geraint H. Jenkins, *The Foundations of Modern Wales* (Oxford, 1987), p. 245.

46. Jenkins, *The Foundations of Modern Wales*, p. 246, citing NLW MS 6146B, pp. 179–86 (at 179).

47. Jeremy Brown, *New Heavens and a New Earth: The Jewish Reception of Copernican Thought* (Oxford, 2013).

48. Samuel Pike, *Philosophia Sacra: Or, the Principles of Natural Philosophy. Extracted from Divine Revelation* (London, 1753), p. 43.

49. Quoted in Sarah Prescott, *Eighteenth-Century Writing From Wales: Bards and Britons* (Cardiff, 2008), p. 46. The poem has a complex context, part of which is its commemoration of 'Merlin's Cave', one of two '"associational" pavilions' erected in Richmond for Queen Caroline, wife of George II; the other pavilion being 'an Hermitage' containing 'busts of Newton, Locke, Samuel Clarke, William Wollaston and Robert Boyle,' some of whom had benefitted from Caroline's patronage. See discussion of the poem pp. 39–47.

50. For details of the important attack on Vigo, see N. A. M. Rodger, *The Command of the Ocean* (2004; London, 2005), pp. 166–7.

51. See *ODNB* entry for William Jones Sr.

52. Wolfgang Steinicke, 'William Herschel, Flamsteed Numbers and Harris's Star Maps', *Journal of the History of Astronomy*, 45/3 (2014), 287–303.

53. For a full discussion of the transits, see Andrea Wulf, *Chasing Venus* (2012; London, 2013).

54. The original manuscript can be seen at NLW MS 17529C entitled 'An Account of the Late Transit of Venus over the Sun, as it was observed at Trevecka in Breconshire by Joseph Harris'.

55. For a full transcript of Harris's observations as sent to the Royal Society, together with other related letter extracts, see Jennifer Stanesby Moody, 'The Trefeca Meridian and Transit of Venus', *Brycheiniog*, 41 (2010), 51–64. See also Martin Griffiths, 'Joseph Harris of Trevecka, Scientist, Artisan, Servant of the Crown', *Antiquarian Astronomer*, 6 (2012); Jones, 'Despite the Clouds'; and the American Astronomical Society at *https://aas.org/posts/news/2021/05/month-astronomica l-history-june-2021* (accessed 28 May 2024).

56. Richard Price, 'A Letter from Richard Price, D.D. F.R.S. to Benjamin Franklin, LL.D F.R.S. on the Effect of the Aberration of Light on the Time of a Transit of Venus over the Sun', *Philosophical Transactions of the Royal Society*, 60 (1770), 536–40.

57. See Peach and Thomas, *Correspondence of Richard Price*, I (1983), pp. 93–4.

58. See Frame, *Liberty's Apostle*, p. 60.

59. Referenced under year 1765 (p. 71) in *List of Catalogue of English Book Sales 1676–1900, Now in the British Museum* (London, 1915) with BL reference number S.-C.S. 6. (1.).

60. NLW MS 12418D f. 12.

61. NLW MS 12500B.

62. NLW MS 12500B.

63. NLW MS 12500B.

64. NLW MS 12419D f. 27.

65. NLW MS 12419D.

66. See Edward Pigott, 'Determinations of the Longitudes and Latitudes of some remarkable Places near the Severn', *Philosophical Transactions of the Royal Society*, 80 (1790), 385–90. The Pigott brothers also involved themselves in a triangulation survey of the mouth of the River Severn, which they believed to be shown as too wide on maps of the time. For a discussion of this, see Anita McConnell and Alison Brech, 'Nathaniel and Edward Pigott, Itinerant Astronomers', *Notes and Records: The Royal Society Journal of the History of Science*, 53/3 (Sept 1999), 305–18 (at 310). See also Edward Pigott to Sir Henry Englefield, 'On a well in Glamorganshire whose waters rise and fall in an unusual manner', RSC: L&P VII.227.

67. E. Pigott, 'Account of nebula in Coma Berenices', *Philosophical Transactions of the Royal Society*, 71 (1781), 82–3; Nathaniel Pigott, 'Double stars discovered in 1779 at

Frampton House, Glamorganshire', *Philosophical Transactions of the Royal Society*, 71 (1781), 84–6.

68. For details of Nathaniel and Edward, see McConnell and Brech, 'Nathaniel and Edward Pigott', 305–18.

69. See *The Society for the History of Astronomy, https://shasurvey.wordpress.com/ Wiltshire-observatories/* (accessed 7 November 2021). They suggest that one of his pieces of equipment was 'an excellent 2-foot transit circle (2-inch OG) by Ed. Troughton, London' also used by Greville at Milford.

70. OHSM MS EVANS 18.

71. Lewis Evans, 'Mr Evan's on The Woolwich Observatory', *The Monthly Magazine*, 41 (1 July 1816), 483–4. Edward Troughton (1753–1835, FRS 1810) was a leading instrument maker in London at the time.

72. See *The Society for the History of Astronomy, https://shasurvey.wordpress.com/wales/ anglesey/* (accessed 11 November 2021), and *Y Rhwyd* ('The Net'), north Wales Community Newspaper, 18 (April 1981), p. 11.

73. See Jake E. Bridges, 'A Science Fit for the Chapel: Astronomy in Nineteenth Century Wales' (unpublished MA thesis, University of Alberta, 2018), pp. 39–43 (at p. 39 and p. 40); available online.

12

THE SCIENCE OF US: HEALTH AND SOCIAL WELL-BEING

There is not ... any circumstance of our existence,
or any one of our bodily or mental powers, which
has not in it something to perplex us; and one
of the greatest mysteries to man is man.[1]

Richard Price (1723–91)

In this chapter, we consider those scientists of Wales who attempted to unravel at least some of the perplexity to which Richard Price alludes. We do so by examining their attempts to improve the 'circumstances of our existence' through medical science and social initiatives, specifically, in the case of the latter, the development of life insurance and pensions.

Despite the attraction of Newtonian 'mechanical' medicine in the early part of the long eighteenth century (previously described in Chapter 5), most physicians continued to rely on balancing the four bodily humours – blood, phlegm, yellow bile and black bile – through bloodletting, enemas, induced vomiting and a mainly herbal, though increasingly chemical, pharmacology. Surgical techniques were being developed, but the days of proper anaesthetics were still some way off. Autopsies, to ascertain causes of deaths, were undertaken either in a morgue or at the deceased person's home. However, medical anatomists also relied on the corpses of hanged criminals, sometimes the deceased poor, and the best efforts of the 'resurrectionists' or body snatchers to learn how the body worked.

Medicines and the Search for Relief

Folk remedies, charms and the planetary and astrological influences touted in some almanacs as having medicinal value continued to intrigue and influence people in Wales throughout the long eighteenth century. However, it was plant and occasionally zoological and mineralogical material that, along with bloodletting, provided the main armoury for those seeking or providing medical relief and care. A number of medical recipe books utilising plant and animal ingredients were being used, and written, in the period. Among them were recipes said to be from the Physicians of Myddfai, in Carmarthenshire (see Chapter 1). With origins dating back to the late fourteenth and early fifteenth centuries, the recipes rely on the balancing of humours from herbal material, in alliance with astrological input, bleeding, surgery, urinoscopy and medicinal plasters. The last of the long line of Myddfai physicians, John Jones, is said to have died in 1739.

The *Spiritual Botanology* of Edmund Jones, written in 'Blank Verse, and Rhime' had as a subtitle 'What of God Appears in the Herbs of the Earth; Together with Some of Their Natural Virtue'. As its editor, Adam N. Coward relates, the book is a complex mix of herbal remedies and 'a religious work, treating plants as physical and natural "scriptures" to be "read" and interpreted within God's Creation, reflecting back upon Him to reveal His nature, goodness, and intention'.[2] As Edmund Jones himself says, when talking of the herb *Horehound* (*Marrubium vulgarre*):

> The Same Power wch made
> All other things, Hath made Herbs.[3]

These works, however, often remained in manuscript and so were of limited and probably local use. Most did not appear in print until much later: the recipes of the Physicians of Myddfai remained in manuscript until their 1861 publication by the Welsh Manuscripts Society;[4] Edmund Jones's *Spiritual Botanology* was originally composed around 1771 and 1772, but was not published until the South Wales Record Society and Adam Coward did so in 2017.[5] Still unpublished is *A*

Volume of Herbal Recipes and Remedies (1803), a 289-page herbal written by the author of *Observations on the Snowdon Mountains* (1802), William Williams (1738–1817) of Llandygai.[6]

In print, meanwhile, the bilingual Welsh/English *Pharmacopoeia* of Nathaniel Williams appeared in 1793 (see Figure 20). It contained 'A Select Body of Useful and Elegant Medicines' of the mainly traditional kind, and offered cures for a multitude of ills. These ranged from ringworm and impotency to asthma and the painless extraction of teeth, all accomplished with the aid of a wide variety of plants and herbs, as well as the likes of millipedes, powdered newts and salmon tails.

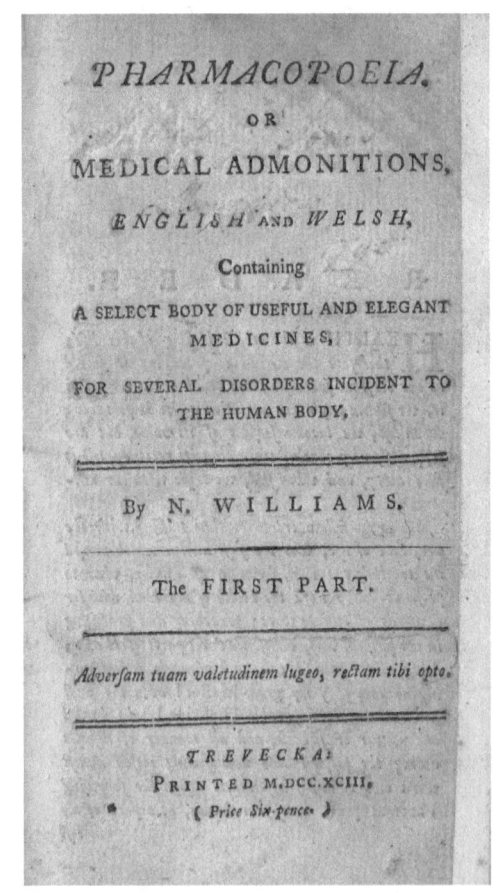

FIGURE 20 Title page to the bilingual (Cymraeg and English) *Pharmacopoeia or Medical Admonitions* of Nathaniel Williams (Trevecka, 1793)

The aim of such 'innocent and harmless' recipes being, as Nathaniel Williams put it:

That health may be recovered with facility, and the knowledge of the powers of medicine be obtained; that all the sick may cease to be patients, and become practical physicians for themselves; every patient his own Doctor, and every man his own Physician.[7]

The hope that everyone could always be their own physician was unrealistic. Nor were the recipes on offer truly science-based. They certainly represent a complete system of medicine, arrived at over many generations, but one resulting from trial-and-error experimentation, rather than a true understanding of the processes at work when creating or using the medicines. Yet, even as these traditional plant and herb-based medicines continued to be prescribed by doctors and apothecaries in Wales, newer medicines were becoming available.

Chemically based medicines and salves had developed out of the sort of iatrochemical experimentation undertaken by the likes of Thomas Vaughan and his wife Rebecca in 1650s London (see Chapter 2). Although the poet and self-declared physician Henry Vaughan (the Wales-based brother of Thomas Vaughan) appears to have preferred traditional galenic or herbal remedies,[8] chemical medicines were, according to Withey (2011), stocked in at least some of the approximately 124 identifiable apothecaries existing in Wales between 1600 and 1762. One early example is identified from the 1665 will-inventory of John Bell of Wrexham, whose apothecary shop contained 'Chymicall oyles' and 'a round box with some chemical medicines', presumably for purchase by early chemical physicians. Withey also notes the presence of English proprietary medicines in Wales, as well as evidence for a 'direct trade in medical goods between south Wales and London by the early eighteenth century'.[9]

One of the most important early eighteenth-century pharmacies in London was established by two Quaker brothers from Swansea. Having probably arrived in the city *c.*1708, and having served a seven-year apprenticeship with an apothecary in Cheapside, Silvanus Bevan

(1691–1765, FRS 1725) opened his pharmacy at 2 Plough Court, off Lombard Street, in 1715 (see Figure 21).

Sometime after 1730 he was joined there by his younger brother, Timothy Bevan (1704–86). Shortly thereafter the brothers published

FIGURE 21 2 Plough Court, off Lombard Street, London; the Pharmacy founded by Silvanus and Timothy Bevan of Swansea in 1715. Engraving from Robert Carruthers (ed.), *Poetical Works of Alexander Pope* (London, 1853). Pope was born in 2 Plough Court.

their *Catalogue of Druggs, and of Chemical and Galenical* [herb-based] *medicines*. It contained remedies using parts of Egyptian mummies for the treatment of consumption and ulcers; pieces of human skull for diseases such as epilepsy, and the use of Bezoar stones (animal gallstones – a calculus or stone-like material formed around a nucleus of a seed or even hair). These were carried as a talisman against poison, fever and plague. The catalogue also listed botanicals, including herbs, roots, barks and leaves, together with preparations derived from chemical and metallic sources:[10]

> Alum, arsenic, antimony and several of its salts, borax, calomel cream of tartar, calamine, liquid ammonia and carbonate, magnesia, mercury and several salts, sugar of lead, sal ammoniac, sulphur, black and precipitated, white vitriol, tartar emetic; the acids were nitric, sulphuric and hydrochloric.

The Plough Court pharmacy prospered mightily. In part this may have been due to advantageous marriages. Silvanus Bevan's first marriage was to Elizabeth Quare, the daughter of Royal clockmaker Daniel Quare. Taking place on 9 November 1715 at the Friends' Meeting House, White Hart Court, Gracechurch Street, in the City of London, their marriage certificate is signed by more than 100 people, including many from Court circles. Timothy Bevan married Elizabeth Barclay, lived at Plough Court, and their third son, Silvanus, became 'one of the pillars of the notable banking family of Barclay'. We might also suppose, however, that the reliability and efficacy of the medicines on offer helped in the prosperity of the Plough Court business, particularly in an age when adulterated and quack medicines were rife. The pharmacy continued under the Bevan family until Timothy Bevan's son, Joseph Gurney Bevan, ended the family interest in 1794 (Timothy Bevan, having become a widower, married his second wife, Hannah Gurney, in 1752). The new owner was William Allen (1770–1843, FLS 1801, FRS 1807, FGS) who had been taken into Plough Court two years before Bevan gave up the business.[11] The firm ultimately became Allen and Hanburys. It continued to flourish and had a presence in various parts of the world by the end of the nineteenth century. It would ultimately become part of Glaxo Laboratories and then GlaxoSmithKline.[12]

How far the laboratory space at Plough Court was used for scientific experimentation, of the sort aimed at developing not only new pharmaceuticals, but also understanding their chemical properties and the how and why of their effectiveness or otherwise, is difficult to determine. Equally elusive is the degree of chemical experimentation among apothecaries within Wales. Withey (2011) records that some individuals in Wales owned stills and 'Alchymy spoons'. He also notes that some wealthy individuals with an interest in science might indulge in chemical experimentation when self-medicating. The MP John Meller, of Erddig Hall, Wrexham, for example, used a notebook to record various 'physical observations' and the results of his experimenting with chemical oils for medical uses.[13] Nevertheless, the full extent of scientific experimentation among apothecaries in Wales in the long eighteenth century remains little known and awaits further research.

So, what can be said of doctors in Wales with regard to scientific experimentation and any accompanying observational skills? It would be possible to fill the remainder of this chapter with examples of doctoral discussions of patient symptoms and diagnoses, and requests being made to other doctors for help in coming to a diagnosis. To take just two early examples: in 1672, Dr Charles Clermont (or Caroli Claromontii MD) produced as part of his *De Aere, Locis, et Aquis Terrae Angliae: deque morbis Anglorum vernaculis* a section titled 'Observationes Medicae Cambro-Britannicae'. Though touted as illustrating his control of some twenty-six diseases in Wales, the work is actually a compendium of fascinating individual case histories, his own diagnoses and his prescribed cures. Clermont has been called a Welsh physician, but he is believed to have come from Lorraine in France. A little later, between 1700 and 1701, John Powell, who appears to have been a doctor in Carmarthen, wrote nine letters to the eminent English surgeon Charles Bernard (1652–1710, FRS 1696). In them, Powell asks for advice on or describes cases of venereal disease, a molar pregnancy (i.e., where a fertilised egg has a problem such that the baby and placenta do not develop properly after conception), and the case of a woman with an ulcer on the outside of her thigh. At age fourteen, this same woman had developed an ulcer on her heel, from which 'several caries bones came out'; an affliction seemingly 'perfectly cured by King Charles'.[14] The royal touch, of course, was principally known as a cure for the tubercular skin disease scrofula, a royal remedy used by Arise Evans (*fl.*1607–60) of north Wales when, having reputedly encountered King Charles II in London's Saint James's Park, he proceeded to rub his afflicted nose onto the king's hand.[15]

Beyond their day-to-day work and various consultations with specialists in London, some doctors in Wales made valuable use of scientific experimentation and observation to try and expose quack remedies and universal cures. They also investigated, in sometimes considerable detail, the properties and effects of particular medicinal ingredients. Thomas Knight (*fl.*1725–49), a doctor born and raised in Caernarfon, and then working there as an Extra-licentiate of the Royal College of Physicians, thought that 'The misapplication of medicines' was 'a subject of no little moment to the welfare of mankind'.[16] In *A Critical Dissertation upon the Manner of the Preparation of Mercurial Medicines, and their Operation on Human Bodies* (1734) he challenged 'the dangerous fashion of taking Argentum Vivum or Quick-Silver [mercury] in all cases and circumstances alike'. The challenge to widely prescribed mercury-based medicines would continue throughout the eighteenth century. As late as 1797 a pamphlet titled *Mercury Stark Naked* appeared, its declared aim being to strip 'that poisonous mineral of its medical pretensions' while also 'excluding it from the practice of medicine' altogether. For a short time in the 1780s there had also been a society 'supporting the substitution of vegetable-based remedies over metallically based ones'. Among its members were the Caerphilly-born deist philosopher and political writer David Williams (see Chapters 4 and 6); his fellow Welshman, James Tilly Matthews (*fl.*1790s), who, along with David Williams, would be involved in Parisian events relating to the French Revolution that began in 1789;[17] and Isaac Swainson (1746–1812), who established a botanical garden in Twickenham. Swainson owned the recipe for the vegetable-based cure, Dr Velnos's vegetable syrup. Described today as 'a nostrum' or quack medicine, numerous pamphlets appeared in the 1780s and 1790s detailing its almost miraculous curative powers, all of these seemingly written by David Williams.[18]

But to return to Thomas Knight in Caernarfon; in 1749 he felt another dangerous 'Panacea, Catholicon or Universal Medicine' had got itself into such esteem with the public 'as to make it almost everyone's physick'. This was 'Ward's Drop or Pill, both in a liquid and solid form'. Not only did Knight believe the product to be 'highly acrimonious, and even poisonous', he set out to prove it scientifically. He did so by performing a series of experiments on the product before publishing his results in *Reflections upon Catholicons or Universal Remedies* (1749). In one experiment he reduced the pill to a powder before putting it on to a red-hot spatula where 'it gave a flashing fulminating flame, and became a whitish antimonial calx'. Another piece, thrown on to melted nitre, 'emitted sparks of flame' and provided a 'convincing demonstration of antimony'. Ward's Pill and Drop, he concludes, were 'not properly innovations, but what had been revived from time to time'; in this particular case, the use of the toxic metallic element antimony.

An antimony revival had already occurred in the mid-seventeenth century when 'antimony cups' were sold in London's Gunpowder Alley by one John Evans (*c*.1594–1659). An Oxford-educated astrologer, originally from Llangelynin in west Wales, his *Universal Medicine, or the Virtues of my Magneticall or Antimoniall Cup* had appeared in 1634. Despite being censured by the Royal College of Physicians, and having William Laud, the Archbishop of Canterbury, ordering *Universal Medicine* to be burnt, the work continued to be reprinted until 1651 at the earliest. When filled with wine, the acid in the liquid oxidised some of the antimony of the cup. This produced symptoms similar to mild arsenic poisoning and induced a desired vomiting – a sovereign remedy in restoring the balance of the bodily humours. Despite Evans seeing the cups as 'health procuring, health preserving and health restoring', the cure occasionally proved fatal. Antimony cups, though, remained available well into the eighteenth century, with one being owned by Captain James Cook. Antimony, like mercury, would continue to be used as a medicinal element into the nineteenth century.[19]

Aside from exhibiting the experimental side of a scientist, Thomas Knight also possessed good recording and observational skills. These were on display in a letter he sent from Caernarfon on 20 February 1737 to the President of the Royal Society, Sir Hans Sloane. Subsequently published in the *Philosophical Transactions* as an 'Account of hair voided by urine' it describes in technical detail this particular episode; the hair being provided to Sloane in a pillbox accompanying Knight's letter. At the end of the letter Knight notes how the patient drank cow's milk early in the morning without it having been put through a *Colatorium* (a sieve). He makes no comment other than to say that some of 'the downy hair about the udder might get along with the milk into the primæ viæ ['intestinal tract']'.[20] As Jones (1962) has noted, the likely cause of this urinary event was the rupturing of a dermoid cyst. According to Shih and Char (1937), such cysts are found 'in various positions with respect to the bladder' and may be 'in no way distinctive until rupture into the urinary bladder has occurred. These cysts always contain hair and the passage of hair in the urine, pilimication, then constitutes a major symptom'.[21]

In the second part of his *Reflections upon Catholicons or Universal Remedies*, Knight reflected on yet another urinary tract problem. Titled 'Sundry Experiments and Observations made upon the Human Calculus', this second part concerned an issue that generated reams of correspondence, published works and quack remedies throughout the long eighteenth century – kidney and bladder stones. One remedy considered by Knight was the recipe for dissolving stones that had been created by Joanna Stephens, and for which Parliament had paid £5,000 in 1739.[22]

One of the more comprehensive studies of the stone to be published in the eighteenth century came from the Wales-born, London-based doctor Nicholas Robinson (1697–1775). Between graduating as a doctor of medicine at Rheims in 1718, and being made a licentiate of London's Royal College of Physicians in 1727, he published *A Compleat Treatise of the Gravel and Stone with all their Causes, Symptoms and Cures, accounted for* (1721). It has been described in modern scholarship as 'perhaps the most extensive natural philosophical account of human stones' written in the eighteenth century.[23] The term 'natural philosophical' relating to the fact that besides ruminating on aspects of natural philosophy – specifically geology (such as the weight of water acting on sediments) – the initial part of Robinson's treatise contains a section on 'the Original, Cause, and Generation of Stones in Man's Body; and the Affinity they have to those generated in Rivers'. Robinson, like Martin Lister before him, and many others in the early 1700s, saw the formation of human stones as comparable to petrifaction, the method of stone formation in nature.[24] Beyond this, the treatise presents a comprehensive discussion of the contemporary understanding of the nature and causes of the stone in humans, as well as detailing possible cures, weighing up the evidence concerning them, providing examples of prescribed medicines and the results reportedly achieved. As the subtitle to Robinson's treatise indicates – 'Propositions demonstrating that the STONE may safely be dissolv'd, without any detriment to the body' drawn from 'Reason, Experiments, and Anatomical Observations' – his preferred method of treatment, like that of Joanna Stephens, was to dissolve the stones using tinctures that he had devised himself.

Unsurprisingly, dissolving the stones became a popular remedy by the mid-eighteenth century given the excruciating pain involved in the surgical method of extraction. First, the stone was properly positioned within the bladder, by an instrument inserted through the penis. An incision was then made along the line of the perineum, the line between the anus and the scrotum. Pincers were then inserted to grab and, one hoped in an age of little anaesthetic, quickly remove the stone or stones. In 1820, a description of being cut for the stone covered five pages of *Cofiant Neu Hanes bywyd a marwolaeth y Parch. Thomas Jones ...* the biography of the Revd Thomas Jones (1756–1820) of Denbigh. Others afflicted by the illness were William Williams Pantycelyn, and Richard Price who traded remedies and pamphlets on the issue with his friend Benjamin Franklin. Both Price and Franklin would eventually die from the consequences of the complaint.

Besides writing about the stone, Nicholas Robinson produced a number of other medical texts, including *A New Method of Treating Consumptions* (1727) and *A New System of the Spleen, Vapours, and*

Hypochondriack Melancholy (1729). Also appearing in 1729 was *A Discourse on the Nature and Causes of Sudden Death*, in which he rejected the use of bleeding for some cases of apoplexy, and describes, from personal observation, the appearance of the brain following opium poisoning.[25] Twenty-nine years earlier, the then Chancellor of Llandaff Cathedral in Cardiff, and licentiate of the College of Physicians in 1677, John Jones (1645–1709), followed his 1683 Latin work on abstruse and intermittent fevers, *Novarum dissertationum de morbis abstrusioribus tractatus primus*, and his *De morbis Hibernorum: speciatim vero de dysentaria Hibernica* of 1698, with *The Mysteries of Opium Revealed* (1700). Although once described as 'extraordinary and perfectly unintelligible', Lucy Inglis, in her *Milk of Paradise: A History of Opium* (2018), suggests that while the *Mysteries of Opium* is a 'slightly eccentric' work, Jones's observations on the effects of opium are 'clear and accurate'. Not only that, as one of the first 'to include the feeling of transportation, a dreamlike removal from the ordinary world' Jones 'shows an intimate familiarity with the mental effects of the drug'.[26] He also describes the various types of opium grown (his preferred one being Egyptian), as well as the effects of the drug when used internally or externally, and in large or moderate doses. The general effects of the drug he sees as mimicking those of excessive drinking. Older people might also suffer a stoppage of the urine, though his sovereign cure for this was to place the scrotum on the edge of a cold chamber pot. He warns that the consequences of leaving off crude opium after long and lavish use could be dangerous, and that excessively large doses might lead to 'venereal fury', alienation of the mind, convulsions and occasionally death. Interestingly, for a religious age, and given his position at Llandaff Cathedral, Jones is remarkably non-judgemental on the use of the drug. As his scientific treatise reveals, he intended not only to reveal the drug's noxious nature, but 'how to render it safe and a noble panacea whereof, its palliative and curative use'.

Opium and the tincture derived from it – laudanum – were remedies for many ills in the long eighteenth century including another of the period's mainstay illnesses: gout, a severe joint-based and painful form of arthritis. Going through ten editions in two years, and being published in Dublin, Philadelphia and Boston, *A Dissertation on the Gout, and all Chronic Diseases, jointly considered, as proceeding from the Same Causes; What those causes are; and a Rational and Natural Method of Curing Proposed. Addressed to all Invalids* (1771) was written by sixty-year-old William Cadogan (1711–97, FRS 1752). Though noted as London-born in some biographical sources, contemporary reports and recent research confirm his family ties to south Wales: his grandfather came from Trostre, his father was known as 'Roger of Usk', and his mother came from Glamorgan. Cadogan himself is now believed to have been born in either Usk or Cowbridge, with the latter considered the most likely by his biographers.[27] Having attended Oxford University, Cadogan studied medicine in Leiden and then practised in Bristol, where he also became Physician to the Royal Infirmary. He later moved to London. In contrast to the opinion of his time, which generally saw gout as hereditary, Cadogan considered it to be a consequence of diet and indolence, as well as possible infection and trauma. As a curative, he recommended activity, food that was mild, soft and taken in moderation, and with wine best avoided. In an age of overindulgence, at least on the part of some, his treatise was heavily criticised for its joy-denying and puritanical ideas, not least in a 2,000-word anonymous tract *The Doctor dissected, or Willy Cadogan in the Kitchen*.[28] Cadogan, meanwhile, made clear what the consequences of inaction would be for the glutton and gout sufferer: 'With a few journies to Bath, he drags on, till, in spite of all doctors he has consulted, and the infallible quack medicines he has taken, lamenting that none have been lucky enough to hit his case, he sinks below opium and dies long before his time.'[29] The journey to Bath, of course, reflects the developing interest in spas and their curative mineral waters; waters that a number of scientists of Wales would investigate and analyse in the course of the eighteenth century. In his *Vindication of a late Essay on the Transmutation of Blood* (1731), Thomas Knight of Caernarfon had included 'A Dissertation Concerning the Operation of Chalybeate Medicines in Human Bodies' based on 'experimental observations and demonstrable principles', these being 'the sole original and fundamental of true knowledge, which our senses are witness to'.

Sometimes referred to as ferruginous waters, the so-called 'chalybeate medicines' were mineral waters impregnated with iron salts, and numerous analyses of them were produced, not least by a colourful émigré from Hemmerde, in the Westphalia region of today's Germany, Dr Diederick Wessel Linden (*fl*.1745–68). In 1748 he published a general treatise on mineral waters, which included St Winefride's Well, in Holywell, Flintshire. Later, having visited Llandrindod Wells, in 1754, he produced a *Treatise on the three medicinal mineral waters at Llandrindod, in Radnorshire* (1756). The waters he describes were sourced from a chalybeate or rock-water spring, a saline spring, and a sulphur spring. Linden considered local geology in relation to these springs and discussed the effects of 'mineral and fossil mixtures in their native veins and beds'. His efforts though were not well received by everyone; both Lewis Morris and Richard Morris were particularly critical of his analyses, Lewis even going so far as to suggest that Linden had written 'a pompous book' so that 'after we have read it, we are never the wiser, or know how to use the waters no more than if he had not wrote'.[30]

Lewis himself spent six days at Llandrindod in August 1760 drinking from the various types of waters and writing an extensive description of their situation, appearance, taste and medicinal and curative properties – the chief use of one spring being 'for evacuation: which is the cure for abundance of diseases'. Having taken the sulphurous waters at meals and with milk at night, he found that:

> I now long for it. I am much better in health, and even my asthma and cough is easier. I can put on my shoes and stockings which I have not been able to do this six months, and have mounted my horse without a horse block which I have not been able to do this two years. These are surprising effects![31]

Linden would produce further works on springs in Britain and Wales. 'An Account of a mineral water at Llangyba' appeared in the *Gentleman's Magazine* of 1766, and *An experimental and practical enquiry into the ophthalmic, antiscrophulous, and nervous properties of the mineral waters of Llangybi, in Caernarvonshire* was published a year later.[32] Although Ireland-based Dr John Rutty, who William Morris of Anglesey had met on his trip to Dublin (see Chapter 10), used his *Philosophical Transactions* paper *On the Vitriolic Waters of Amlwch in the Isle of Anglesey* (1759) to describe his experiments on those particularly strong chalybeate waters, detailed chemical analysis of mineral waters would not properly take place until the nineteenth century. One study, comprising experiments using reagents to determine the true chemical and mineralogical composition of the medicinal waters and their associated soils and rocks, was published by Richard Williams of Aberystwyth as *An Analysis of the Medicinal Waters of Llandrindod* (1817).

Having now considered experimentation and observation with regard to medicines, we turn to a consideration of the part played by medical scientists of Wales in developing new ideas and techniques within medicine.

Medical Ideas and Techniques

One sentiment that comes across in a number of the medical texts mentioned in the previous section is a respect for the medical profession and the need to care for its reputation, a consideration sometimes expressed in terms resonant with the concerns of our own time. For example, Nathaniel Williams, in a section on 'Rules for the Preservation of Health and Life' in part two of his *Pharmacopoeia* (1796), was of the opinion that physicians would be more successful if they were simpler 'in their prescriptions'. He notes too how some patients 'never approve a physician, who does not plunder a *Shop* at a prescription!!' and he thought 'it would be a good policy to depend on simple medicines with a proper Regimen'.[33] Nicholas Robinson, meanwhile, was keen to emphasise that medicine could not cure everything, and that the 'Noble art of Physick does not oblige to impossibilities'. Yet he also believed that 'every man should exert his best endeavours for the benefit of that society he is a member of'; and among the arts and sciences most conducive to the support of mankind:

> I know none more deservedly honoured, than this of Physic; whether we consider the Dignity of the Subject it has for the peculiar object of its Studies, which is the Contemplation of the

Structure and Mechanism of that noble creature MAN; or the end it proposes, *viz.* To restore his lost or impair'd Health.[34]

Teaching provided one way for medical scientists to not only express their care for the profession but to also pass on established or new ideas and techniques aimed at restoring lost or impaired health. The situation within Wales, in this regard, is another area ripe for research. Outside Wales, a number of Welsh doctors and surgeons certainly undertook the task.

For example, the Neath-born, London-based physician to George III, Sir Noah Thomas (1720–92, FRS 1753), had been educated at St John's College, Cambridge, made an FRS in 1753, and a Fellow of the Royal College of Physicians in 1757, yet, though renowned for his learning, he never published any book or paper. He did, though, deliver the Royal College of Physicians' Goulstonian lecture in 1759 and, before that, lectured at St Thomas's Hospital and at Cook's Court, Carey Street in London. There, in 1753, Erasmus Darwin heard him lecture. Like any good student, Darwin took notes during the lectures on 'salivation and poisons, both acrimonious and narcotic'. These notes run to fifty pages and are today in the Wellcome Institute in London awaiting an intrepid interpreter of Darwin's shorthand.[35] In 1761, Noah Thomas would be one of the sponsors of Erasmus Darwin for Fellowship of the Royal Society. From 'his house in the Royal College of Physicians', Nicholas Robinson also gave lectures, and he published twenty-five of them in *A General Scheme for a Course of Medical Lectures Intended for the Improvement of Young Physicians, and Gentlemen* (n.d.). Born in Pembrokeshire, George Rees (1776–1846) received his medical education at St Thomas's and Guy's hospitals before qualifying MD at Glasgow in 1801. Formerly the House Surgeon at the Lock Hospital in London and a licentiate of the College of Physicians in 1808, Rees published on disorders of the stomach in *Observations on Spasms of the Stomach* (1810), which included an early description of cirrhosis of the liver related to alcohol. A *Treatise on Hemoptysis* ('coughing up blood') followed in 1813. Rees's lectures were given from the site of his practice at 2 Soho Square in London, and, like Robinson, he published a course of twelve of them. They appeared as an addition to his *Treatise on the Primary Symptoms of Lues Venerea* (1802), a work dedicated to physician and entomologist John Coakley Lettsom FRS, founder of the Medical Society in 1773.

New ideas of course can become the subject of controversy, and this was certainly the case when *An Enquiry into the State of Medicine on the Principles of Inductive Philosophy* (1781) appeared under the name of Robert Jones 'of Carmarthen'. The work passionately advocates the theories outlined by Scottish physician John Brown (1735–88) in his *Elementae Medicinae* (1780, translated into English by the author and published as *The Elements of Medicine* in 1788). What later came to be called the Brunonian system of medicine asserted that all illness resulted from the over or under-stimulation of the patient. The resulting imbalance then needed correcting, usually by large doses of laudanum – the tincture of opium – taken in alcohol. Brown practised in Edinburgh at the time Jones was a medical student there. They certainly knew each other, for at some point the pair appear to have attempted to subvert the medical care of a fellow student of Jones by bribing an attending nurse to administer their prescribed laudanum doses; all this without the knowledge of the already attending physicians. Those attending physicians were Andrew Duncan (1744–1828) and Alexander Monro (1733–1817). These two pillars of Scottish medicine were also criticised in Jones's *An Inquiry into the State of Medicine* for their method of treating the young patient. Duncan fought back for the sake of his reputation and in *A Letter to Dr Robert Jones of Caermarthenshire* (1782) he denounced the Brunonian content of *An Inquiry into the State of Medicine*, and noted that Jones had 'retired beyond the reach of the laws of this part of Britain [Scotland]; and at present, I have neither the leisure nor inclination to follow you into Wales'.[36] Duncan went on to note with regard to *An Inquiry into the State of Medicine* that 'an opinion is very commonly entertained here [Edinburgh], that Dr Brown himself ought in reality to be considered as the author of it'. A suggestion later confirmed by John Brown's own daughter, Elizabeth.[37]

While we might justifiably criticise Jones for his part in subverting the medical care of a fellow student, and for disparaging those providing that care, we should not dismiss out of hand his enthusiasm for Brown's ideas. Although largely ignored today, Brown's work gained significant traction at the time. Thomas Beddoes later translated *Elementae Medicinae* into English, and Benjamin Rush, one of the signatories of the American Declaration of Independence, would take up its ideas in America, and it would become influential in Germany. More importantly, from the perspective of Jones, the ideas were a challenge to prevailing orthodoxy. There was an increasing realisation that the long-established bleeding and purgation regimen, used for almost all illnesses, were less than truly effective methods of treatment. New ideas were clearly needed. In one case study given in *Inquiry into the State of Medicine*, Jones/Brown strongly 'criticized the repeated use of emetics and purgatives' for a patient already suffering from vomiting and diarrhoea. Such challenges to established orthodoxy clearly appealed to the young Dr Jones.

The treatment of mental health was another area where new ideas were sorely needed. An attempt had been made in 1754 to regulate private madhouses, but the Royal College of Physicians considered the proposal that they should vet and regulate such institutions as unworkable. It would not be until 1774 that an Act of Parliament provided for five commissioners, appointed by the Royal College of Physicians, to inspect madhouses within seven miles of London and Westminster. In the same year as he was appointed a censor at the Royal College of Physicians (1781), Neath-born Noah Thomas (see above) became one of the 'Commissioners for granting licences to persons for keeping houses for the reception of lunatics'.[38] Of the other medical scientists we met in the previous section, George Rees established a private lunatic asylum in Hackney. He also spent some time as a medical superintendent in Cornwall's lunatic asylum in Bodmin.

Horror stories concerning the treatment of madness in the long eighteenth century abound. They include the display of inmates for the titillation of the public at the Bethlem Hospital in London (a.k.a. Bedlam), a place designed originally in 1676 by Robert Hooke of the Royal Society as a 'palace beautiful'. One self-declared Welshman (Welsh father and French mother) destined to end up in Bedlam was

James Tilly Matthews, who, in late 1792, travelled into revolutionary France with Caerphilly-born, London-based political writer and educationalist David Williams (see above). Having been made an honorary French citizen in August 1792, along with Tom Paine and Joseph Priestley, Williams had been invited to Paris in November to help draft a new constitution for the emerging French state. However, following the execution of the French king in February 1793, war now loomed between France and Britain. Williams left Paris and returned to Britain carrying a message for the London Government. Tilly Matthews remained in Paris and, as Constantine (2013) records, waged 'his rogue pacifist mission of diplomacy back and forth across the Channel, completely confusing both sides as to the true intention of the other'. Eventually arrested as a spy, and imprisoned in Plessis prison near the Sorbonne, Matthews continued to bombard the French government with letters (these are preserved in the Archives des Affaires Étrangères). Released in February 1796 on grounds of insanity, Matthews somehow made his way home. There he claimed to have fallen under the control of the criminals of the 'Air Loom Gang'. Its members were controlling and influencing his mind, and those of various government ministers, by thought waves coming from the Air Loom machine located underground in Moorfields. Having accused Lord Liverpool of 'Treason' from the gallery of the House of Commons, Matthews was committed for the rest of his life to Bedlam. His case, as described in detail in *Illustrations of Madness* (1810) by the doctor John Haslam, is considered to be 'the first fully documented case of paranoid schizophrenia'.[39]

We might at least hope that *Treatise on Insanity* (1806) by Carmarthenshire-born, Sheffield-based David Daniel Davis (1777– 1841) provided some compassionate guidance to such institutions in later years. Davis's work was a translation into English of the more humane approach to madness adopted by Philippe Pinel in his *Traité médico-philosophique sur l'aliénation mentale; ou la manie* (1801) – Pinel opposed the keeping of patients in chains, for example. Only in 1828 would a Madhouses Act come into force in Britain to 'regulate the Care and Treatment of Insane Persons'.[40]

To develop and advance new medical ideas it was of course necessary to have knowledge of the internal structure and workings of the human body. Aside from the availability of bodies of hanged criminals and those resulting from body snatching, autopsy seems to have also played its part and it has been said that 'until the twentieth century most autopsies were carried out by GPs either at a patient's home or in a morgue'.[41] From Wales, we have Perrot Williams of Haverfordwest reporting his 'Observations upon dissecting the body of a person troubled with the stone' in a brief letter to the *Philosophical Transactions* in June 1723. And when John Powell wrote his letter to the surgeon Charles Bernard seeking advice about a molar pregnancy (see above), he felt his diagnosis had already been confirmed by the autopsy undertaken on the patient by his son. Whether Powell's son was qualified to do an autopsy is not known. In London, we have the example of Richard Price. Following his death on 19 April 1791 from the effects of a urinary tract infection and kidney stones, his body was autopsied at his home in St Thomas's Square, Hackney, in front of a few invited friends. As the minister and teacher Thomas Belsham wrote: 'When opening him up, his viscera were found in a very diseased state – one kidney quite gone – the other very unsound – a small stone in one of the Ureters – and a fleshy excrescence at the neck of the bladder.' We can only agree with Belsham in his closing comments: 'From such miseries, how happy the release'.[42]

Whether advanced through autopsy and dissection or not, new ideas and methods were certainly appearing in surgery, with a particularly notable contribution coming from Griffith Rowlands (1761–1828). In presenting 'A Case of an Un-united fracture of the Thigh, cured by sawing off the ends of the bone' in the *Medico-Chirurgical Transactions* in 1811, he was quite possibly the first to report on this operation in Britain, and was certainly one of the first persons in Europe to perform it (perhaps the third). Born in Llanfair, near Harlech, and practising surgery in the Infirmary at Chester for more than forty years, Rowlands would publish another paper on amputation as well as one on lithotomy (cutting for the removal of a stone). In 1802, he helped remove a two-and-a-quarter-ounce stone from the gall bladder of Thomas Jones of

Denbigh. In this pre-anaesthetic age Jones was no doubt relieved that Rowlands was considered to be 'a bold and dextrous operator'.

Griffith Rowlands is also credited with being the 1798 instigator of the Chester Lying-in Charity.[43] The introduction of ideas and techniques concerning the care of expectant mothers and of children is another contribution made by medical scientists of Wales. An early example came in *Magia Adamica* (1650), when the alchemist Thomas Vaughan discussed child development. Given its early date and psychological approach it is worth quoting in full (the spelling and italicisation has been modernised here):

> For little children before ever they can speak, will stare upon any thing, that is strange to them; they will cry, and are restless till they get it into their hands, that they may feel it, and look upon it, that is to say, that they may know what it is in some degree, and according to their capacity. Now some ignorant nurse will think they do all this, out of a desire to play with what they see, but they themselves tell us the contrary; for when they are past infants, and begin to make use of language, if any new thing appears, they will not desire to play with it, but they will ask you, what it is? for they desire to know; and this is plain out of their actions; for if you put any rattle into their hands, they will view it, and study it for some short time, and when they can know no more, then they will play with it. It is well known, that if you hold a candle near to a little Child, he will (if you prevent him not) put his finger into the flame, for he desire to know what it is, that shines so bright; but there is some thing more than all this, for even these infants desire to improve their knowledge. Thus when they look upon any thing, if their sight informs them not sufficiently, they will, if they can, get it into their hands, that they may feel it: but if the touch also doth not satisfy, they will put it into their mouths to taste it, as if they would examine things by more senses than one. Now this desire to know is born with them, and it is the best, and most mysterious part of their nature. It is to be observed, that when men come to their full age, and are serious in their disquisitions, they

are ashamed to ere, because it is the property of their nature, to know. Thus we see that a philosopher being taken at a fault in his discourse, will blush, as if he had committed something unworthy of himself, and truly the very sense of this disgrace prevails so far with some, they had rather persist in their error, and defend it against the truth, than acknowledge their infirmities.[44]

Later, in another text, Vaughan adds:

but least you think, I have only convers'd with children I shall confess, I have convers'd with children and fools too; that is, as I interpret it, with children and men, for these last are not in all things, as wise as the first. A child I suppose, *in puris Naturalibus*, before education alters, and ferments him, is a subject hath not been much consider'd, for men respect him not, till he is company for them, and then indeed they spoil him. Notwithstanding I should think, by what I have read, that the natural disposition of children, before it is corrupted with customs and manners, is one of those things, about which the ancient philosophers have busied themselves even to some curiosity.[45]

The London Foundling Hospital, established through the efforts of Thomas Coram, received its first 'exposed and deserted young children' in 1741. In June 1748, Mr Waple, one of the hospital's Governors, received a letter concerning the 'nursing and management of children'. He laid this before the General Working Committee of the hospital, who recommended its publication. *An Essay upon Nursing, and the Management of Children, from their Birth to Three Years of Age* appeared 'by order of the General Committee' later in 1748 (see Figure 22). It went through at least ten editions by 1772 and was translated into French in 1752 and 1768.

Although originally published anonymously, the essay was written by William Cadogan, whose writing on gout we have already discussed. His aim in this earlier work on childcare was to be 'intelligible and useful' and for the hospital to use the work as it wished:

FIGURE 22 Title page to *An Essay upon Nursing, and the Management of Children* by William Cadogan (1748; London, 1752). Courtesy of Special Collections, University of Bristol Library.

If you think it may be of Use to publish this Letter, I am not unwilling it should appear; if not, do with it what you please. I deliver it up as a *Foundling* to be disposed of as you think proper. I shall only add by way of Persuasive to those who may be inclined to make Trial of the Method I recommend, that I am a Father, and have already practiced it with the most desirable success.[46]

Cadogan used the London Bills of Mortality to illustrate the wrongs in the current treatment of children. As he notes: 'almost half of those who fill up that black list die under five years of age.' To that end, he

heavily criticises the continuing practice of swaddling babies and recommends instead loose clothing for both babies and children. He also
recommends fresh air and the changing of babies often, so as to free
them from 'stinks and sourness'. As his biographers note, this disagreed
with those who taught that washing babies 'robbed them of their natural juices'. He discussed the methods of feeding babies and the food
to be provided as well as the ill effects of teething and worms. He also
wanted to introduce 'a more reasonable and more natural method of
nursing', which included improving the quality and provision of wet
nurses and midwives: 'Children in general are over-cloath'd and over-
fed, and fed and cloath'd improperly. To these Causes I impute almost
all their diseases.'[47] On 28 June 1749, Cadogan was elected a Governor
of the Foundling Hospital. He was elected to Fellowship of the Royal
Society in 1752. As a result of his publication, he has been dubbed 'The
Father of Child Care'.[48]

Other works related to children and their mothers followed. In 1778,
the surgeon John Evans (1736–1846), son of mapmaker John Evans (see
Chapter 9), produced a thesis at Edinburgh titled *Disputatio physiologica
inauguralis, de foetus humani nutrimento* ('An essay on the nutrition of the
human foetus'). In 1802, surgeon Griffith Rowlands wrote *Observations
on the hydrocele* ('fluid collection and consequent swellings in the scrotum'; often common in newborns), and in 1813 David Daniel Davis (see
above) would be elected as physician accoucheur (an assistant at child-
birth and often an expert in obstetrics, and also known in less refined
circles as a man-midwife) at the Queen Charlotte Lying-in Hospital
in London. He later adopted the same role as one of the attending
physicians at the birth of Queen Victoria in 1819. In 1827 he became
the first professor of midwifery at the newly founded (1826) University
of London and he went on to publish *Elements of Operative Midwifery*
(1825), and the three-volume *Principles and Practice of Obstetric Medicine
in a series of systematic dissertations on the diseases of women and children*
(1836). In the last work, according to Fay Bower (2003), Davis also
'discovered the pathology of the phlegmasia dolens [a form of deep vein
thrombosis common in mothers after childbirth], and its dependence
on phlebitis'.[49]

Finally, our medical scientists of Wales played their part in another significant advance for both children and adults: the eighteenth-century development of vaccination as a means of giving immunity to a particularly dreadful disease.

'Buying the Smallpox'

On 8 May 1980, the World Health Organization finally declared that the disfiguring and all too often deadly disease of smallpox was finally eradicated. A major step on the road to this magnificent scientific and social achievement came in 1796. Having noticed how a previous infection with the less potent cowpox virus gave immunity from the more virulent smallpox, the Gloucestershire-born Edward Jenner (1749–1823) successfully vaccinated an eight-year-old boy, James Phipps, against smallpox.[50] In c.1803 the English conchologist, William Turton (see Chapter 10) published from his 'private press' in Swansea a *Treatise on cold and hot baths* [and] *their application in various diseases.* To this he added a letter *on the introduction of the cow pock in the principality of Wales.* There is, though, an even earlier and less well-known history to these developments, one in which scientists of Wales played a part.

In 1721, Lady Mary Wortley Montagu (1689–1762) brought into England a technique of smallpox inoculation that she had witnessed and had cause to employ when resident with her diplomat husband in Turkey. The method – variolation – involved deliberately introducing puss or scab material from a smallpox infected person into a healthy individual, who had not previously had the disease, through a scratch or cut made on their skin. The method usually produced a milder episode of the illness and subsequent immunity from the disease. The seemingly 'new' technique produced a flurry of papers in the *Philosophical Transactions* and a correspondence with the then secretary of the Royal Society, James Jurin. Among the letters written to Jurin, and others, which later appeared in the *Philosophical Transactions*, were those from two Haverfordwest doctors: Richard Wright, a surgeon, who had written on the subject of smallpox and inoculation to the apothecary Silvanus Bevan;[51] and Perrot Williams of Haverfordwest, who had written on the issue to Samuel Brady (1680–1748), a physician at the Portsmouth garrison.[52]

By January 1723, Jurin had obtained Perrot Williams's letters to Samuel Brady. In them, Williams detailed the Welsh practice of 'buying the smallpox'; a method whereby scabs from an infected person were *bought* and, as in the Turkish method, used to infect a healthy person by variolation so as to produce immunity. In a letter of January 1723, Jurin notes how the account given by Williams 'of ye Practice among Boys at School of buying ye Small Pox as they call it, is so remarkable, that it will be highly worth ye Notice of ye Royal Society before whom I design to lay it at their next meeting'. The letters from Williams to Brady were duly read at the Society that same month. Furthermore, since Jurin was in the process of 'drawing up an account of ye danger of ye Natural Small Pox compared wth that of ye same disease by innoculation', he requested Williams to provide further information. From that point on Williams sent a series of letters to Jurin between 1723 and 1728 giving details of smallpox mortality in his area, as well as details of other inoculations carried out there.[53]

Significantly, Wright and Williams asserted in their letters that the method of 'buying the smallpox' had been an 'Immemorial Custom' in their part of Wales. Indeed, from the evidence presented by Wright it can be suggested that the technique was known there from at least 1600.[54] It soon became clear too that the method was well known in many other countries. Needham (1954) notes how similar practices had been first reported from Germany in 1671 and 1674, Denmark in 1673, Poland in 1677 and Scotland in 1715. The method was also known in the Baltic states, Asia (especially China) and Africa.[55] It did not, though, appear to have been known in England before Emanuel Timoni mentioned it in the *Philosophical Transactions* of 1714, or before Lady Montagu introduced the method in 1721. This omission would instigate a nineteenth-century debate on the subject into which a degree of national and class consciousness crept. As Samuel Miller, in America, put it in 1803: 'How shall we account … for this art being confined chiefly to the common people, or the less civilized part of mankind, while the learned were ignorant of it?' May it not be admitted, he suggests, 'as proof of the *great antiquity* of the practice, that precisely that portion of the community, whose habits, in every country, are in general most simple, uniform, and stationary, were found to retain a practice which the more polished had lost?' In 1857, a British government report presented *Papers Relating to the History and Practice of Vaccination*. It noted that variolation was known among the Chinese and the Brahmins, and in Persia, Armenia and Georgia, and was said to be 'in vogue, and to have spread as a popular custom, not only about the shores of the Mediterranean, but even to those of the Baltic, to Scotland, and still less accountably to Wales'.[56] Why 'still less accountably to Wales'? Today it is possible to at least suggest how the technique might have come to be known there. Nabil Matar (1999) has noted some of the 'English' merchants living and trading in Morocco in the sixteenth century. Among them, in 1572, at a time when a Welsh identity would very likely have been subsumed into the broad category 'English', we find one 'Thomas Owen', who spoke Arabic, and, in 1589, a 'M. Richard Evans'.[57] It has also been noted elsewhere that in 1669 a 'Mr. William Richards "from Wales" was appointed chaplain to India under the East India Company',[58] India being a country where variolation was known. It is not hard to believe that such travellers could have brought the method back to Wales. Though this does not lessen the mystery of why it remained unknown in England until the early eighteenth century.

There were, of course, dangers in using variolation. It did not always work in providing protection; it might even cause severe illness and, in some cases, death. There was also little awareness that the inoculated themselves could be infectious for a time. Unsurprisingly, therefore, there were those who vehemently opposed its use. In 1722 Edmund Massey gave *A Sermon against the Dangerous and Sinful Practice of Inoculation* in Holborn, London. Diseases, he declared, 'are sent, if not for the trial of faith, for the punishment of our sins'. So, to inoculate could be seen as refusing God's will. Meanwhile, in Wales itself, even though Perrot Williams 'never thought fit to perswade [*sic*] any to make the experiment on 'em selves', he found that what he had already done in the affair had been 'strangely misrepresented' by 'very unjustifiable methods'. There was also disagreement within the medical profession. In a letter written to Jurin in 1725, Williams made a passionate appeal for the medical establishment to reach a decision on inoculation, and to make their position public:

> For the Benefit of Mankind (which I am satisfy'd is y^r. chief concern in the Affair of Inoculation) I heartily wish, the College of Physicians wou'd vouchsafe to be so publick Spirited, as a Society, to Oblige the World with their Opinion of that Practice; that if they judge it conducive to the End proposed, it may be establish'd as such; and if otherwise, suppressed. For the Approbation of such a Learned Body, procures sufficient Authority and weight to a worthy Design; and what's in such Cases above all, effectually supports it against the side of popular Prejudice, and also avoids the Odium and Envy, which commonly attends and but too often bears down single Attempts, how reasonable so ever in 'emselves. And this wou'd conduce much more to their Honour, as the Publick Guardians of Health, than by vilifying one another (of which practice I cou'd give not a few late Instances) to ruin the Reputation of, I believe I may venture to say, the most useful Profession in the World, Divinity excepted.[59]

The Royal College of Physicians declined to endorse inoculation until 1755.[60]

Despite opposition, both Jurin and Williams passionately defended the method. Jurin's immediate response to the revelation of the Welsh method had been a scientific one. He established, as Needham notes, with the aid of information from the likes of Williams 'the first statistical study of the technique's value'. Jurin would also defend the 'Welsh method' against its opponents and have three of his own daughters inoculated by variolation in 1728. Williams too advanced a scientific defence based on observation and experimentation when inoculating his two sons. To prove 'that Inoculation, is a Sufficient Preservative', he caused his '2 boys, not only to see but even to handle a child, dying of the most Malignant sort of small-pox'. The boys, he thanked God, 'continue in perfect health'.[61] When hoping for the approval of the Royal Society for the method, Williams also defended it on the basis of reason and experience. As he wrote to Jurin:

> For my part, I am so fully convinc'd, both from Reason and Experience, of the safety of this Method; whatever some Physicians think fit to advance, in order to make it appear Dangerous, and some Divines unlawful; that I am confident, if countenanced by y[r]. Learn'd Society, 'twill prove a very Advantageous Discovery to Mankind, and nowise inconsistent with the Christian Religion; which no where forbids us the use of any Rational means to prevent our falling a Sacrifice to one of the worst of Diseases we are obnoxious to.[62]

Ensuring Life by Insuring Lives

The fear of being unable to provide for oneself and others as a result of diseases like smallpox, as well as accident or unexpected death, was an ever-present one for people living in a time with few social benefits or other safety nets. Although various forms of insurance and mutual aid had been developed since the Middle Ages, these were not generally based on the sound statistical and actuarial science necessary for the provision of long-term life insurance or pensions.

The move towards such provision began in the sphere of gaming, with attempts to understand mathematically the idea of chance – the relative frequency of a particular number or numbers being thrown with dice. Although begun in Italy during the Renaissance, and developed by the likes of Blaise Pascal and Pierre de Fermat in the seventeenth century, the stimulus for such enquiry in Britain came from the treatise *De Ratiocinilis in Ludo Aleae* (1657). Written by the Dutchman Christiaan Huygens (1629–95) it applied mathematical theorems to problems related to games of chance. The next step was to convert the relative frequencies derived from dice throwing into a proper understanding of the operations of chance. In short, to develop a mathematical and statistically based theory of probability that could then be applied to life expectancy in different age groups. A major contribution to this came from the mathematician and astronomer Edmond Halley (1646–1742) and the Huguenot refugee to Britain, Abraham de Moivre (1667–1754).[63]

In taking the matter further, several scientists of Wales made import-ant and lasting contributions. We previously met the Anglesey-born, London-based mathematician William Jones Sr publishing work by Isaac Newton and his own efforts on navigation and pure mathematics, including, of course, his introduction of the symbol π. Jones was the 'Intimate friend' of de Moivre, and he also came to know Halley, some-what later, when they served together on the Royal Society's Newton/Leibnitz committee in 1712, (see Chapter 5). Halley had published life tables in the *Philosophical Transactions* in 1693 based on demographic data from Breslau (today's Wrocław in Poland). Abraham de Moivre had made key contributions to probability theory in a series of works: *Doctrine of Chances; or, A Method of Calculating the Probability of events in Play* (1718), *Annuities Upon Lives* (1725) and *Miscellanea Analytica* (1737). To the last of these Jones Sr had subscribed and, in a preface to the first, *Doctrine of Chances*, de Moivre described Jones as 'my Intimate friend and a very skilful mathematician'.[64] Jones had clearly used this skill when reading de Moivre's *Annuities Upon Lives* for, having done so, he wondered why de Moivre had neglected to demonstrate a particular theorem, one that Jones considered 'the most curious of them all'. The result was a letter to Jones Sr from de Moivre giving his reasons for the omission: 'as the demonstration depended upon a principle which was not commonly known, I was afraid the publishing of it would have swell'd the book too much'. He then provided Jones Sr with the omitted demonstration of a theorem for *A short method of calculating the value of annuities on lives, from tables of observations*. The letter was subsequently published in the *Philosophical Transactions* in 1744.[65]

Although Jones Sr did not publish a book on probability or annui-ties he did produce a manuscript titled *The Practice of Interest*. Passed around among his friends, it centred, as Ogborn (1949) has written, on a discussion of compound and simple interest, together with the nature of the relationship between five quantities of significance in life insurance and annuity provision. These were: 'the amount of an annuity, the period and the rate of interest, the present value and the accumulated sum of the payments.'[66] Ogborn also notes that Jones seems to have given permission for others to publish elements of the work. As William Gardiner records in the preface to his *Tables of Logarithms for all numbers 1 to 102100* (1741), to which Jones subscribed:

> The Explication of the tables, with the most necessary rules, precepts and examples, such as might in some measure, shew the extensive use of logarithms, I have collected wholly from the papers of W. Jones, Esq; F. R. S. who, with his usual freedom, and readiness to assist in the promoting of all useful knowledge, gave me leave to dispose them as I thought best.[67]

The Jones material includes a section on 'Various rules and problems, wherein some of the uses of Logarithms are shewn', together with others on the value of logarithms in calculating compound interest and their uses in proportion, trigonometry, and latitude and longitude. Jones's work on compound interest also appears in the 'Introduction' to *The Anti-logarithmic Canon* (1742) by James Dodson (*c.*1705/10–57, FRS 1755).[68]

Dodson was the first person to show 'that life assurance would be practicable with premiums properly calculated on actuarial principles according to age, and the type of assurance'. In this regard, he also became the earliest and principal promoter of what became, in 1762, the *Equitable Society for Assurances on Lives and Survivorships*.[69] Dodson would not live to see the actual formation of the Society, but it prospered mightily until its ignominious demise in 2000, by which time it was known as the Equitable Life Assurance Society.

For a substantial part of the early history of the Equitable its three Presidents between 1773 and 1875 were linked to Wales through marriage or birth.[70] At the same time, the two Actuaries in office from 1775 to 1870 were Bridgend-born, London-based William Morgan (1750–1833, FRS 1790, Copley Medal 1789) and his son Arthur Morgan (1801–70, FRS 1835). William Morgan was in office between 1775 and 1830 and Arthur Morgan between 1830 and 1870. Furthermore, for some fifteen years from 1768, William Morgan's uncle, Llangeinor-born, London-based Richard Price, acted as an unpaid adviser to the Society.

It would be Richard Price who made a major and lasting contribution to the development of probability theory when 'An Essay towards solving a Problem in the Doctrine of Chances by the late Revd Mr. Bayes F.R.S. communicated by Mr. Price, in a letter to John Canton A.M.F.R.S.' (1763) was read at the Royal Society and published in the *Philosophical Transactions*.[71] As the title indicates, the work derived from papers that Price had found among those of his recently deceased friend, the mathematician and Dissenting minister of Tunbridge Wells, Thomas Bayes (*c*.1701–61, FRS 1742). Recognising the importance to probability theory of theorems contained in the Bayes papers, Price undertook to edit them for publication. In doing so he may actually have contributed up to 52 per cent of the published paper (Tucker, 2017). On that basis alone the work qualifies, by any modern standard, as being of joint authorship. In 1765, Price contributed a second paper to the *Philosophical Transactions* in which he proposed new theorems and improved the calculation method to be employed.[72] At the risk of gross oversimplification, we can say that their concern is with how the probability of future events happening can be predicted by calculating the frequency of their previous occurrence. Suffice to say: 'the papers Price created are landmarks in the history of probability and, as their legacy grows, in modern science'. Indeed the *New York Times* has reported that what *we* would call Bayesian/Price statistics 'are rippling through everything from physics to cancer research, ecology to psychology'.[73] A comprehensive and comprehensible discussion of the content of these essays can be found in Tucker (2017).

Besides an understanding of the workings of probability and compound interest, the development of reliable actuarial science, life assurance, pensions and annuities also depend on the availability of accurate statistics; in particular those pertaining to population numbers, and mortality numbers within different age groups. To these Price contributed through his magnum opus: *Observations on Reversionary payments, on Schemes for providing Annuities for Widows and Persons of Old Age; on the Method of Calculating the Values and Assurances on Lives; and on the National Debt* (see Figure 23).

First published in 1771, and with seven editions to follow, the work was described by Benjamin Franklin (who also worked on insurance issues), as 'the foremost production of human understanding that the century has afforded us'. Tucker (2017) notes that the book contains a rigorous, critical and comprehensive account of all aspects of actuarial science and a significant part of its value lies in the life tables that Price

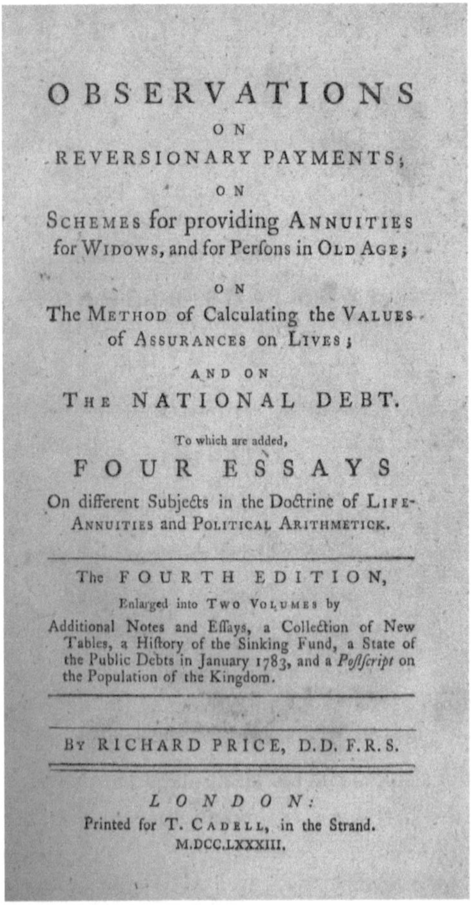

OBSERVATIONS
ON
REVERSIONARY PAYMENTS;
ON
SCHEMES for providing ANNUITIES
for WIDOWS, and for Persons in OLD AGE;
ON
The METHOD of Calculating the VALUES
of ASSURANCES on LIVES;
AND ON
THE NATIONAL DEBT.

To which are added,
FOUR ESSAYS
On different Subjects in the Doctrine of LIFE-ANNUITIES and POLITICAL ARITHMETICK.

The FOURTH EDITION,
Enlarged into TWO VOLUMES by
Additional Notes and Essays, a Collection of New Tables, a History of the Sinking Fund, a State of the Public Debts in January 1783, and a *Postscript* on the Population of the Kingdom.

BY RICHARD PRICE, D.D. F.R.S.

LONDON:
Printed for T. CADELL, in the Strand.
M.DCC.LXXXIII.

FIGURE 23 Title page to *Observations on Reversionary Payments* by Richard Price (1771; London, 1783)

developed; a life table being 'a collection of data for evaluating and predicting life in a group or location. It gives mean probabilities that tell how long people live on average at every age'.[74] Price undertook thousands of calculations to construct such tables for various places at home and abroad. Although subject to later criticism, those he developed for Northampton would become famous and a standard in the insurance industry for more than fifty years.

Population statistics were crucial to these developments and in 1779 Price appended to the first publication by his nephew William Morgan – *The Doctrine of Annuities and Assurances on Lives and Survivorships, stated and explained* (1779) – an essay on the population of England and Wales. Price would be rightly criticised by his contemporaries for concluding that the population was falling, but, at the same time, he was aware of the difficulty of obtaining accurate population statistics. As a result, he became a powerful advocate for the instigation of a regular census; though this would not occur in Britain until 1801, ten years after his death in 1791.

Meanwhile, William Morgan became actuary at the Equitable in 1775 and built on the work of his uncle, who in turn presented some of his nephews' contributions at the Royal Society. As Bennetts (2020) has discussed, Morgan, in his first paper – 'On the Probabilities of Survivorship between Two Persons of Any Given Ages, and the Method Determining the Value of Reversions Depending on Those Survivorships' (1788)[75] – took to task aspects of the work of de Moivre and Dodson before concentrating on 'specialist actuarial mathematics'. A second paper followed in 1789, 'On the Method of Determining, from the Real Probabilities of Life, the Value of Contingent Reversions in Which Three Lives are Involved in the Survivorship'. For his work on these two papers Morgan was awarded the Copley Medal of the Royal Society in 1789, a year before his being elected a Fellow.[76] Morgan published further papers on survivorships in the *Philosophical Transactions* in 1791, 1794 and 1800, and in 1821 he published *The Principles and Doctrine of Assurances on Lives, Annuities on Lives, and Contingent Reversions, stated and explained*, a revision of his earlier volume of 1779 (see Figure 24).

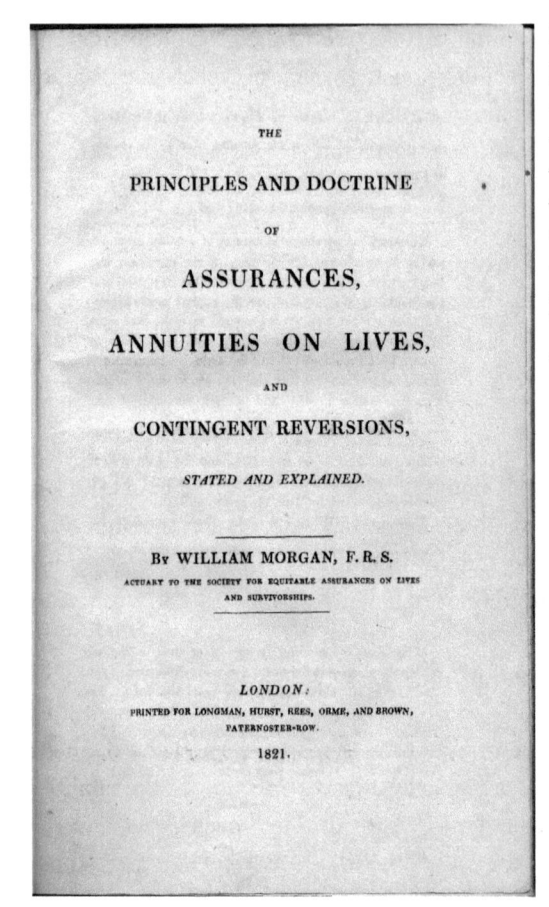

FIGURE 24 Title page to *The Principles and Doctrine of Assurances, Annuities on Lives, and Contingent Reversions, Stated and Explained* by William Morgan (London, 1821).

Within this overall corpus of works William Morgan's contribution is considered pioneering, for reasons outlined by the distinguished actuary Sir William Elderton in the 1930s:

He was the first to show how to work out complicated benefits, involving several lives, from any mortality table; the first to value the liabilities of a life assurance company and appreciate the meaning of the result; the first to see that, with the valuations in use, a margin of surplus had to be carried forward to prevent his bonus system from breaking down; the first to set down the available sources of profit and obtain measures for them; the first to keep record of the mortality of a life assurance office, and to notice that there was such a thing as 'select' mortality. Further than this, he was the first practical administrator of life assurance and a successful business man.[77]

Today, Morgan is widely referred to as the 'father of the actuarial profession', a legacy carried on into the nineteenth century through the work of his son, Arthur Morgan, at the Equitable and, elsewhere, by the likes of the quarryman, mathematician and teacher, Griffith Davies (1788–1855, FRS 1831). Largely self-taught, Davies became actuary at The Guardian Assurance Company for thirty-two years while making further contributions to actuarial science.

Today, Morgan, Price and Griffith Davies have modern biographies devoted to their life and work.[78] It is hoped that William Jones Sr will eventually have one within the Scientists of Wales series. Their various contributions to mathematics, statistics, probability theory and the development of annuities and life assurance constitute just one aspect of their overall contribution. Nevertheless, it is one of the major contributions to science, and to the social and industrial life of Britain made by scientists of Wales in our long eighteenth century – a contribution as relevant in today's world as it was in theirs. Indeed, the truism spoken by Andrew Kippis at Richard Price's graveside in 1791, that 'The blessings of those who might otherwise have perished came upon him; and he has given cause to many a widow's heart, that knew him not, to sing for joy',[79] and the words of the *Morning Chronicle* obituary for William Morgan and his fifty-six years at the Equitable Assurance Society, 'diffusing its benefits to thousands of families, and securing them in the enjoyment of comforts of which they would otherwise have been rendered destitute by the death of their friends and relations',[80] illustrate just some of the many benefits to result from their combined efforts.

Notes

1. William Morgan (ed.), *Sermons on Various Subjects, by the Late Dr. Richard Price, D.D. F.R.S.* (London, 1816); Sermon IX, p. 165.

2. Adam N. Coward (ed.), *A Spiritual Botanology by S. Lucilius Verus (Edmund Jones)*, (Newport, 2017), p. 1.

3. Coward (ed.), *A Spiritual Botanology*, p. 158.

4. John Pughe (trans.) and Revd John Williams (ed.), *The Physicians of Myddfai; Meddygon Myddfai* (Llandovery, 1861); NLW MS 17166C; Diana Luft, 'Introduction: The Nature of the Corpus', in *Medieval Welsh Medical Texts, Volume 1: The Recipes* (Cardiff, 2020), pp. 3–10.

5. Coward (ed.), *A Spiritual Botanology*.

6. William Williams, *A Volume of Herbal Recipes and Remedies* (1803) is held at Bangor University Archives in north Wales.

7. Nathaniel Williams, *Pharmacopoeia, or Medical Admonitions in English and Welsh* (Trevecka, 1793), p. 3.

8. A conclusion based on evidence presented in Donald R. Dickson, 'Henry Vaughan's Medical Library', *Scintilla*, 9 (2005).

9. Alun Withey, *Physick and the Family* (Manchester, 2011), pp. 110, 114.

10. Desmond Chapman-Huston and Ernest C. Cripps, *Through a City Archway: The Story of Allen and Hanburys 1715–1954*, (London, 1954), pp. 17–8 and Appendix 3, pp. 276–7.

11. Allen founded the Askesian Society in 1796. A society for debating scientific matters, its members used the laboratory at Plough Court for experiments. It folded in 1807 but some of its members, including William Allen, went on to found a Mineralogical Society (at Plough Court on 2 April 1799) and the Geological Society (1807). See Chapman-Huston and Cripps, *Through a City Archway*, p. 287. Also, H. S. Torrens, 'Dissenting Science: the Quakers among the founding fathers', in C. L. E. Lewis and S. J. Knell (eds) *The Making of the Geological Society of London* (London, 2009), Special Publications, pp. 317, 129–44.

12. For the Bevan and the later Allen and Hanbury years, see Chapman-Huston and Cripps, *Through a City Archway*; Ernest C. Cripps, *Plough Court: The Story of a Notable Pharmacy 1715–1927* (London, 1927); and *ODNB* entry for Silvanus Bevan.

13. See Withey, *Physick and the Family*, p. 102.

14. BL Add. MS Sloane 1786.

15. *DWB* entry.

16. For Knight and his Welsh roots in Caernarfon, see G. Penrhyn Jones, 'Some Aspects of the Medical History of Caernarvonshire', *Caernarvonshire Historical Transactions*, 23 (1962), 67–91 (at 77–9).

17. See Mary-Ann Constantine, 'The Welsh in Revolutionary Paris', in Mary-Ann Constantine and Dafydd Johnston (eds) *'Footsteps of Liberty & Revolt': Essays on Wales and the French Revolution* (Cardiff, 2013).

18. See J. Dybikowski, *On Burning Ground: An Examination of the Ideas, Projects and Life of David Williams* (Oxford, 1993), pp. 310–11 (for the pamphlets), pp. 278–9 (for the society).

19. See Institution of Engineering and Technology, *https://ietarchivesblog.org/2020/03/35/rare-book-john-evanss-the-universal-medicine-or-the-virtues-of-my-magneticall-or-antimoniall-cup/* (accessed 23 June 2022); also, Royal College of Physicians,

London, *https://history.rcplondon.ac.uk/blog/may-god-protect-us-such-drugs-and-physicians-17th-century-antimony-cup* (accessed 23 June 2022); R. I. McCallum, 'Captain James Cook's Antimony Cup', *Vesallius*, 7/2 (2001), 62–4.

20. T. Knight, 'A Letter from Mr. T. Knight to Sir Hans Sloane … concerning Hair Voided by Urine', *Philosophical Transactions of the Royal Society*, 41 (1739–41), 705–7.

21. See Penrhyn Jones, 'Medical History of Caernarvonshire', 79–80; also H. E. Shih and G. Y. Char, 'Dermoid Cyst Ruptured into the Urinary Bladder', *Journal of Urology*, 38/2 (1937), 165–72; Wm. H. Ranking and John Henry Walsh (eds), 'On Pilimication, or passing of hairs with the urine', *Provincial Medical and Surgical Journal* (London, 1852), 150.

22. See 'Joanna Stephens, 18th century quack or medical pioneer?', at The British Association of Urological Surgeons, *www.baus.org.uk/museum/1438/joanna_stephens* (accessed 22 June 2022). Her fame was widespread in Europe as well as Britain.

23. Philip K. Wilson, 'Acquiring surgical know-how: occupational and lay instruction in early eighteenth-century London', in Roy Porter (ed.), *The Popularization of Medicine 1650–1850* (Abingdon, 1992), pp. 42–71 (at p. 58).

24. For Lister's views, see Anna Maria Roos, *Web of Nature: Martin Lister (1639–1712), the First Arachnologist* (Leiden, 2011), p. 209 onwards.

25. For all details of Robinson see *ODNB* entry and Royal College of Physicians website.

26. Lucy Inglis, *Milk of Paradise: A History of Opium* (2018; London, 2019), pp. 145–6.

27. For all details see Morwenna and John Rendle-Short, *The Father of Child Care: Life of William Cadogan (1711–1797)* (Bristol, 1966), p. 7.

28. See Rendle-Short, *The Father of Child Care*, pp. 47–8.

29. Rendle-Short, *The Father of Child Care*, p. 46.

30. *ML*-II, p. 234.

31. *ML*-II, pp. 234–7 (at 237).

32. For all details of Linden, see R. C. B. Oliver, 'Dietrich Wessel Linden', *NLWJ*, 18/3 (1974), 241–67.

33. N. Williams, *Pharmacopoeia, or Medical Admonitions, English and Welsh* (Part II, Trevecka, 1796), p. 6.

34. Nicholas Robinson, *A Compleat Treatise of the Gravel and Stone* (London, 1721), p. 252 and Preface (n.p).

35. Desmond King-Hele, *Doctor of Revolution, the Life and Genius of Erasmus Darwin* (London, 1977), pp. 28–9, 49. Also, *https://wellcomecollection.org/works/d26c899v* (accessed 25 July 2022); and Wellcome manuscript MS 2043; Darwin used the shorthand created in *Brachygraphy*, a manual published by Thomas Gurney in 1750.

36. Andrew Duncan, *A Letter to Dr Robert Jones of Caermarthenshire* (Edinburgh, 1782), p. 2.

37. For details of the affair, see Robert Jones, *An Inquiry into the State of Medicine* (Edinburgh, 1781); Duncan, *A Letter to Dr Robert Jones*. See Guenter B. Risse, 'Explaining Brunonianism: A Biography of Edinburgh's Master of Conviviality', p. 38 n. 93 for Elizabeth Brown's confirmation of Brown's authorship of Jones's

book, *www.researchgate.net/publication/344483633_EXPLAINING_BRUNONIANISM_A_BIOGRAPHY_OF_EDINBURGH'S_MASTER_OF_CONVIVIALITY* (accessed 24 May 2024).

38. See *The Universal Magazine of Knowledge and Pleasure*, 68/9 (1781), 218.

39. For all details see Constantine, 'Welsh in Revolutionary Paris', in Constantine and Johnston *'Footsteps of Liberty & Revolt'*, pp. 78–84.

40. Catherine Arnold, *Bedlam, London and its Mad* (2008; London, 2009), p. 177.

41. See *https://bathmedicalmuseum.org/dissections-and-autopsies/* (accessed 28 May 2024).

42. See Tony Rail, 'A Previously Unpublished Letter from Thomas Belsham to Samuel Fawcett, 21 April 1791', *Enlightenment and Dissent*, 30 (2015), 136–49 (at 147); available online.

43. For all details see *DWB* entry; also, Joseph Hemingway, *History of the City of Chester*, 2 vols (Chester, 1831), II, pp. 364–6.

44. *WTV(MA)*, p. 144.

45. *WTV(E)*, p. 521.

46. Anonymous [William Cadogan], *An Essay upon Nursing, and the Management of Children, from their birth to three years of age* (London, 1748), p. 36; full text reproduced in Rendle-Short, *Father of Childcare*.

47. Anonymous [William Cadogan], *An Essay upon Nursing, and the Management*, p. 9; full text reproduced in Rendle-Short, *Father of Childcare*.

48. All details from Rendle-Short, *Father of Childcare*.

49. See Fay Bower, 'David D. Davis's Obstetric Textbook and Atlas', *Australian and New Zealand Journal of Obstetrics and Gynaecology*, 43 (2003), 338–40, available online, and references therein for Davis.

50. One of the first to use the Jenner method was Dr William Woodville (1752–1805, FLS 1791). Born in Cumberland, he was a physician in Denbigh between 1776/7 until 1782. After leaving Denbigh he would publish a number of works on smallpox and cowpox and, in 1800, co-authored with Jenner, *A Comparative Statement of Facts and Observations Relative to the Cow-pox*.

51. Richard Wright, 'A Letter on the Same Subject, from Mr. Richard Wright, Surgeon at Haverford West, to Mr. Sylvanus Bevan, Apothecary in London', *Philosophical Transactions of the Royal Society*, 32 (1772/3), 267–9.

52. Perrot Williams, 'Part of Two Letters concerning a method of procuring the small pox, used in South Wales. From Perrot Williams MD Physician at Haverford West, to Dr Samuel Brady, Physician to the Garrison at Portsmouth', *Philosophical Transactions of the Royal Society*, 32 (1723), 262–4.

53. For details of the development of variolation and vaccination, see Gareth Williams, *Angel of Death: The Story of Smallpox* (2010; Basingstoke, 2011). All the letters by Wright and Williams can be found online. They are also calendared, and many printed in full, in Andrea Rusnock (ed.), *The Correspondence of James Jurin (1684–1750)* (Amsterdam and Atlanta GA 1996).

54. Wright, 'A Letter on the Same Subject … to Mr Sylvanus Bevan, Apothecary in London', 267–9.

55. Joseph Needham, *Science and Civilisation in China* (Cambridge, 1954), 6, p. 167.

56. General Board of Health, *Papers Relating to the History and Practice of Vaccination* (London, 1857), p. vii.
57. Nabil Matar, *Turks, Moors, and Englishmen in the Age of Discovery* (New York, 1999), p. 64.
58. Mary Clement, *Correspondence and Records of the S.P.G. Relating to Wales (1701–1750)*, (Cardiff, 1973), p. 3.
59. See Rusnock, *Correspondence of James Jurin*, p. 313.
60. See Rusnock, *Correspondence of James Jurin*, p. 314 n. 2.
61. See Rusnock, *Correspondence of James Jurin*, p. 130. Also noted in Perrot Williams, *Some Remarks upon Dr. Wagstaff's Letter Against Inoculating the Smallpox ...* (London, 1725), 'Appendix by F. Slare', pp. 26–7. Jenner would later employ a similar method of exposing previously vaccinated people to those currently in the throes of the disease to prove the efficacy of inoculation with the cowpox vaccine.
62. See Rusnock, *Correspondence of James Jurin*, p. 130.
63. See Maurice Edward Ogborn, *Equitable Assurances* (London, 1962), pp. 19–27 (at pp. 22–3). For a detailed discussion of the early history of probability, see Anders Hald, *A History of Probability and Statistics and their Application Before 1750* (New York, 1990).
64. D. R. Bellhouse, E. M. Renouf, R. Raut and M. A. Bauer, 'De Moivre's Knowledge Community: An Analysis of the Subscription List in the *Miscellanea Analytica*', *Notes and Records: The Royal Society Journal of the History of Science*, 63 (2009), 137–62 (at 145).
65. 'A Letter from Mr Abraham De Moivre F.R.S. to William Jones, Esquire, F.R.S. concerning the easiest method for calculating the value of annuities upon lives, from tables of observations', *Philosophical Transactions of the Royal Society*, 43 (1744), 65–78.
66. For all details, see M. E. Ogborn, 'The Theory of Simple and Compound Interest: An Eighteenth-Century Manuscript', *Journal of the Institute of Actuaries* (1886–1994), 75/1 (June 1949), 73–4.
67. William Gardiner, *Tables of Logarithms for All Numbers from 1 to 102100* (London, 1742), Preface n.p.; see also the Subscribers list. For 'The Explication and Use of the Logarithmic Tables' taken from Jones, see pp. 1–14.
68. James Dodson, *The Anti-Logarithmic Canon* (London, 1742), p. viii.
69. Ogborn, *Equitable Assurances*, pp. 21, 261.
70. These were Sir Charles Gould (1726–1806, President 1773–1806), who married into the Morgan family of Tredegar and took the name Sir Charles Morgan, Bart. in 1792; his son Sir Charles Morgan, Bart. (1760–1846, President 1806–46); and Sir Charles Morgan Robinson Morgan, Bart. 1st Baron Tredegar (1792–1875, President 1846–75).
71. *Philosophical Transactions of the Royal Society*, 53 (1763), 370–418.
72. Richard Price, 'A Demonstration of the Second Rule in the Essay toward the Solution of a Problem in the Doctrine of Chances', *Philosophical Transactions of the Royal Society*, 54 (1764), 296–325.

73. John Horgan, 'Bayes's Theorem: What's the Big Deal?', 4 January 2016, *https://blogs.scientificamerican.com/cross-check/bayes-s-theorem-what-s-the-big-deal/* (accessed 9 June 2022).

74. John V. Tucker, 'Richard Price and History of Science', *THSC*, 23 (2017), 84.

75. *Philosophical Transactions of the Royal Society*, 78 (1788), 331–49.

76. For all details see Nicola Bruton Bennetts, *William Morgan: Eighteenth-Century Actuary, Mathematician and Radical* (Cardiff, 2020).

77. See William Elderton, 'William Morgan', *Transactions of the Faculty of Actuaries*, 14 (1931–4), 15; Bennetts, *William Morgan*, p. 60.

78. See Bennetts, *William Morgan*; Paul Frame, *Liberty's Apostle, Richard Price his Life and Times* (Cardiff 2015); Haydn E. Edwards, *Griffith Davies – Arloeswr a Chymwynaswr* (Cardiff, 2023).

79. Andrew Kippis, *An Address, Delivered at the Interment of the Late Rev. Dr, Richard Price, on the twenty-sixth of April, 1791* (London, 1791), p. 21.

80. Bennetts, *William Morgan*, p. 203.

13

THE SCIENCE OF US: NATURE AND ECONOMICS

To Richard Crawshay, Esq;

Who in applying the Materials of Nature to the Purposes
of Life and the Uses of Society, has best answered the
Ends of Science, and advanced its Interests.[1]

With these words the Swansea-based, English conchologist, William Turton, dedicated his English-language translation of the complete *Systema Naturae* by Linnaeus (see Chapter 10) to Richard Crawshay (1739–1810), owner, by the early nineteenth century, of the largest ironworks in the world – the Cyfarthfa works, in Merthyr Tydfil.

Aside from human input through work, three things were necessary to apply the materials of nature to the purposes of life and the uses of society: i) a knowledge and understanding of the material resources available, together with their possible uses; ii) a sound economic framework; and iii) the technical knowledge necessary to develop such resources. Knowledge of the resources available has largely been the preoccupation of earlier chapters in this book, which have detailed the collecting, cataloguing and illustration of material by scientists of Wales. The understanding of the availability and possible use of those materials, together with their economic development, are the subjects of this chapter. Technical know-how is then considered in the chapters on artisan science and innovation and invention that follow this one.

Nature's Bounty

Scientists of Wales were never averse to utilising natural resources as a small-scale aid to their financial security. For example, in 1708, while working as a librarian at the Ashmolean Museum in Oxford, Alban Thomas, whose later position, as assistant secretary at the Royal Society, Methusalem Bowen attempted to obtain (see Chapter 7), took out a four-page advert in the *Philosophical Transactions*.[2] It advertised for sale, at one guinea, a collection of some fifty-two mineral and fossil specimens, with multiple specimens available of some of the material. The specimens were named in accord with Edward Lhwyd's *Lithophylacii Britannici Ichnographia* (1699), and the place where they were collected would be made available to any purchaser. As Thomas (2018) notes: 'This may well rank Alban Thomas as the earliest known dealer in mineral specimens.'[3] We have also seen Lewis Morris busy collecting cregyn ('shells') in the Dyfi Estuary in order to earn a little money by selling them to gentry and aristocratic collectors. According to Edmund Jones in his 1779 work *A Geographical, Historical, and religious Account of the Parish of Aberystruth*, 'The Earth itself produces no treasure but minerals, and stones of divers sorts and uses', and he records seeing 'Small stones of Crystalline Colour, which are sometimes sold at shops in *Abergavenny*, but little is paid for them'.[4] While Edmund Jones and the customers in Abergavenny may not have seen much value in the products of the earth, others certainly did.

When, in September 1665, the Secretary to the Royal Society, Henry Oldenburg, wrote to Sir Robert Moray concerning Moray's proposed 'incursion into Wales' (see Chapter 10), Oldenburg's initial desire was to learn about Welsh husbandry practices. However, his interest in the country actually went much further, as the continuation of his letter to Moray reveals:

> You will also, I am confidant, make inquiry after ye curiosity and variety, observable in ye country; of their mines, and especially ye leaden-ore, which, I often heard, the Dutch, not many years since, have made great advantage off, by extracting considerable quantities

of silver of it: of their springs, ye ebb and flow; St. Winifreds Well, as they call it, ye British Alpes, Snowdon-hills, in Caernarvonshire, if it be in yr way; ye Pearls, said to be fished[?] in great abundance in the same shire, and if I mistake not, in ye river of Conwey.[5]

Whatever the value to Oldenburg and the Royal Society of topographical descriptions and river pearls, it is the enquiry concerning mines and ores that is of immediate interest here.

Prior to leaving for Wales, Robert Moray had settled in Oxford in December 1665 so as to avoid the plague then ravaging London. It was there, as we saw in Chapter 2, that he began dabbling in experiments at the house of Samuel Kem, assisted by Welsh alchemist Thomas Vaughan. Moray employed himself in extracting 'from lead ore all the metal it contains with one wash, great ease, and small charge', and he did 'the same in extracting silver out of lead with the same advantages'. It was probably while assisting with these experiments that Thomas Vaughan accidentally inhaled mercury fumes and met his end in February 1666.

Having organised Vaughan's funeral, we find Moray in April 1666 describing his work on 'the ore in Cardiganshire that is called silver or rich ore'. When worked on by the usual methods of extraction this yielded '40lb of lead', but, presumably as a result of his experimentation, Moray had managed to get '69 out of it'. According to Stevenson (2016) it was after this that Moray visited Wales, where he hoped to reopen a silver mine in Cardiganshire with the help of unnamed friends including, perhaps, Christopher Wren. Wren had supplied Moray with a 'long account of trials made on lead ore from a Welsh mine' in 1664. Following his visit to Wales Moray knew there were great riches underground there, but he was forced to abandon plans to move to the country to develop them when the King refused him permission to go. Instead, we find Moray presenting 'several Mineralls and Marscasites out of Wales' to a meeting of the Royal Society in July 1666.[6]

Henry Oldenburg hailed from Germany, long a centre of metal production and metallurgical expertise, which may explain his interest in Welsh ores. Moray was a Scot, and when Porter discussed in *The Making of Geology* (2008) what was 'characteristic and distinctive about Scottish Earth science' by the eighteenth century, he concluded that interest in the subject reflected a 'deep and continuing concern for utility'. He also suggested this was a response to an 'all-pervasive need to modernize the backward Scottish economy'. By contrast, 'At no point were leading English naturalists so genuinely concerned with the economic aspects of Earth Science as their Scottish counterparts'.[7] Porter makes no mention of Welsh attitudes, but in referencing *The Natural History of the Mineral Kingdom* (1789), by Wales-born, Scottish-based geologist John Williams, as a source for his conclusion regarding the Scottish position, Porter confirms what can be deduced from other evidence – that Welsh attitudes mirrored those of the Scots.

The desire of gentry landowners in Wales to develop the minerals found on their estates is evident from well before the opening of our period in 1650. In *c.*1579, Sir John Wynn (1553–1627), of Gwydir, in north Wales, witnessed, in company with Lord Burghley and Sir Francis Walsingham, the chief ministers of Elizabeth I, an experiment undertaken in Anglesey in which a 'Mr Medley' had boiled a quantity of iron in vitriolic water. Out of this experiment alum and copperas were produced, both being of commercial interest at the time since they were used as mordents and fixatives in the cloth-dyeing industry.[8] Some twenty-eight years later, in November 1607, we find the entrepreneur Sir Thomas Myddelton (1550–1631) writing to John Wynn having been made aware, by a gentleman 'lately come out of Germany', that mineral waters were to be found on Sir John's estate. Wynn, in turn, notes how trials had been made 'out of alum and copperas both in the country and in London'. However, to take these trials further a skilled worker was needed, but since Sir John found some 'uncertainty in alchemy' he 'never durst adventure so great a work himself',[9] a hesitancy perhaps rendering John Wynn more of a committed exploiter than a true scientist of Wales.

By contrast, his son, Sir Owen Wynne (1592–1660), of Gwydir, who took over the lands of his father at the start of our long eighteenth century, possessed a scientific as well as a commercial attitude, and had far less uncertainty about alchemical experimentation than his father.

In 1629, for example, he purchased in Duck Lane, London, a large collection of manuscripts and books 'germane' to his 'alchemical and metallurgical interests'.[10] Being keen to develop the minerals on his lands, Owen Wynne visited his lead mines in Cardiganshire in the 1650s. Once there, he planned to:

> Enquire if there be amongst the miners at the lead works any outlandish man, (as a Dutchman or High Dutchman) that hath skill in mines and in how to find out and discover the lead mines in the ground where lead ore is found in several places thereof.[11]

More interesting from the perspective of science was his desire to find 'an expert salt petre boyler, not a salt petre finder' who might be 'drawne into ... the country [north Wales] to trie an experiment'.[12] Whether Owen Wynne proposed to undertake this experiment in some form of laboratory is currently unknown.

Throughout the long eighteenth century other scientists of Wales showed an interest in the productive nature of their country. In 1684, Edward Lhwyd described in a letter to his then boss at the Ashmolean Museum in Oxford, Robert Plot, the nature of the asbestos found in a northern part of Anglesey. Later published in the Royal Society's *Philosophical Transactions*, the letter gave a detailed description of asbestos in outcrop, and related some of the experiments that Lhwyd had made on a sample of the material. These included putting 'a small quantity of the lint in the fire, which grew red hot; but though it remained there ¼ of an hour, I could not perceive that it was anything consumed'. Having satisfied himself that the material was incombustible, Lhwyd remembered that Plot had suggested in one of his 'Chymical Lectures in the Natural History Schoole' that paper might be made of the material. To this end Lhwyd pounded a quantity of the asbestos in a stone mortar, 'till it became a downy substance and seem'd very fit for that purpose'. Following continued refining he brought a quantity of the prepared material to a paper mill 'and desired the workmen to proceed with it in their usual method of making paper'. This they did, and though the 'Paper made of it proved very coarse and too apt to tear' he had 'some reasons to believe it may be much improved', whence 'it would make good [non-flammable] writing-paper'.[13] Later, in 1702, we find Lhwyd writing to his aunt in Wales asking her to take care of Johan Angerstein (*c*.1672–1720), a visiting Swedish mineralogist who made a tour of Wales and England in 1702–3. Angerstein later donated a collection of some fifty or so minerals to the Ashmolean Museum.[14]

Resource interest is evident too among the surveyors and map makers of Wales. The map of south Wales produced by Emanuel Bowen as early as 1729 includes 'coal pits', while the later 1795 map of north Wales by John Evans includes the position of slate quarries. When Lewis Morris published his coastal survey, *Plans of Harbours, Bars, Bays and Roads in St George's-Channel*, in 1748 he made sure to include a lengthy section containing *Observations Relating to the Improvements that might be made in the Harbours whose Plans are here given: together with some account of the Natural Commodities and Trade of Those Places.* In the sections on 'natural commodities' that accompanied each of his maps, he supplements local knowledge with his scientific understanding as a natural historian and geologist; all with the aim of highlighting commercial possibilities. For example, at Holyhead he notes 'a plant growing on the Sea Rocks, called by the Natives *Gwymmon*, (in English *Tang*) of which they make great Profit, by burning it into a Kind of Salt, called KELP; one of the ingredients in making Glass, and used also in Allum works'.[15] *Gwymmon*, or gwymon today, is Welsh for 'seaweed', but *tang* appears to have been a traditional Scottish (and perhaps northern English) name for seaweed gathered for agricultural and industrial use. Such seaweed was extensively collected in the poorer coastal communities of Wales, Scotland and Ireland. Though we today use the word *kelp* to describe a particular type of seaweed, the name originally applied to the alkali produced by burning seaweed. As Morris records, this 'kelp' was then used in the eighteenth century in both the glass and alum industries.

We also learn from Morris that aside from the excellent fishing and wool manufactures at Barmouth there are, on the Afon Mawddach, in Merionethshire:

some Lead and Copper Veins discovered up this River, which were never worked to any Purpose: And here is one of the greatest Veins of Marcasite I ever met with, containing some small Strings of yellow Copper Ore.

All the rocks of the Coast of *Merionethshire* abound with *Pyrites Tesselatæ*, a mineral which is generally a Follower of Copper and richer Metals.

Up this River, near Dolgellau, there is Iron Stone in abundance; some Attempts have been made to make Iron of it, but they have not succeeded, perhaps for want of Skill: It is probable, if it was mixed with more fusible Iron Ore, such as that of *Lancashire*, it might answer.[16]

In an entry for Dulas Bay, in Anglesey, he suggests how earth science might come to the aid of art. At 'Mynydd Paris' ('Parys Mountain'), not far from Dulas Bay, 'there is plenty of a reddish Okery Earth, something like Spanish-brown, but a far better bodied Colour for Painting'.[17] As Thomas Pennant remarks in his 'Preface' to *British Zoology*: 'the permanency of colors depends on the goodness of the pigments; but the various animal, vegetable, and fossil substances (out of which they are made) can only be known by repeated trials.'[18]

In his already noted *Natural History of the Mineral Kingdom* (1789), Welsh geologist John Williams stressed another important pairing – that of education in aspects of earth science with economic development:

It is a particular loss to the increase of knowledge in the natural history of the mineral kingdom, that this branch of science is neglected in our public schools. Mineralogy is taught in the Universities abroad. I believe, that what may be called fossilogy, or the arrangement and description of mineral fossils, is taught in some of our public schools; but their instructions are founded upon small detached samples, the collections of the cabinet, which leave the country gentleman and the young miner as much in the dark as before, with respect to knowledge of nature and of real mineral appearances, which are the true sources of useful knowledge in these matters; and this species of knowledge is of great importance. No country in the world depends so much upon the productions of the mineral kingdom, for the means of comfortable accommodation, wealth, and power, as the island of Britain.[19]

As a passionate geologist Iolo Morganwg saw potential in the geological resources of Wales. Sometime in the 1780s he had taken out a lease on a quarry. Though this had not prospered, he began a second attempt at development in 1800, with the possibility of mining lead for use in pencils.[20] In writing about this possibility in a November 1799 letter to Owen Jones ('Owain Myfyr'), he also detailed wider geological possibilities in Wales:

> There are in many parts of Wales stones of various sorts that would be worth sending to London, as several kinds of good marble, freestone, pavements, much better than that of Yorkshire or Purbeck stones; filtering stones &c. with ochres [earthy materials rich in iron oxides], boles [fine compacted or unctuous clay with iron oxide], &c. There is in many places in Northwales and Cardiganshire a very fine potters clay, fit for pipes and the finest wares. This, if worked, would employ many hands and bring money into the country.[21]

Finally, at least in the earth science stakes, we have John Lloyd of Wigfair corresponding in 1813 with the geologist and mathematician John Farey (1766–1826) concerning a proposed, though not undertaken, mineralogical survey of north Wales.[22]

The principal scientist of Wales to see commercial and industrial potential in zoology was Thomas Pennant. As he writes in the 'Preface' to *British Zoology*: 'Why ... should we neglect inquiring into the various benefits that result from these instances of the wisdom of our Creator, which his divine munificence has so liberally, and so immediately placed before us.'[23] Important branches of commerce, he suggests, 'may be enlarged by discovering new properties in animals, or by the further cultivation of those already discovered'. To achieve this 'The science of zoology is requisite'. One example of Welsh commerce being enlarged by the properties of one particular animal is to be found in the case of the leech, which was widely used in medicine as an aid to bloodletting. For example, in 1659 advice was being given to Sir Owen Wynne of Gwydir concerning the application of leeches to his haemorrhoids as an aid to relief.[24] By the eighteenth century, production of the necessary leeches was a well-established enterprise on Marloes Marsh, near Dale in Pembrokeshire. With their reputation established in Wales and England, Marloes leeches were exported in bulk to France and to Ireland from Haverfordwest.[25]

Botanical science also had a long history of commercial application, particularly through the apothecaries and doctors who used plants and herbs, along with chemical means to produce the medicines and treatments for which they might charge.

There were also, of course, less scientific considerations at play, as Thomas Pennant also asserted in his *British Zoology*: 'if we reflect but a little on the unwearied diligence of our rivals the French, we should attend to every sister science that may any ways preserve our superiority in manufactures and commerce.'[26] To preserve that superiority, sound economic science was another prerequisite for the development of the natural resources available.

On Economics

In 1755, Welshman 'C. B.' writing under the pseudonym 'Giraldus Cambrensis' produced his *Proposals for Enriching the Principality of Wales* (see Figure 25). The proposals were addressed primarily to the 'Nobility, Gentry, Clergy and Yeomen' of the country, and the proposals themselves were essentially a plea for improved landed estates and commerce from various agricultural innovations.

These innovations included cultivating turnips and making changes to the fattening of livestock before their move to market. In the letter Iolo Morganwg had written to Owen Jones detailing the geological possibilities of Wales (see above), Iolo had highlighted one of the problems in moving not only livestock, but any other commercial product

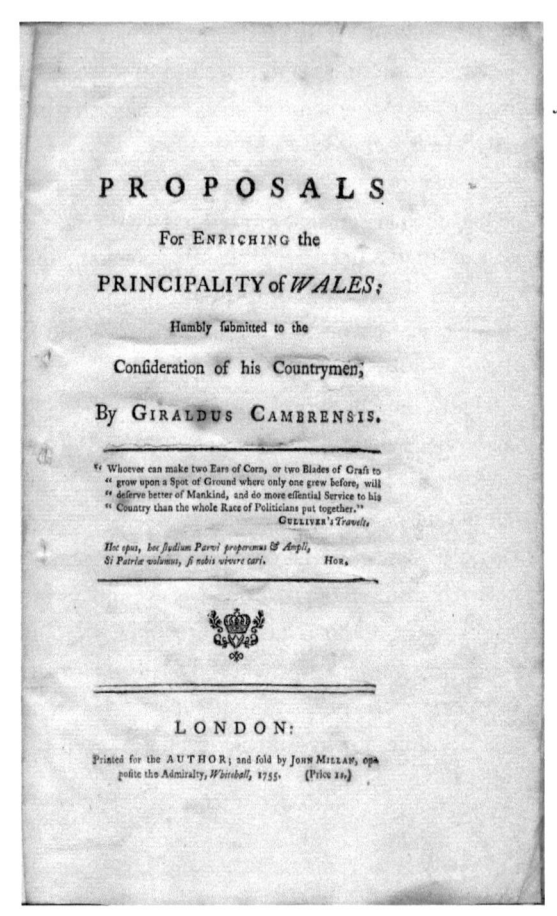

FIGURE 25 Title page to *Proposals for Enriching the Principality of Wales* by 'Giraldus Cambrensis' (London, 1755)

within Wales at this time – the limited and uneven infrastructure development in the country. As he notes with his usual south Wales bias and somewhat ironic tone:

> About 30 years ago Southwales boasted, and at that time it might justly do so, of having the best roads in the kingdom, but in this very important improvement it must now yield to Northwales which has left far behind even the vicinities of the metropolis where it might naturally be supposed the best roads might be found. North Wales is deficient in natural advantages; Caermarthen, Pembroke

and Cardiganshires have more than double portions of these; and yet with the exception of a few – how very few! – spots, are as much in the savage state as they were two thousand years ago. Pinkerton [who we will meet in Volume 2] has been I fear in Carmarthenshire &c., whence he derives his idea that the Celts are radically and irreclaimably savages. The counties that I have named may properly be termed the British Negroland.[27]

Proposals for Enriching the Principality of Wales contains little by way of real economic theory. By contrast, the Scottish economist Adam Smith (1723–90) represents the doyen of eighteenth-century and most-modern economic theorists. Through his publication of *An Enquiry into the Nature and Causes of the Wealth of Nations* (1776), with its advocacy of economic progress through the pursuit of self-interest, the division of labour and freedom of trade, Smith has achieved a place in economics and business akin to that of Newton in physics. However, as with Newton in recent years, there has been a reappraisal of Smith in relation to the wider contexts of his times; one such context being the influence of earlier writers on his ideas. From that perspective, it is of note that Smith's library contained publications by three scientists of Wales who wrote on economics.

The first is the London-based Assay Master, Joseph Harris. In 1757, Harris published anonymously the first part of his *Essay on Money and Coins*; the second part appeared in 1758, again anonymously. A third part, in the form of an unpublished manuscript, would be found much later in a copy of *Essay on Money and Coins* sold at auction in 1991. This third part appears to have been written by Stanesby Alchorne (d.1800). In 1757, he became a deputy to Harris during the latter's tenure as King's Assay Master at the Royal Mint, then situated in the Tower of London. In a preface to the manuscript, Alchorne records Harris's authorship of parts one and two, and that the manuscript of part three had been composed after Harris's death in 1764; Alchorne having perused, during seven years close intimacy with Mr Harris 'the original manuscript, and hearing the several parts repeatedly explained and enlarged upon'. The content of the various parts can be discerned from their subtitles: part one concerned *The Theories of Commerce, Money and Coins*; part two the fact that *the established standard of money should not be violated or altered, under any pretence whatsoever*; and the third *Humbly Proposing some regulations for remedying the present base state of our coin, and for obviating all cause of complaint about our money for the future*.[28] The Harvard University political economist Joseph Schumpeter (1883–1950) described Harris's work as 'the standard English work on money' of the eighteenth century, and *An Essay on Coins* (Parts One and Two) certainly features in the library of Adam Smith. According to the compiler of a catalogue of Smith's books, Hiroshi Mizuta, 'it seems to be one of the works of political economy most frequently used by Smith'.[29]

It was said of our second economics writer, Josiah Tucker (1713–99), that 'Almost all public events bring him forth'. With interests in economics, politics and religion, Tucker produced more than forty works.[30] Of these, eleven are to be found in Smith's library, of which three concerned economics: *The Elements of Commerce and Theory of Taxes* (1755, though only privately circulated and only three copies are known), *Instructions for Travellers* (Dublin, 1758), and *A Treatise concerning Civil Government* (London, 1781).[31] According to Schuyler (2021), *The Elements of Commerce and Theory of Taxes* represents:

> a pioneer work, written before the publication of any of the important products of the French physiocrats [French economists, such as Mirabeau, who believed in nature's bounty, the inherent goodness of men, and *laissez-faire* economics] and more than twenty years before the appearance of the *Wealth of Nations*. But as it was not published it did not influence the subsequent course of economic thought. Tucker distributed copies of it among his friends for criticism, but only three of these copies are known to be in existence. It justifies the opinion that if the larger treatise [that Tucker planned] had been completed, the author would have gained a position of eminence in the history of economics.[32]

Born in Laugharne, Carmarthenshire, Tucker lived near Aberystwyth as a boy before education in Ruthin, Denbighshire, and then St John's

College, Oxford. He later spent time in Bristol before becoming Dean of Gloucester Cathedral in 1758.[33] According to a number of writers, 'The similarity of some of the basic economic ideas expounded in the *Wealth of Nations* to those which Tucker had previously set forth would make it reasonable to suppose that Adam Smith was influenced by his older contemporary'. According to Oslington (2017), 'Ideas and even specific language' from Tucker's works 'appear in Smith's works and early 19th century works of political economy'. This even includes the description of England as a nation of shopkeepers, which later appeared in Smith's writing before being taken up by Napoleon and many others.[34]

Both Tucker and Smith believed in self-love or self-interest as an economic force, with Tucker stating that 'every individual (whether he intends it or not) will be promoting the good of his country, and of mankind in general, while he is pursuing his own private interest'. He also believed that 'the Laws of Commerce, when rightly understood, do perfectly coincide with the Laws of Morality, both originating from the same good Being, whose mercies are over all his works'. What perhaps distinguishes Tucker from Smith is the former's belief that the market cannot always reconcile the public and the private, and that government is needed in order to do so. As Oslington concludes: 'In Tucker we have a clear and powerful statement of the providential harmony between self-love and the common good. Tucker though is no proponent of laissez-faire, and government must direct self-love to its proper end.'[35]

Tucker also involved himself in arguments with David Hume, with whom he corresponded, over how far trade might equalise wealth between poor and rich countries, and he exchanged views with Edmund Burke over the American Revolution. Later, two of Tucker's works would be translated into French by Anne-Robert-Jacques Turgot, the French Finance Minister in the years running up to the French Revolution of 1789. Coincidentally, and somewhat remarkably, our third Welsh writer with books in Smith's library also entered into a debate with David Hume (on philosophy), was confronted by Edmund Burke (over the French rather than the American Revolution), and would correspond with Turgot. This was Richard Price.

Price certainly knew Smith. Both were members of the Royal Society and, according to Benjamin Franklin, as reported in Cone (1952), when Smith was in London between 1773 and 1776 working on the *Wealth of Nations* he:

> was in the habit of bringing chapter after chapter as he composed it to [Franklin], Dr. Price, and others of the literati; then patiently hear their observations and profit by their discussions and criticisms, sometimes submitting to write whole chapters anew, and even to reverse some of his propositions.[36]

Price generally wrote respectfully of Smith, but in a 1785 letter to William Eden, who Price was debating with about population numbers in England, Smith declared that 'Price's speculations cannot fail to sink into the neglect that they have always deserved. I have always considered him as a factious citizen, a most superficial philosopher, and by no means an able calculator'.[37] Cone (1952) notes that:

> Smith's opinion of Price was not clear; he respected Price's judgment on some matters and rather nastily disagreed on others. There is evidence that Smith liked Price, and there is evidence to the contrary. In any case, Smith was not always a pleasant person; his affability on some occasions was matched by rudeness on others.[38]

To prove the point, when Samuel Rogers, a neighbour of Price's at Newington Green, visited Smith in 1789, he took with him a letter of recommendation from Price, together with one from Price's friend and fellow Dissenting minister Andrew Kippis. These 'won for Rogers the kindest possible reception'.[39]

Smith's library contained five works by Price but only one centred on economics, and that related to what would be an economic fixation for much of Price's career – the National Debt. As well as including discussion of the topic in his books on other subjects, Price published in 1772 *An Appeal to the Public on the Subject of the National Debt*. For its second edition, which appeared that same year, he included both the mathematical calculations and the illustrative tables necessary to show how different kinds of a so-called 'sinking fund' could be used to progressively redeem the debt. In 1783, as the terms for peace with revolutionary America were being finalised in London and Paris, he returned to the debt issue with *The State of the Public Debts and Finances at signing the Preliminary Articles of Peace in January 1783. With a Plan for Raising Money by Public Loans, and for Redeeming the Public Debts.* Having next been asked, in 1786, by the Prime Minister, William Pitt the Younger, to submit a plan for redeeming the National Debt, Price produced not one but three plans in a manuscript submission titled 'Three Plans for shewing the progress and effect during 40 years of a Fund consisting of a million per ann. as is therein express'd and apply'd to the redemption of the public debts'. These three plans would later be published by William Morgan in 1792, the year following Price's death in 1791, as *A Review of Dr Price's Writings, on the Subject of the Finances of the Kingdom: to which are Added the Three Plans communicated by him to Mr Pitt in the year 1786, for Redeeming the National Debt.* Morgan published the work as a way of defending his uncle against criticism of the plan of redemption adopted by Pitt – and which Price had considered the weakest of the three he had suggested.

William Morgan's writing on economics would not appear in Smith's library as he did not publish on the subject until after Smith's death in 1790. Like his uncle, Morgan would concentrate on the issue of national debt redemption. Both his *Facts addressed to the Serious Attention of the People of Great Britain respecting the Expence [sic] of the War, and the State of the National Debt* as well as *Additional Facts* (on the same subject) were published in 1796, with each running to four editions. The presentation of *facts*, however, does not appear to have been enough, so in 1797 came the more direct *Appeal to the People of Great Britain, on the Present Alarming State of the Public Finances, and of Public Credit.*

William Morgan lived well into the nineteenth century (he died in 1833) in a country with a growing economy and an expanding empire. Both Price and Tucker were anti-imperialists and criticised the imperial developments in their published work. Underpinning this economic and imperial expansion, beyond the 'benefits' accruing from slavery, a trade which both Price and Tucker criticised, was the linkage between scientific knowledge and the advances in technology seen in the long eighteenth century, and particularly from the 1750s onward – a linkage facilitated by artisan science and the innovation and invention that are the subject of our next two chapters.

Notes

1. William Turton (trans.), *A General System of Nature* (Swansea, 1800), I, (n.p.); available online at Hathi Trust Digital Library, *babel.hathitrust.org/cgi/pt?id= chi.27986221&view=1up&seq=5* (accessed 18 July 2023).
2. Alban Thomas, 'Advertisement', *Philosophical Transactions of the Royal Society*, 26 (1708), pp. 77–80.
3. See Wendell E. Thomas (2018), 'Mineralogical Record Biographical Archive', *www.mineralogicalrecord.com* (accessed 24 May 2024).
4. Edmund Jones, *A Geographical, Historical, and religious Account of the Parish of Aberystruth* (Trevecka, 1779), p. 38.
5. RSC EL/OB/36a.
6. For details, see David Stevenson (ed.), *Letters of Sir Robert Moray to the Earl of Kincardine* (Aldershot, 2016).
7. Roy Porter, *The Making of Geology: Earth Science in Britain 1660–1815* (1977; Cambridge, 2008), p. 153.
8. *CWP* pp. 77–8; items 470, 471. Also J. R. Harris, *The Copper King: Thomas Williams of Llanidan* (1964; Ashbourne, 2003), pp. 18–25. Even as late as 1759 we find one Ambrose Morris sending vitriolic waters from Amlwch, which is near Parys Mountain, across to William Morris's friend in Ireland (see previous chapter), Dr Rutty; the results of Rutty's analysis being published in the *Philosophical Transactions*.
9. *CWP*, p. 75; items 455, 456.
10. For details of this purchase and Owen Wynne's metallurgical and bibliophile interests, see Owen Morris, *The 'Chymick Bookes' of Sir Owen Wynne of Gwydir: An Annotated Catalogue* (Cambridge, 1997).
11. *CWP*, p. 339; item 2071.
12. *CWP*, p. 359; item 2248.
13. *Philosophical Transactions of the Royal Society*, 14 (1684), 823–84. Also, *Lhwyd*, p. 53.
14. For details, see Arthur MacGregor, 'The Ashmolean as a Museum of Natural History, 1683–1860', *Journal of the History of Collections*, 13/2 (2001), 125–44; Gunther, pp. iv, 450. In 1754, an R. R. Angerstein would travel in Wales and keep a wonderfully illustrated travel diary depicting all manner of industrial processes, machinery and facilities he witnessed, see Torsten Berg and Peter Berg (trans.), *R. R. Angerstein's Illustrated Travel Diary, 1753–1755: Industry in England and Wales from a Swedish Perspective* (London, 2001).

15. *Lewis Morris*, 'Observations Relating to the Improvements that might be made in the Harbours whose Plans are here given: together with some account of the Natural Commodities and Trade of Those Places', in Lewis Morris *Plans of Harbours, Bars, Bays and Roads in St. George's Channel* (1748; Beaumaris, 1987), p. 4.

16. Morris, 'Observations Relating to the Improvements', in *Morris, Plans of Harbours*, p. 9.

17. Morris, 'Observations Relating to the Improvements', in *Morris, Plans of Harbours*, p. 3.

18. Thomas Pennant, *British Zoology* (London, 1768), p. xi.

19. John Williams, *The Natural History of the Mineral Kingdom*, 2 vols (Edinburgh, 1789), I, p. ix.

20. *CIM*-I, p. 200 and n. 2.

21. *CIM*-II, p. 245.

22. NLW MS 12420D.

23. Pennant, *British Zoology*, p. ix.

24. *CWP*, p. 353; item 2194.

25. See Gareth Williams, *Angel of Death: The Story of Smallpox* (2010; Basingstoke, 2011), pp. 52–3.

26. Pennant, *British Zoology*, p. xvi.

27. *CIM*-II, p. 245. The development of the roads in north Wales is discussed in A. H. Dodd, *The Industrial Revolution in North Wales* (Cardiff, 1971). The influence of John Pinkerton, who in 1788 had challenged the authenticity of some of the Celtic materials being published as original by the likes of Iolo Morganwg can be found in Philstor [John Pinkerton], 'Letters to the People of Great Britain, on the Cultivation of their National History', *Gentleman's Magazine*, 68/1 (1788), 500; and Moira Dearnley, '"Mad Ned" and the "Smatter-Dasher": Iolo Morganwg and Edward "Celtic" Davies', in Geraint H. Jenkins (ed.), *A Rattleskull Genius: The Many Faces of Iolo Morganwg* (Cardiff, 2005), pp. 425–42 (at pp. 425, 426, 435).

28. See Marvin Lessen, 'Harris, Alchorne and an Essay', *British Numismatic Journal*, 62 (1992), 196–7; available online.

29. See Hiroshi Mizuta, *Adam Smith's Library: A Catalogue* (Oxford, 2000); the list of Smith's books can also be found in Daniel B. Klein and Andrew G. Humphries, 'Foreword and Supplement to "Adam Smith's Library: General Check-List and Index"', *Econ Journal Watch*, 16/2 (September 2019), 374–474, or *www.econjwatch. org/articles/adam-smith-s-library-general-check-list-and-index* (accessed 24 May 2024). Note, the two quotes are taken from an excellent essay by Peter Moody, *http://joseph-harris.org/archives/documents/a-new-review-of-joseph-harris-essay-upon-money-and-coins* (accessed 22 September 2016; sadly this site no longer seems to be available).

30. For a listing see Robert Livingston Schuyler, *Josiah Tucker: A Selection from his Economic and Political Writings* (Carmel IN, 2021), pp. 555–8.

31. See note 29 – Mizuta, and Klein and Humphries.

32. Schuyler, 'Introduction', in *Josiah Tucker*, p. 12.

33. See *DWB* entry.

34. See Schuyler, *Josiah Tucker*, p. 16; Paul Oslington, 'Anglican Social Thought and the Shaping of Political Economy in Britain: Joseph Butler, Josiah Tucker, William Paley and Edmund Burke', *History of Economics Review*, 67/1 (2017), 26–45, *https://d3nr8uzk0yq0qe.cloudfront.net/media/upload/research/Butler_Tucker_Paley_Burke.pdf* (accessed 1 August 2022).

35. Oslington, 'Anglican Social Thought'.

36. See Carl Cone, *Torchbearer of Freedom: The Influence of Richard Price on 18th Century Thought* (Lexington KY, 1952), p. 59. Cone cites for the Franklin comment: 'Quoted in John Rae, *Life of Adam Smith* (London, 1895), pp. 264–65'.

37. See Cone, *Torchbearer of Freedom*, pp. 59–60. Cone cites: 'Quoted in John Rae, *Life of Adam Smith* (London, 1895), p. 398'.

38. Cone, *Torchbearer of Freedom*, p. 59.

39. Cone, *Torchbearer of Freedom*, p. 59. Cone cites: 'Quoted in John Rae, *Life of Adam Smith* (London, 1895), p. 416'.

14

SCIENCE AND TECHNOLOGY: ARTISAN SCIENCE

Nothing tends so much to the advancement of
knowledge as the application of a new instrument.[1]

Humphry Davy, *Elements of Chemical Philosophy* (1812)

The dictionary definition of an artisan, or *crefftwr* in Cymraeg/
Welsh, is of someone trained to practise a manual art, a skilled
craftsman or workman – a *gweithiwr medrus*. Artisans are the unsung
heroes of science whose fabrications usually require, along with basic
labour, a considerable degree of scientific knowledge – an understanding of the correct temperature of a furnace, for example, or the use of
mathematics and the rules of physics and chemistry. The word artisan
is used here in a wide sense to incorporate not only those practising
hands-on skills, but more cerebral ones too. Included in this latter group
are astronomical observers, mathematical calculators, and translators
of scientific works. We begin, though, with instrument makers and
the presence of their instruments within places of scientific endeavour
associated with scientists of Wales.

Equipped for Science

In 1634, the biographer, antiquarian and future Fellow of the Royal
Society, John Aubrey, wrote how his Welsh paternal great-grandfather,
William Aubrey, dabbled in alchemical experimentation with his friend
John Dee:

In the house of Dr Dee [at Mortlake, just outside London], I have heard, they used to distil eggshells and other revolting ingredients: menstrual blood, human hair, clouts, chalk, shit and clay. The children were frightened because they thought Dr Dee was a conjuror of evil spirits.

I do not think I would have been frightened of him.[2]

Both John Dee and William Aubrey fall outside our period but in 1651, at the start of our long eighteenth century, and twenty years after John Aubrey wrote his description, the alchemist Thomas Vaughan also saw his calling as one in which some practitioners '*mistake* their own *Excrements* for that *Matter* out of which *Heaven* and *Earth* were made. Hence they *drudge*, and labour in *Urine*, and such filthie dirty stuffe which is not *fit* to be *nam'd*.[3] Apart from a strong stomach, the wringing of new alchemical knowledge from 'such filthie dirty stuffe', or any other stuff, required furnaces and stills, crucibles, files, weighing scales, pestles and mortars, alchemy spoons, knives, axes, and all manner of glass containers including simple vials, retorts, alembics for distilling and even glass stills[4] – just the sort of equipment the Glamorgan alchemist Bassett Jones illustrated and described in an appendix to his lengthy alchemical poem *Lithochymicus* of *c*.1650.[5] Among his illustrations is that of an alchemical furnace, reproduced here as Figure 26.

At the same time in the seventeenth century more complex and precise instrumentation emerged, including microscopes, telescopes, air pumps and thermometers. While Dutchman Antonie van Leeuwenhoek (1632–1723, FRS 1680), one of the founders of microscopy, used a simple microscope comprising a single lens he had ground himself, the first compound microscope (one with two lenses: an objective lens above the specimen examined and an eyepiece lens), had appeared in Holland *c*.1600. It was this form of microscope that Robert Hooke later used to reveal the hidden worlds of creation via the stunning illustrations in his *Micrographia* of 1665. At the end of the book, Hooke included observations of the stars and the moon made through a telescope; invented in Holland in 1608 and developed by Galileo in 1609, the first example of a 'Dutch Trunk' arrived in Wales

FIGURE 26 An alchemical furnace, as drawn by Bassett Jones in his seventeenth-century manuscript 'Lithochymicus or A Discourse of a Chymic Stone'. Source: BL MS Sloane 315, fol. 27r, © The British Library.

from England in 1610 (see Chapter 1 n. 15).[6] By 1668 Isaac Newton had invented the first reflecting telescope. Shortly before publishing *New Experiments*, in 1660, Robert Boyle commissioned the making of an air pump or pneumatic engine, an instrument invented by Otto von Guericke in 1650. In Boyle's later work, *New Experiments and Observations Touching Cold* (1665), he provides the first description of a graduated thermometer scale, or what he called a thermoscope.

Some of the above almost certainly formed part of the contents of those early laboratories associated with scientists of Wales for which we have but brief glimpses. For example, in May 1664, the Welshpool antiquary, lawyer and chymist, Meredith Lloyd (*fl.*1655–77),[7] who acted as something of a middleman/fixer for Sir Thomas Myddelton (1586–1666), began buying equipment in London for a laboratory to be established in a small house near Myddelton's home at Chirk Castle in Denbighshire, the castle itself having been rendered temporarily unliveable due to the depredations of the Civil War. As Lloyd wrote to Sir Thomas, from London:

> I have not called for any money to Mr Midleton[8], nor shall unless you order me to buy such materials & vessels as are requisite to try the conclusions you are upon, which will not amount to any considerable sum & you having coals and bricklayers and workmen for iron vessels. You will in the compass of three months in some small house of your own find more experience in metallurgy and physic than I will hear express; and for the furnaces they may be made by my direction by your own bricklayers & any one of your poor neighbours may attend the fires & labour. But if you desire a refiner for greater works I will bring one down.[9]

John Aubrey, who knew Meredith Lloyd, described him as a collector of alchemical and chymical books and 'an able chymist'. With the Chirk laboratory designed to enquire into 'metallurgy and physic', Lloyd clearly intended to play his part: 'I shall god willing come suddenly down provided I have a lodging in or near to the laboratory & a good room for my books which I presume are more in number & more choice than any study in Northwales contains.'[10] In June 1664, in another letter to Myddelton, Lloyd declared himself 'well satisfied wi[th the] place you intend for me & my books', and he was convinced:

> that all the materials crucibles, glasses & other instruments will be safe at Chirk Castle, at a reasonable rate & under £15 ... If God be pleased to lengthen your days and mine forty years longer we should not want any thing suitable to our purposes save glasses which are cheap. As for a refiner, the [As]say master of the Tower [John Woodward Jr., d.1665] upon my letters will come for a month or 6 weeks to stay with us in the summer or autumn.[11]

Outside Wales, Samuel Pepys made a visit in 1669 to the King's laboratory at Whitehall, where Sir Thomas Williams (c.1621–1712) worked on chemical experiments with Charles II and Sir Robert Moray. Williams, who came from a Breconshire family at Llangasty Tal-y-llyn, rose to prominence in 1665 as one of the signatories on a petition to the King advocating the establishment of a *Society of Chemical Physicians*. Although the King refused the petition, he continued to encourage the development of chemical medicine. On 'hearing of the extraordinary learning and skill' that Williams had shown 'in compounding and inventing medicines', some of which had been prepared in the King's presence, he appointed Williams his Chemical Physician in 1667. Williams took full advantage of his royal situation. By 1674, he was drawing down £1,000 a year for laboratory equipment and expenses at Whitehall.[12] Unfortunately, Pepys, on his visit to Whitehall, simply records entering 'the king's little laboratory under his closet, a pretty place' where he saw 'a great many chemical glasses and things, but understood none of them'.[13] Edward Chamberlayne proved equally succinct when, in 1684, he visited the basement laboratory of the recently opened (1683) Ashmolean Museum in Oxford, where Edward Lhwyd would spend much time after being made 'Register of the Chymical Courses of ye laboratories at Oxford'.[14] Chamberlayne simply described the laboratory as 'perchance one of the most beautiful and useful in the world; furnished with all sorts of furnaces, and all the necessary materials, in order to use and practise'.[15]

By the mid-eighteenth century we have a better idea of the type and complexity of instruments then owned by scientists of Wales thanks, in large part, to catalogues detailing the contents of estate sales. In 1765, Samuel Baker of Covent Garden sold the instruments and library of Joseph Harris, King's Assay Master at the Royal Mint in the Tower of London (examples of the equipment and instruments auctioned are given in Table 4). Among the buyers at the sale were the electrical experimenter John Canton, Joseph Banks of the Royal Society, and the instrument maker Jesse Ramsden.[16]

In the late eighteenth century the various scientific instruments made for the Carmarthen Academy, under the auspices of Richard Price, included a set of magnets, a planetarium, an orrery and various electrical machines.[17] Latterly, Price himself possessed a number of philosophical instruments, including a telescope presented to him by the Equitable Life for his otherwise unpaid services to that institution. At his London

TABLE 4 Examples of the equipment and instruments
auctioned from the estate of Joseph Harris in 1765

---

compasses

brass rules

levels and plummers

a copper hydrometer

magic lanterns with show boxes

multiple telescopes and microscopes including an 18″ reflector by Short. This
telescope has different metals for shifting, to make it either a Newtonian, a
Gregorian, or a Caffegraine; also a 2ft speculum, and a small ditto by Bird

theodolites

a brass apparatus, of a new construction, in a mahogany case, for drawing the
Elliptic and other Geometrical curves

horizontals and dipping needles

piles of Troy weights

an air-pump and condensing engine, both in one, of a new construction,
with mercurial gauges, and a variety of receivers and other appurtenances for
experiments; with a large copper vessel; either for weighing air, or to serve
alone as a fountain, having various jets. The whole in a deal cupboard: being a
most compleat apparatus of the kind, and quite new, by Nairne £35.

---

home in Newington Green (see Figure 27) there was enough equipment
to produce a regulus of various minerals and ores for his scientifically
minded friends and acquaintances; a regulus being: 'the purer mass of a
metal that sinks to the bottom when an ore is being smelted'.

As Price wrote in a 1783 letter to the ex-clerk of the Royal Society
and author of *The Natural History of Fossils* (1757), Emanuel Mendes
da Costa (1717–91):

> Dr Price presents his compliments with the enclosed specimens to
> Mr DaCosta. He is sorry that his stock at present constrains him
> to make them so small especially that of the Reg:[ulus] of Nickel
> which however is half of what he had by him which he thought
> sufficiently pure. He has not yet made any more Reg:[ulus] of
> manganese but when he does, he will request Mr DC's accept-
> ance of some.[18]

FIGURE 27 Richard Price's home at Newington Green, north London, where he was visited by the likes of Joseph Priestley and Benjamin Franklin. The house has been identified as the one with the squared off top. In 2023 (the tercentenary of Price's birth) it was marked with an English Heritage Blue Plaque.

We previously noted how a wide range of laboratory equipment, as well as multiple telescopes, were sold from John Lloyd of Wigfair's estate in 1815 (see Chapter 8), and how George Cadogan Morgan, Richard Price's nephew, kept his philosophical apparatus, electrical machines, globes and telescope in his study. In a similar vein the landowner Sir John Nicoll, of the Merthyr Mawr estate in the Vale of Glamorgan, appears to have used his 'microscopes, a camera obscura, a pair of globes, a pantograph and much else' while surrounded by 3,600 books in his favourite room – his library.[19] Beyond these essentially private spaces, there also developed, in the course of the eighteenth century, laboratories attached to commercial and industrial concerns, such as the one at the copper works established at Llangyfelach in Swansea, in 1717, although little seems to be known of its actual contents.[20]

So, what part did artisan scientists of Wales play in the field of scientific instruments in the long eighteenth century? The subject is an under-researched one, and the answer that follows must be regarded as no more than a basic introduction to this subject.

Tool and Instrument Makers

One of the earliest mentions of a mechanical clock in European litera-ture is made by Dafydd ap Gwilym in his appropriately titled poem *Y Cloc*. Written in Welsh in strict metre (between 1340 and 1370) it tells how the clock's ropes, wheels and 'stupid balls which are its weights' rouse the poet from his sojourn in a 'girl's encircling arms, / folded between the breasts of Deifr's form' (Deifr or Dyfr being one of the golden-haired maidens at King Arthur's Court).[21] Four hundred years later the ticktock of the clock truly began its autocratic rule. As Pryce and Davies (1985) show in their meticulous and comprehensive study of clockmaker Samuel Roberts (*c.*1720–1800), the number of makers in Wales expanded from just nine or ten before 1700, to seventy-two by 1750, and 238 by 1800. Although these numbers are small in compari-son to Scotland (sixty-four by 1700) and England (571 in Cumberland alone by 1800), they still reflect the remarkable growth of early con-sumer related industries in eighteenth-century Wales.[22]

Given the commercial nature of clockmaking, it is easy to dismiss, as having little to do with science, the skills of artisan makers like Samuel Roberts in his small town of Llanfair Caereinion, Montgomeryshire. Yet, as Pryce and Davies indicate, while Roberts's use of iron over brass in his early clocks suggests his initial training could have been as a blacksmith, he may also have undergone the usual seven-year appren-ticeship of a clockmaker, followed by a period as a journeyman. In addition, the creation of his more complex clocks, including repeat-ers, chimers, pendulum clocks, nine-day clocks and thirty-day clocks, involved him in a practical knowledge of arithmetic and elementary mathematics, and a degree of experimentation, which Roberts diligently recorded. Astronomical and other knowledge might also be needed when constructing so-called astronomical and tidal clocks. One of the former, built by Philip Lloyd of Llawhaden in Pembrokeshire, bears the inscription:

AN ASTRONOMICAL CLOCK
Showing the apparent Daily motions of the Sun, Moon and Stars
with the times of their Rising, Southing and Setting; the places of
the Sun and Moon in the Ecliptic and the Age and Phase of the
Moon for every Day of the Year.[23]

The astronomical and tidal clocks made by Maurice Thomas of
Caernarfon showed 'the apparent diurnal motions of the sun and moon,
the age and phases of the moon, with the time of her coming to the
meridian, and the times of high and low water', the tidal indicator being
adjustable for any place on the coast.[24] Tidal clocks were particularly
important in the main ports, with clocks being made, for example, to
show tide times at places such as Aberthaw (made by Henry Williams
of Llancarfan in c.1780), Chepstow (by William Charles of Chepstow
in 1755) and Bristol (made by Henry Williams of Llancarfan in 1780).[25]
For the making of sundials, and the accurate placing of their central
indicator gnomon, an understanding of more than basic geometry was
also needed. Beyond their commercial importance, the artisan skills of
Samuel Roberts and local makers like him would also have been relied
on by those scientists of Wales, such as John Lloyd of Wigfair, who not
only collected timekeepers requiring occasional repair, but who wanted
accurate clocks for their astronomical and other scientific observations.

Clocks figure prominently in the lives and archives of a number
of Welsh astronomers. An accurate timekeeper was among the instruments Joseph Harris sent from London to his family home at Trefeca in
Breconshire, prior to his arrival there to observe the Transit of Venus in
1761, even though the one he sent gave him some problems by gaining
fifty-three seconds in a twenty-four-hour period. Harris might have
played a role too in establishing a four-faced clock at Trefeca, one face
of which survives there to this day.[26] In 1819, in a newsletter discussing a recent comet and a telescope made by Herschel, the astronomical
observer Abraham Robertson (1751–1826, FRS 1795) wrote from the
Radcliffe Observatory in Oxford (in operation from 1773 to 1934),
commending the Caerleon-born mathematician and astronomer Lewis
Evans: 'The care you have taken to amend and correct the [observatory]

clock deserves and has my most sincere thanks.'[27] Clocks were also the subject of letters between Lewis Evans and Nevil Maskelyne, the Astronomer Royal, with one letter including a discussion of the work of English watchmaker Thomas Earnshaw (1749–1829), the inventor of a new form of spring escapement.[28] Earnshaw had been associated (and possibly apprenticed) with William Hughes of 119 High Holborn in London. Hughes, who arrived in London before 1755, appears to have come from Llanfflewin in Anglesey. During his career he sold a cabin clock to Captain Cook (possibly the one today owned by the National Museum of New Zealand),[29] a watch to Captain Bligh, and a musical clock destined for the Emperor of China (see Volume 2). He also acted as vice-president of the then eight-year-old Honourable Society of Cymmrodorion in 1759.[30]

Pryce and Davies also indicate in their study of Samuel Roberts how it is 'generally understood that until the 1780s virtually every clock maker designed and made clocks by assembling all the parts which he had made himself'. It is a comment that can be made about other scientists of Wales who acted as their own artisan-scientists by making the equipment or instruments they needed. Just as Samuel Roberts may have been apprenticed to a blacksmith early in his career, so Joseph Harris of the Royal Mint had been apprenticed to his blacksmithing uncle, Thomas Powell, in Breconshire. The skills Harris acquired, together with his youthful love of making small instruments and tinkering with mechanical things, almost certainly proved valuable during his time at the Mint. In 1745, he oversaw the melting-down in the Mint's furnaces of £800,000 of silver bullion taken from two French ships by British privateers. In 1741, he informed the vice-president, and soon-to-be president of the Royal Society, Martin Folkes, that he had assayed 'with great care a small ingot of gold out of which Mr Tanner proposes to make the medals for your society'.[31] The medal in question was the gold Copley medal, the oldest award given by the Royal Society. The model, die and proofs of the medal were made by John Sigismund Tanner, chief engraver at the mint from 1741.[32]

The practical skills of Lewis Morris in making cabinets of curiosity in the 1750s have been previously noted. Slightly earlier, on 4 July 1737, we find him in Beaumaris, on Anglesey, preparing to undertake a survey of the island. Having spent the next day in Holyhead gathering his various instruments, he spent 6 July making the wooden part of a 'waywiser'. Resembling a unicycle in shape, but formed of a sizeable wooden wheel and frame with an attached meter, it was an essential surveyor's instrument for measuring distances walked, and thereby measured.[33]

Standard measures of length, volume and weight are, of course, crucial for accurate experimentation, the production of scientific instruments, and for the development of science as a whole. However, attempts to achieve such standards by the eighteenth century had met with only variable degrees of success. Prior to the Edwardian Conquest of 1282, Wales had its own system of weights and measures – the Venedotian scheme in the north and the Dimetian in the south. These had been codified by, among others, Hywel Dda (Hywel the Good, d.950).[34] Post the conquest of 1282, English measures were more widely adopted but, as Lewis Morris reveals in his eighteenth-century manuscript compilations concerning measures of length, volume and weight, complete standardisation still had some way to go. Weights in Anglesey, he records (possibly in the 1740s), are 'the same as all over Britain', but in measures of length 'the Anglesey yard for measuring cloth &c., is 3 foot 3 inches and hath been so time immemorial'.[35] Furthermore, even though the length of a mile had been standardised at 1,760 yards or 5,280 feet in 1538, during the reign of Elizabeth I, eighteenth-century travellers to Wales commented frequently on the fact that the standard Welsh mile remained three times as long as the English one. But, according to Lewis Morris: 'Thou' the old Welsh mile was but 5000 foot, yet each foot being 13 Inches Engl. Makes ye old Welsh mile still 135 foot 4 inches longer than Queen Elizabeth's mile.'[36] In 1758, the Westminster Government established the Carysfort Commission on Weights and Measures. The commission soon revealed that significant discrepancies existed between the various standards already in existence. New, accurate standards were thus demanded. In his capacity as Assay Master at the Royal Mint, in the Tower of London, Joseph Harris produced a series of weights and volume measures for the Commission, some of which are preserved today in the Science Museum in London.[37]

Also in 1758, Harris made a standard one-pound weight. This later formed part of the 'Imperial Standards' established by law in 1824 and 1825, with the standard yard length being made by the London instrument maker John Bird in 1760.[38] As Pasley (1834) records, the 1824/5 legislation decreed that the 'Standard Yard of 1760', by Bird, should be the only legal standard of lineal measure, when used at 62 degrees Fahrenheit, and 'That the brass weight made by Harris [in 1758] ... of one pound of Troy weight, should be the only standard of weight from which all others should be derived'.[39] Both the one-pound Imperial Troy weight and the Imperial standard yard were subsequently committed to the care of the Clerk of the House of Commons. Unfortunately, they suffered significant damage during the fire at the Palace of Westminster in 1834, with the one-pound Troy weight completely destroyed. The yards of 1760 survived, as do two other 1758 weights – the 16 lb and 32 lb – both of which would, in all probability, have been made under the supervision of Harris. Made from 'gun-metal with the surfaces a little oxidised by the flames; they are not signed by the maker' (see Figure 28). Today they are in the care of the Clerk's Department in the

FIGURE 28 One of the standard Troy weights made under the auspices of Joseph Harris at the Royal Mint in the Tower of London. Photograph courtesy of Dr Mark Collins, Estates Historian, UK Parliament.

Palace of Westminster.[40] Following his death in 1765, Joseph Harris's artisan skills were also made apparent in the sale catalogue of his estate. Not only did the auctioneer, Samuel Baker, advertise the 'Large collection of Optical, Astronomical and other Mathematical Instruments' (made by John Bird) that were being offered for sale, but that the principle ones were 'either of a New Construction, or greatly improved by the late Mr Harris'.

Seeing More

Precisely what improvements Harris made to his optical and astronomical instruments is not known, but he certainly appears to have possessed enough artisan skills to make 'a new form of azimuth compass able to read magnetic variation to one degree' and to design 'a more accurate fore-staff for navigational purposes'.[41] His desire to impart practical skills as well as theoretical knowledge is also apparent from the history of his *Treatise of Optics*. In 1742, Harris initially began collecting material for a proposed *Treatise on Microscopes* and, to that end, published that year his three page 'Proposals for printing by Subscription a Treatise upon Microscopes, containing a compleat account of all sorts of microscopes, both as to their theory and mechanism. With several new improvements'. Written in French, and published in the *Bibliothèque britannique*, he intended the *Treatise* to consider not only the limits of vision and provide descriptions of micrometry, the camera obscura and microscopical discoveries, but also to detail the building of simple, double, reflection and refracting microscopes.[42] As a consequence of the volume of material collected, he then hoped to expand the proposed work on microscopes into a broader *Treatise on Optics*. Sadly, his efforts towards this, as well as his optical and light experiments undertaken at the Tower, were interrupted by pressures of work at the Mint and by ill health. Consequently, the *Treatise on Optics*, edited and published in 1775, ten years after his death, essentially comprises the two books 'containing the elementary parts of this science' that he had already substantially prepared. 'Upon this foundation', according to its editors in 1775, he had 'proposed to explain the theory and mechanism of optical instruments, in a third book, under the general head of Telescopes and Microscopes'; this was never completed.[43]

Even though the proposed work on microscopes did not come to fruition, Harris's initial proposal on the subject was timely for two reasons. First, little had been published on the subject since Leeuwenhoek's seventeenth-century works were republished in the early eighteenth century,[44] and much of the material published since had been more concerned with microscopic observation than the practicalities of microscope construction. Second, there was a commercial opportunity since, as Ratcliff (2016) observes: 'After insects began to provide a shared microscopical object in the 1730s, the demand for microscopes began to rise, leading to a sharp increase in their manufacture in the early 1740s.'[45] With the absence of Harris's proposed 1742 work on microscopes, the principal book to popularise the subject in Britain after Leeuwenhoek's time was *Microscope made Easy* (1742) by naturalist Henry Baker (1698–1774, FRS 1740, Copley Medal 1744). One artisan scientist of Wales clearly aware of Baker's work was Lewis Morris. Writing from London to his brother William, in Anglesey, on 10 September 1755, Lewis notes:

> I promised you in my last a specimen of a letter to ye President of ye Cymmrodorion Society on ye subjects they propose to examine, and I cannot do it better than in ye inclosed copy or an extract of a letter from my acquaintance, Mr Henry Baker, to ye Royal Society, about his microscopical observations in which he hath exceeded all that Society of which he is a member.[46]

The President of the Cymmrodorion, which had been established in London in 1751, was its principal founder Richard Morris, the London-based brother of Lewis Morris. Baker made many contributions to the Royal Society's *Philosophical Transactions* and the work to which Lewis refers is uncertain. He may actually be referencing Baker's *Employment for the Microscope* (1753), although this was not directly addressed to the Royal Society. In another letter, written ten days before the one quoted above, Lewis Morris had effused to his brother William on the general subject of microscopes. It is a wonderful example of the joy of science and the opening up to one individual of a new world:

> The chiefest pleasure I have had was an opportunity of making microscopical observations, which I might never have looked into, if I had not this leisure time. And I have also made some improvements (which hath not been seen before) in the structure of my microscope, having an opportunity here [London] to get a thing done according to direction, and I insist upon it that my microscope exceeds everything of the kind ever yet publishd. It is plainer, more useful, more natural, and not a quarter of ye price of some of them. If I was once at home and in quiet I could make you a microscope that would serve as well as ye grandest made in London. This is the most amusing study in the world, and it is impossible to make any progress in natural philosophy without microscopes; it is amazing, it is beyond conception, and beyond description – a new world! The microscope for opaque objects being imperfect, checked [hindered] my observations in ye country [presumably at home in Wales], so that I was quite tired ... till I had an opportunity of setting about it here [London and presumably his various modifications], which hath given me vast satisfaction, but I have not the objects here, as I can come at in ye country. Would not you wonder to see millions of animals larger than elephants? And is not ye wonder greater to see millions of animals of which a hundred of 'em would not make the thickness of ye hair of one's head, and all in as great or greater perfection of beauty, etc., than an elephant? Rhyfeddodau annhraethadwy! [Unspeakable wonders!][47]

Unfortunately, Lewis did not detail the improvements he had made to his own microscope, but he did, two years later, in September 1757, inform his brother William:

> I have been very busy lately in finishing a microscope for you of my own making and contrivance, chiefly adapted for sea plants and

the use of your cabinet, but will serve for anything else. It hath 8 lenses and is a curious thing in its kind. The worst of it is I have very few tools except mynawyd ['bradawl'], twcca ['tuck-knife'], ffeil ['file'], morthwyl ['hammer'], a[nd] phlyers ['plyers'], not so much as a handvice or an oilstone, and yet you'll say when you see this that Robinson Crusoe could not have done the like.[48]

Lewis sent the finished article to William as 'a microscope of my own making from top to toe', an instrument that their London-based brother, Richard, had tried out and who was 'so delighted with it that I must make him another'.[49] Perhaps taking a further cue from Henry Baker, who had won the Copley medal of the Royal Society in 1744 for his microscopical observations on the crystallising of saline particles, Lewis Morris later (1760) employed his own microscope in an investigation of the various spring waters at Llandrindod Wells. The salts of the so-called 'pump water', he observes, have sharp spiculae. By contrast, the spiculae of the 'sulphureous water', when seen by microscope 'in the configuration made over a gentle heat', were blunter than those of the pump water: 'its configurations as well as crystals resembling those of Sal Ammoniac (see Baker's *Employments for the Microscope* [1753])'. Finally, salts observed in the 'Rock water' were 'no more sharp than the sulphureous well, but the configurations most beautifully variegated with figures with blunt points'.[50]

Around the time Lewis Morris made his microscopical examinations at Llandrindod, another scientist of Wales composed a text on microscopes. Francis Watkins (*bap.* 1723–91) was born in Newchurch, Radnorshire and apprenticed to Nathaniel Adams, a spectacle maker of London, in 1737. Watkins eventually set up in business at the sign of *Sir Isaac Newton's Head* in Charing Cross, where he made spectacles that contained lenses of Brazil Pebble, Crystal Glass or fine Venetian Green. He later held a Royal Warrant and made all manner of other optical instruments, including reflecting and refracting telescopes of all lengths and sizes, single, double and solar microscopes, barometers, thermometers, hydrometers, hydraulic balances, air pumps and 'all sorts of Mathematical Instruments finished with the greatest accuracy'.

Examples of his work are to be found today in museum collections in the United Kingdom, Portugal, Germany and the United States (see Figure 29).[51]

Acknowledged as being inspired by Baker's earlier *Microscope Made Easy* (1743), Watkins published in 1754 his *L'Exercise du Microscope*, a ninety-seven-page effort, written in French, and with an eye to a possibly lucrative market among the French settlers in London, one of whom may have been his Huguenot wife Clarinda; she may also have helped in the pamphlet's production.[52] 'I could not do better' he wrote, 'than to give the explanation in the most universally understood language: having been much encouraged to do so, by the realisation that

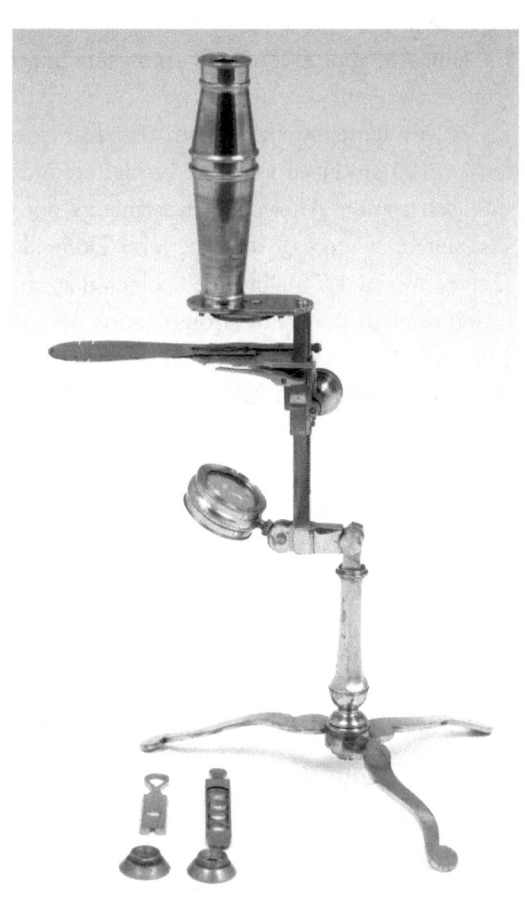

FIGURE 29
Compound
Microscope by
Francis Watkins,
London, *c.*1760.
Courtesy of the
Museum of the
History of Science,
Oxford University
(Inventory
number 47695).

there is nothing substantive in the French language on this subject'. According to Gee et al. (2014): 'The text was an invaluable contribution to the diffusion of the subject because it set out a number of obser-vational exercises in a graded fashion. It was also a practical guide to handling and adjustment of the instrument.'[53]

In 1758, Watkins became involved/embroiled in one of the major eighteenth-century inventions relating to the telescope – the devel-opment of the achromatic lens. In the eighteenth century there were numerous problems with lenses, and the viewing of objects through them. As Lewis Morris informed his brother William: 'The costly appa-ratus of the double microscopes makes y_m intolerable, besides some inconveniences attending y^e instrument of inverting objects and mak-ing concaves appear convex.'[54] Alongside image distortions produced from uneven lens curvature – spherical aberration – one of the biggest issues with lenses was 'the prismatic effect close to the perimeter of a lens', which produced 'a coloured fringe around a bright object' – chro-matic aberration. After many experiments, one of the leading London instrument makers of the day, John Dolland (1707–61, FRS 1761, Copley Medal 1758), 'invented' a lens that removed chromatic aber-ration through the use of a combination of crown glass and flint glass. However, Dolland lacked the finance necessary to obtain a patent, and to make the lens a commercial possibility. In April 1758, Dolland approached Francis Watkins with the aim of having him obtain a pat-ent (a fairly complex process at a cost to Watkins of nearly £50) and provide capital (initially £200) for commercial development of the lens and associated instruments. In return Watkins obtained a partnership and began selling telescopes containing the new lens. The ensuing events are long and complex. The central question was whether John Dolland had actually invented the lens, or whether priority for it lay with earlier researchers into the problem of chromatic aberration. Also important was whether Dolland had knowingly misled Watkins over his 'invention' of the lens during the patent application. These questions resulted in a number of complex court cases between Watkins and John Dolland's son Peter, John Dolland having died in 1761. They resulted in the dis-solution of the partnership. The issues surrounding the case, together with a fascinating discussion of the life and work of Watkins and his later partnerships, are comprehensively discussed for the first time in Gee et al. (2014).[55]

Far less controversial were efforts by other artisan scientists of Wales to produce their own instruments, using lenses of polished glass, or specula made from polished metal. An example of the latter can be seen today in the telescope at Trefeca which Joseph Harris used for his observation of the Transit of Venus in 1761.[56] Among artisan makers from Wales were John Lloyd of Wigfair, and the mathematician and astronomer Lewis Evans. Writing to his mother from Bewdley, in Worcestershire, on 18 September 1769, John Lloyd first recorded his observations of the Great Comet, which had made its closest approach to Earth on 10 September that year, before going on to discuss his work on optics:

> The weather has been remarkable cloudy and unfavourable for making observations on this comet, notwithstanding which I have made four curious observations on it together with a friend of mine (for which he is) who is usually curious and hath all kinds of instruments, he says the tail is larger by much than that of 1744 but not so bright yet. It is by our measuring 37 million miles in length. It will make a glorious appearance in its return from the sun in a few days. My leisure hours have been imployed in the study of optics and my curious friend has supplied me with proper books and instructions, and I am as intimate with him as you can conceive. I have now made some proficiency in that curious science. I can now make a telescope that will magnify 150 times from end to end except the glasses and what in London would cost 20 or 30 guineas I can make for so many shillings.[57]

Lloyd's optically curious friend could be either William Herschel, with whom he corresponded and may well have visited in Windsor, or Sir George Shuckburgh-Evelyn with whom he had a close scientific friendship. The latter is certainly the more likely since Bewdley is en route from Shuckburgh Hall in Warwickshire back to Lloyd's home in north Wales. Lewis Evans, meanwhile, describes in a letter of 1791 how he has 'been lately engaged in casting and publishing specula for a Gregorian telescope, in which I have succeeded beyond my expectations';[58] the purpose of all such effort being, of course, to see more and to be able to make better observations and calculations.

Observers and Calculators

On opening some of the texts compiled by long-eighteenth-century scientists of Wales, less mathematically inclined readers might feel certain bewilderment, even awe at the degree of mathematical calculation and complexity they contain (as Figure 30, taken from a work on annuities by William Morgan helps illustrate).

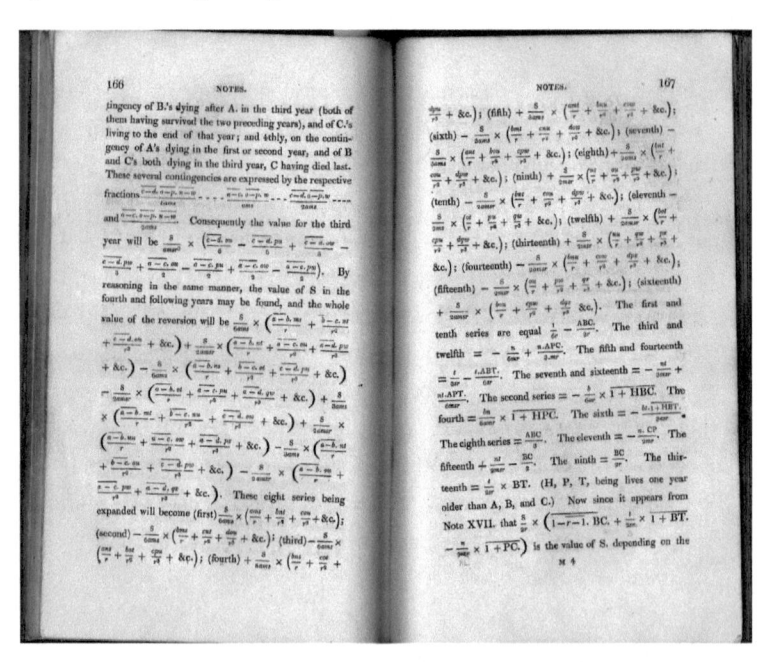

FIGURE 30 Part of the complex calculations given in William Morgan's *The Principles and Doctrine of Assurances, Annuities on Lives, and Contingent Reversions* that were required to answer Problem XXVII: 'To Determine the value of a given sum payable on the contingency of C. surviving B., provided the life of A. shall be then extinct.'

The enormous amount of computational effort undertaken by William Morgan, as well as his uncle Richard Price, in relation to their foundational work on annuities, pensions, life assurance and population studies has been noted in Chapter 12. In our age of push-button computation, it is hard to conceive of the physical and mental stamina needed to undertake the thousands of complicated calculations and observations such productions required, all done with little more than brainpower, paper, quill and much use of candlelight. There could, of course, be consequences of such effort. William Morgan noted how the hair of his uncle, Richard Price, 'which was naturally black, became changed in different parts of his head into spots of perfect white' as he grappled with a mathematical issue.[59] According to the compilers and editors of his posthumously published *Treatise of Optics*, which contains mathematical material related to focal lengths and refraction angles etc., Joseph Harris, having been interrupted in his initial work on it, by 'an extraordinary demand of duty in His Majesty's Mint', other public works, and a degree of ill health, later resumed 'his application ... with an earnestness that probably hastened his dissolution'.

The range of computational topics that an interest in science might provoke can be gauged from the content headings of a notebook compiled by William Jones Sr and today held in the collections of the Royal Society (see Table 5).

Much of this sort of computational effort remained in the manuscript collections of the scientists concerned; sometimes destined to be revealed, sometimes not. Despite the importance later claimed for them in the long eighteenth century, parts of William Jones's oeuvre would not see the light of day until sometime after his death in 1749. In the early 1770s, for example, the then librarian of the Royal Society, John Robertson, communicated to the Society papers written by Jones on the subject of logarithms and the properties of conic sections. The 'familiar manner' in which the nature of the former were explained 'and the great art with which he obtains the modes of computation, not being exceeded, if equalled, by any writer on this subject'. The work on conic sections, and the properties of their curves, proved interesting because of their similarity 'to those which are described by the

TABLE 5 Contents List from a Notebook of William Jones Sr

Contents List from a Notebook of William Jones Sr – RSC MS 42
Composition and Resolution of Forces
The Algorithm of Fuxions
Method of Drawing Tangents
The Doctrine De Maximus et Minimus
Of Infinite Series
The Inverse Method of Fluxions
The Quadrature of Curves
The Composition of Brigg's Logarithms
The Mensuration of Solids
Of the Surfaces of Curvilinear Solids
To Find the Circumference of a Circle
The Method of Reversing Series
The Centre of Gravity of Bodies
Problems Concerning Gravity
Of the Motion of Pendulums
Of the Density of Air
Of the Resistance of Bodies in Fluids
Of the Attraction of a Prolate Sphaeroid

motions of the celestial bodies in the Solar System'. In his paper, Jones had described these curves in a manner 'very different from what he gave in his *Synopsis Palmariorum Matheseos*' of 1706. Jones 'having laid down a very simple method of describing these curves; seems to have been desirous of arriving at their properties in as expeditious a way as he could contrive' by using the algebraic method to reduce his equations and, on occasion, using fluxions 'to deduce some properties relating to tangents'. Furthermore, 'by a judicious use of these, he has very much abridged the steps which otherwise he must have taken, to have deduced the great variety of relations he has obtained'. Both papers appeared in the *Philosophical Transactions*.[60] The remainder of William Jones's surviving mathematical calculations remained locked away in the library of the earls of Macclesfield at Shirburn Castle, Oxfordshire, to the second earl of which, and his former pupil, Jones had bequeathed them on his death. They reappeared in 2004 when they were sold, along with Jones's papers relating to Isaac Newton, to Cambridge University Library, where they await an enquiring soul.[61]

More fortunate scientists were able to employ others to undertake a part, or even the entirety of any computational grind. In 1796, the Astronomer Royal, Nevil Maskelyne, wrote to Lewis Evans saying he found so appealing the description Evans had provided of his son, Thomas Simpson Evans (1777–1818, FLS 1805), that Maskelyne proposed to 'take him upon trial as soon as he can handsomely and with propriety quit his present situation'.[62] Thomas Evans shared his father's mathematical interests, but life as an astronomical observer and calculator under Maskelyne, at the Greenwich Observatory, proved to be no easy position. As an assistant he was expected, among several other duties, to be on stand-by to observe the heavens every day of the week between 7 a.m. and 10 p.m., and to undertake computations up to 7 p.m. or 8 p.m., when not otherwise observing. Computations involved correcting the transit time observations of lunar and stellar bodies for 'instrumental error, aberration, precession, refraction parallax and nutation' before these were 'converted into celestial latitudes and longitudes'. In addition, other calculations could be derived from the observational data collected. These included 'mean right ascensions of the Sun, Moon, stars and planets … calculating heliocentric and geocentric longitudes and latitudes' and noting instrumental errors. Although he endured it for two years, four months and seventeen days, one of the longest terms of the twenty-four assistants that Maskelyne had between 1765 and 1811, Thomas Evans was not enamoured of this regime.[63] As he wrote:

> Nothing can exceed the tediousness and ennui of the life the assistant leads in this place, excluded from all society, except, perhaps, that of a poor mouse which may occasionally sally forth from a hole in the wall, to seek crumbs of bread dropt by his lonely companion at his last meal. This, of course, must tend very much to impede his acquiring astronomical information, and damp his ardour for those researches which conversation with scientific men never fails to inspire. Here forlorn, he spends his days, weeks, months, in the same long wearisome computations, without a friend to shorten the tedious hours, or a soul with whom he can converse. He is also frequently up three or four times in the night

(an hour or two each time), and always one week in the month when the moon souths in the night time, with the owls perched on the fir-trees in the park below, screaming by way of answer to him when he opens the sliding shutters, in the roof of the building, to make his observations.[64]

Among the rules imposed by Maskelyne on his assistants was his refusal to allow them to submit answers to problems set in the *Ladies' Diary*; or, to give its full title, *The Ladies' Diary or Woman's Almanack, Containing New Improvements in Arts and Sciences, and many entertaining particulars: designed for the use and diversion of the fair sex*. Founded in 1703, it included mathematical problems to which its reading public could submit answers. One of those who provided mathematical questions for the journal was Thomas Simpson Evans's father, Lewis Evans. His manuscript archive in the Oxford Museum of Science contains a number of letters addressed to the mathematician, surveyor and calculator of the density of the earth, Charles Hutton (1737–1823, FRS 1774), editor of the journal between 1773 and 1818. While Thomas Simpson Evans may have been excluded from contributing answers to the *Ladies' Diary* he did, as a mathematician and teacher in later life, publish a number of mathematical papers, and detail his experiments on prismatic rays and terrestrial magnetism in different latitudes, in the *Philosophical Magazine*. He also translated a number of scientific works, including *Trigonometria piana e spherica* ('Plane and Spherical Trigonometry') by the Italian astronomer and mathematician Antonio Cagnoli (1743–1816).

The Translators

Roy Porter (2008) described translation 'as the foreign service of popularization'. As we have seen in previous chapters, it is a service to which a number of long-eighteenth-century scientists of Wales contributed. Among them, John Davies of Kidwelly, who translated some twenty-five works from French to English, including *Reflections upon Monsieur Des Cartes Discourse* and *Another Collection of Philosophical Conferences of the French Virtuosi*; John Evans translating Wesley's hugely popular

Primitive Physick into Welsh; Nathaniel Williams with his bilingual Welsh/English *Pharmacopoeia*; and David Daniel Davis translating work on mental health by the French writer Philippe Pinel. At the same time, works by Welsh writers such as Josiah Tucker and Richard Price were translated into French, Dutch and German. To these can be added two more names – the mathematician Thomas Salusbury (*c*.1620–*c*.65) and the chemist Thomas Henry (1734–1816, FRS 1775, APS 1786).

The inclusion of Thomas Salusbury as a scientist of Wales is based on currently available biographical literature, where the consensus seems to be that he was legitimately or illegitimately part of, or in some way connected to, the Salusbury families of either Bachygraig, Rug or Llewenni in Denbighshire, with the Llewenni branch being the more likely. Salusbury translated a number of works from Italian and Latin into English, but his importance rests principally on the publication of the two volumes of his *Mathematical Collection and Translations* in 1661 and 1665. Their publication history is extremely complex and beyond our scope here, but their contents are of particular importance. Among the contents of Volume 1 are the first published vernacular translations of Galileo's *Dialogue Concerning the Two Chief Systems of the World – Ptolemaic and Copernican* (1632), which was banned by the Inquisition, the *Letter to the Grand Duchess Christina* (1615) and Castelli's *Of the Mensuration of Running Waters*, together with a short work by Kepler. Among the contents of Volume 2 are translations of Galileo's *Discourse* and a *Discourses of the Mechanicks* by Descartes, described by Stillman Drake as a true scientific first edition and 'the first edition of Descarte's *Mechanicks* to appear in any language'. Volume 2 also contained another crucial text *Galilaeus Galilaeus His Life: In Five Books* – the first major biography of Galileo in any language. All copies bar one of this last treasure were subsequently destroyed in the Great Fire of London in 1666, perhaps through being stored in St Paul's Cathedral, where many of the printers stored their products, wrongly thinking them to be safe there from the flames. The one copy known to have survived did so in the extensive library of Welsh mathematician, William Jones Sr. Following his death in 1749 it entered the library of the second earl of Macclesfield, where it remained hidden until sold to a private buyer during the auction of the Macclesfield library between 2004 and 2007. It is hoped that further research into the life of Salusbury, who included in his publication some of his own scientific work, will eventually determine his true origins, whether they be confirmed as Welsh, or otherwise.[65]

Another translator with more certain origins is the chemist Thomas Henry. Born in Wrexham, where he received his early training as a surgeon-apothecary, he later moved to Manchester. There he became a founder member of the Manchester Literary and Philosophical Society (President from 1807 to 1816) and, in 1775, a Fellow of the Royal Society with Joseph Priestley as one of his sponsors. In 1776 he published *Essays Physical and Chemical by M. Lavoisier*, an English translation of the original work by the famous French chemist. In doing so Henry, even with his 'very short acquaintance with the French language', was unafraid, with the aid of 'Mr. Aikin of Warrington', to correct mistakes in Lavoisier's 'account of the discoveries of our great English Philosopher [Priestley]', which 'from an insufficient acquaintance with our language, sometimes misrepresented that author's meaning'. Where this was the case Henry either altered the text by restoring Priestley's own words or, where that could not be conveniently done, 'pointed out and corrected the mistakes by notes'. Added to the text too were a recent memoir by Lavoisier 'on the nature of the principle which combines with metals during their calcination and increases their weight'; and the publication of Priestley's second volume on the subject of air. The latter enabled Henry, with Priestley's approval, 'not only to give his sentiments on the nature of that principle, but also his ideas of the constitution of common air'. As we shall see in the next chapter, Henry would establish himself at Manchester as an educationalist and a chemical innovator.

Notes

1. Quoted in Brian Gee with Anita McConnell and A. D. Morrison-Low (eds), *Francis Watkins and the Dollond Telescope Patent Controversy* (Farnham, 2014), p. xiii; citing Humphry Davy, *Elements of Chemical Philosophy* (1812).
2. Quoted in Ruth Scurr, *John Aubrey: My Own Life* (2015; London, 2016), p. 21.
3. *WTV(LL)*, p. 346.

4. For a fascinating illustrated account of alchemy and glass, see Valeria Montana, 'Glass and Alchemy', *https://valeriamontana.com/glass-and-alchemy/*.

5. See Robert M. Schuler, *Alchemical Poetry 1575–1700* (1995; Abingdon, 2014), pp. 349–56.

6. Bryn Jones, 'Despite the Clouds: A History of Wales and Astronomy', *Antiquarian Astronomer*, 8 (2014), 66–96. See also, Allan Chapman, *Stargazers: Copernicus, Galileo, the Telescope and the Church* (London, 2014), p. 270.

7. Lloyd was also an important antiquary, particularly with regard to Welsh manuscripts and monuments. He is noted a number of times by John Aubrey, whose neighbour he was in Fleet Street for a time. See Scurr (ed.), *John Aubrey*, pp. 104, 253, 386, 392.

8. 'Mr Midleton' is possibly John Middleton, who may have been related to Sir Thomas, and was a 'Grocer at the Starre on Fishstreet Hill', which was then the main route from the City of London to London Bridge.

9. Nesta Lloyd, 'Meredith Lloyd', *Journal of the Welsh Bibliographical Society*, 11 (1975–6), 133–92 (at 170–1).

10. Lloyd, 'Meredith Lloyd', 171.

11. Lloyd, 'Meredith Lloyd', 171–2.

12. For Thomas Williams see *ODNB* entry. For the *Society of Chemical Physicians* see its *ODNB* entry; also *https://practitioners.exeter.ac.uk/sample-data* (accessed 2 May 2019).

13. John Warrington (ed.), *The Diary of Samuel Pepys*, Everyman Library No. 55 (London, 1960), p. 335.

14. Gunther, p. 4.

15. Gunther, p. 5.

16. Samuel Baker, *A Catalogue of the Genuine and Valuable Library of Joseph Harris Esq., His Majesty's Assay Master of the Mint*, Samuel Baker Catalogues. 11 February 1765–16 December 1766, no. 6, listed in *List of Catalogues of English Book Sales 1676–1900 Now in the British Museum* (London, 1915) with BL reference number: S.-C.S. 6. (1.).

17. See Gwyn Walters, 'Richard Price and the Carmarthen Academy', *The Price-Priestley Newsletter*, 4 (1980), 69, citing *The Cambrian Magazine; or Useful Repository, of Science and Entertainment*, 1 (June 1773), 29–30.

18. BL Add MS 28541 f. 40; see also Paul Frame, 'A Further Seven Uncollected Letters of Richard Price', *Enlightenment and Dissent*, 27 (2011), 154, available online.

19. *CIM*-III, Iolo Morganwg to Sir John Nicholl (?1816), p. 387 at n. 2.

20. Chris Evans and Louise Miskell, *Swansea Copper: A Global History* (Baltimore MD, 2020), p. 44.

21. Rachel Bromwich (ed.), *Dafydd ap Gwilym Poems* (1982; Llandysul, 1993), pp. 110–3, 123–4.

22. W. T. R. Pryce and T. Alun Davies, *Samuel Roberts Clock Maker: An Eighteenth-Century Craftsman in a Welsh Rural Community* (Cardiff, 1985), pp. 13–15.

23. William Linnard, *Wales, Clocks and Clockmakers* (Ashbourne, 2003), pp. 128–30.

24. Linnard, *Wales, Clocks and Clockmakers*, p. 130.

25. Linnard, *Wales, Clocks and Clockmakers*, pp. 111–9.

26. See Martin Griffiths, 'Joseph Harris of Trevecka; Scientist, Artisan, Servant of the Crown', *The Antiquarian Astronomer*, 6 (January 2012), 19–33.

27. OHSM, Lewis Evans Papers, MS 18 Part 3 Letters, Abraham Robertson to Lewis Evans 27 September 1819.

28. OHSM, Lewis Evans Papers, MS 18 Part 3 Letters, Nevil Maskelyne to Lewis Evans 26 January 1796.

29. See *https://collections.tepapa.govt.nz/object/59577* (accessed 25 May 2024).

30. As identified by Iorwerth Peate, *Clock and Watch Makers in Wales* (1945; Cardiff, 1960), William Hughes was a native of Llanfflewin in Anglesey who went to London before 1755 (pp. 18–9 and n. 1 and 2). He is listed in the Cymmrodorion members list of 1755 as William Hughes of High Holborn and born in Anglesey (the present author has a 1755 member list that is also inscribed with the note 'watchmaker'). He also appears in the Cymmrodorion members list for 1759, as a watchmaker (he also acted as 'Islywydd' – vice-president of the Society), and in 1762. This data seems conclusive as to his origins. Others have suggested (including *www.kelvinhughes. com* (accessed 25 May 2024) and the Wikipedia entry for Kelvin Hughes), a possible descent from the clockmaker Thomas Hughes Sr, who was a member of the Worshipful Company of Clockmakers in 1712, at the age of twenty-six, and Master of the Worshipful Company in 1742, and his son Thomas Hughes Jr, who was Master in 1762. It is notable that above the entry for William Hughes in the Cymmrodorion members list of 1755 is an entry 'Thomas Hughes, *dead* of Clerkenwell Green'. Thomas Hughes Sr would have been sixty-nine in 1755, though he is noted in the entry as being born in Cardiganshire, not Anglesey. Whatever the relationship may be, the firm Kelvin Hughes (today Hensoldt UK), long-time makers of navigation equipment, is ultimately believed to be descended from William Hughes through a possible son, Joseph Hughes (b.1781), and then through Joseph's son, Henry Hughes (b.1816), who founded Henry Hughes and Sons in 1838.

31. BL, *Royal Society Papers*, vol. X, BL Add MS 4441, f. 80.

32. For the history of the Copley Medal, see M. Yakup Bektas and Maurice Crosland, 'The Copley Medal: The Establishment of a Reward System in the Royal Society, 1731–1839', *Notes and Records: The Royal Society Journal of the History of Science*, 46/1 (January 1992), 43–76; Rebekah F. Higgitt, '"In the Society's Strong Box": A Visual and Material History of the Royal Society's Copley Medal, *c.*1736–1760', *Journal of the Material and Visual History of Science* (in press), text available at *https://kar.kent. ac.uk/73612* (accessed 25 May 2024).

33. NLW Add. MS 607A; see also Hugh Owen, *The Life and Works of Lewis Morris* (Anglesey Antiquarian Society and Field Club, 1951), p. 223.

34. See Aneurin Owen, *Ancient Laws and Institutes of Wales* (London, 1841), Book II, Ch. 17, p. 90. Also Ronald Edward Zupko, *Revolution in Measurement: Western European Weights and Measures since the Age of Science* (Philadelphia PA, 1990).

35. See *ML*-I pp. 108–9; and Owen, *Life and Works of Lewis Morris*, pp. 38–43, citing BL ADD MS 14927, f. 120–4.

36. Owen, *Life and Works of Lewis Morris*, p. 40. Others have made it 136 feet 8 inches.

37. Search the Science Museum website for images of the various sets of Harris weights in its collections. For example, *https://collection.sciencemuseumgroup.org.uk/objects/co57502/nested-gunmetal-troy-counterpoises-troy-cup-weight-standard* (accessed 25 May 2024), by Harris for the Carysfort Commission 1758.

38. Bird provided instruments for the Greenwich Observatory as well as others across Europe and he made the first dividing engine to allow for variations caused by changes in temperature. He also worked for the Board of Longitude, *CHS*-4, p. 528.

39. Charles William Pasley, *Observations on the Expediency and Practicability of Simplifying and Improving the Measures, Weights and Money, used in this Country, without materially altering the present standards* (London, 1834), p. 169.

40. The two 1758 weights and the Bird standard yards are still in the care of the Clerk's department in the Palace of Westminster. I am grateful to the House of Commons Enquiry Service, and Dr Mark Collins, Estates Historian at Westminster, for their help, and for providing information and photographs on this subject.

41. See Griffiths, 'Joseph Harris of Trevecka', 21.

42. Joseph Harris, 'Proposals for printing by Subscription a Treatise upon Microscopes', *Bibliothèque britannique*, 19/1 (1742), 211–3, available online; Marc J. Ratcliff, *The Quest for the Invisible: Microscopy in the Enlightenment* (2009; Abingdon, 2016), pp. 77–81 (at pp. 80–1).

43. Joseph Harris, *A Treatise of Optics* (London, 1775), editor's preface.

44. *CHS*-4, 579 n. 10.

45. Ratcliff, *The Quest for the Invisible*, p. 77.

46. *ML*-I, p. 378.

47. *ML*-I, pp. 374–5.

48. *ML*-II, p. 22.

49. *ML*-II, pp. 22–3.

50. *ML*-II, pp. 235–6.

51. Search the Science Museum website for images of Watkins work in its collections. For example, *https://collection.sciencemuseumgroup.org.uk/objects/co56952/gregorian-reflecting-telescope-of-1-1-4-inch-aperture-by-watkins-telescope-gregorian-telescope-reflecting* (accessed 25 May 2024). Also search the OHSM collections.

52. Gee et al., *Francis Watkins and the Dollond Telescope Patent Controversy*, pp. 69–70 and n. 94.

53. Gee et al., *Francis Watkins and the Dollond Telescope Patent Controversy*, p. 70.

54. *ML*-I, p. 378.

55. See Gee et al., *Francis Watkins and the Dollond Telescope Patent Controversy*, chapters 4 to 6.

56. See Griffiths, 'Joseph Harris of Trevecka', 28, figures 6a, b.

57. NLW MS 12423C, f. 7, letters from John Lloyd.

58. OHSM, Lewis Evans MS 31, Lewis Evans to Charles Hutton, 12 July 1791.

59. William Morgan, *Memoirs of the Life of The Rev. Richard Price*, (London, 1815), p. 39.

60. William Jones, 'Of Logarithms, by the late William Jones, Esq; F.R.S. Communicated by John Robertson, Lib. R. S.', *Philosophical Transactions of the Royal Society*, 61 (1771), 455–61; William Jones, 'Properties of Conic Sections; deduced

by a Compendious Method. Being a Work of the late William Jones, Esq; F.R.S. which he formerly communicated to Mr John Robertson, Libr. R. S. who now addresses it to the Reverend Nevil Maskelyne, F.R.S. Astronomer Royal', *Philosophical Transactions of the Royal Society*, 63 (1773), 340–60.

61. See Paul Quarrie, 'The Scientific Library of the Earls of Macclesfield', *Notes and Records: The Royal Society Journal of the History of Science*, 20 (2006), 5–24.

62. OHSM, Lewis Evans Papers, MS 18 Part 3 Letters, Nevil Maskelyne to Lewis Evans, 26 January 1796.

63. For all details see Mary Croarken, 'Astronomical Labourers': Maskelyne's Assistants at the Royal Observatory, Greenwich, 1765–1811', *Notes and Records: The Royal Society Journal of the History of Science*, 57 (2003), 285–98.

64. Quoted in Croarken, 'Astronomical Labourers', 285, citing John Evans, *Juvenile Tourist* (London, 1810), pp. 333–5.

65. For all details and further information, see Stillman Drake, 'Galileo Gleanings II a Kind Word for Salusbury', *Isis*, 49 (1958); Jacob Zeitlin, 'Thomas Salusbury Discovered', *Isis*, 50 (1959), 455–8; Stillman Drake, 'Introduction', in *Mathematical Collections and Translations … in Facsimile* (London, 1967), pp. 1–27; Quarrie, 'The Scientific Library of the Earls of Macclesfield'; Nick Wilding, 'The Return of Thomas Salusbury's "Life of Galileo"', *The British Journal for the History of Science*, 41 (2008), 241–65; The Archimedes Project (available online).

15

SCIENCE AND TECHNOLOGY: INNOVATORS AND INVENTORS

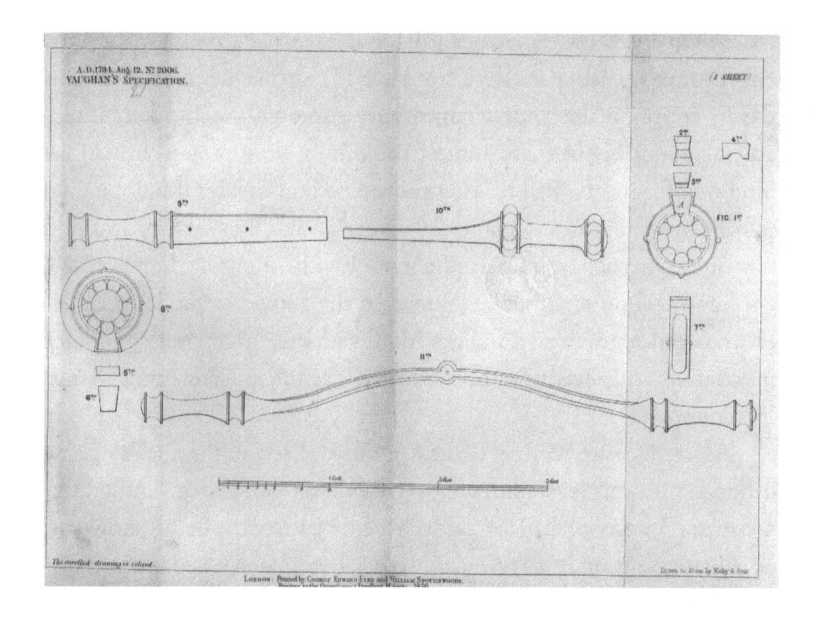

The 'Prologue' to this book opened by referencing an important invention patented in 1794 by Philip Vaughan (bap. 1757–1824) of Carmarthen. The technical drawing made to accompany that patent – the very first for a ball bearing – is reproduced above. Originally from Trellech, in Monmouthshire, Philip Vaughan was part of a family of managers who worked at ironworks with names resonant in the early industrial development of south Wales, the Melingriffith works at Whitchurch, in Cardiff, being among them. By 1790, Vaughan had moved to Carmarthen. There he established himself at 'Vaughan's Iron Foundry' and made and sold his ball bearings. Vaughan's invention for carrying the loads on 'certain Axle Trees, Axle Arms, & Boxes for Carriages' comprised 'a box in a groove, revolving round the axle in rotation'. It also included a dovetail piece of iron in the bearing (at A in the above figure) that could be removed so as to enable the balls to be taken out and replaced when needed. As an advert for the bearing in the *Hereford Journal* of November 1794 related, the bearings were recommended to all owners of wagons and carts, and would prove of great utility to the public by lessening friction; a reduction so notable that 'on average two horses can draw as much as three'. Despite his efforts at advertising, and the trouble and expense he had taken over the invention, it does not seem to have garnered much public interest or financial benefit for him at the time.[1] Nevertheless, in various modified forms, the ball bearing proved pivotal to later developments – from steam engines to planetary rovers, and roller skates to washing machines. As such, it is an example of how the interplay between science and technology that developed in the course of the long eighteenth century provided a foundation for nineteenth century and later innovation and invention. Ball bearings are today alive and well and living on Mars.

As one of the world's earliest industrialised nations, Wales saw other significant developments over the long eighteenth century. On a visit to Pontypool, in 1697, Edward Lhwyd records being shown an excellent invention by the England-born, Pontypool-based industrialist and pioneer of the tinplating industry 'Major Hanbury, for driving hot iron (by the help of a rolling engin[e] mov'd by water) into as thin plates as tin'. The invention of the rolling mill by John Hanbury (1664–1734) would become the basis for the tinning (protecting sheets from rust) and japanning (the application of a black lacquer to the tinplate in imitation of East Asian lacquered wood) industries first developed at Pontypool by another Englishman, Thomas Allgood (*c*.1640–1716).

In the early eighteenth century at Neath, and at the Llangyfelach copper works in Swansea in 1717, coal, instead of traditional wood and charcoal, was used in reverberatory furnaces (those that allowed for the separation of the ore from the actual fuel being used for the smelting). In what became known as the 'Welsh Method' of smelting, the change in fuel resulted in a shift of copper production away from places where the ore and the wood fuel were found, to places rich in coal. The 'Welsh Method' would soon become 'the defining technology of Swansea Copper' with up to a third of the world's smelted copper being produced in the Swansea Valley between the 1770s and 1840s.[2]

In the iron industry, one of the first people to follow Abraham Derby of Ironbridge, on the River Severn, in using coke instead of charcoal for smelting was Charles Lloyd – 'of the stock of Lloyds of Dolobran, the family which gave its name to Lloyds Bank'.[3] He had established his works at Bersham, near Wrexham in north Wales, by 1721. In 1751 the lease of Bersham was bought by Englishman Isaac Wilkinson of Cumbria, whose son, John Wilkinson, eventually became the sole owner of the works. A new method of boring iron cylinders, patented by John Wilkinson, resulted in the works becoming a major canon foundry while also producing cylinders for almost all of James Watts's steam engines.

By the late eighteenth century, thanks to his control of the copper deposits at Mynydd Parys on Anglesey, Thomas Williams (1737–1802), previously an Anglesey solicitor, would become known as the 'Copper King'.[4] With the growth of shipping, in particular, Royal Navy ships sheathed with copper plates as a protection against invasive molluscs and worms, new methods of fixing the plates to ship's hulls were needed; the available iron bolts were seen to corrode rapidly, while those made from copper were too soft. As a result of experiments sponsored by Williams, the cold-rolling of copper was found to produce harder and more lasting bolts; an innovation that secured Williams his fortune.

Since the results of modern research into these developments can be found elsewhere,[5] the focus in the rest of this chapter will be on less-well-known contributions. As was the case in the previous chapter, the subject is a large one and what follows should be considered as no more than a broad introduction to the subject.

Water and Air, Fire and Steam: The Engineers

The distribution of water, so vital to daily life, presented significant challenges in the long eighteenth century, particularly in large metropolitan areas. Not least among the problems was the difficulty of raising water out of rivers or deep wells into associated reservoirs and a distribution system of often leaky wooden pipes. One Welsh entrepreneur who had risen to the challenge in London was Sir Hugh Myddelton (1560–1631), the founder of the New River Company between c.1609 and 1619 (the company existed until 1904). The company managed the New River, a long artificial aqueduct, part funded and built by Myddelton, with its water distributed through a network of wooden (elm) pipes to those who paid. By the early 1720s (see below) other water companies had been established and, away from traditional waterwheels and the use of horse traction, a new method of raising water had appeared – steam power.

The family of Edward Somerset (1601–67), second marquess of Worcester and titular earl of Glamorgan, was based at Raglan Castle where, some sources say, Edward was born (others say in London). In four entries (numbers 68, 98, 99, 100) in his *A Century of the Names and Scantlings of Inventions by me already practised* (1655, published 1663), Somerset relates what is now considered to be one of the earliest known descriptions of a 'Water Commanding Machine' or steam engine. It could raise water to a height of 40 feet and, according to Wolf (1935), was seen by eyewitnesses in 1663 and 1669. However, having apparently obtained, under a parliamentary act, 'rights for ninety-nine years' for his engine, Somerset failed to develop it further.[6] It would be Thomas Savery (c.1650–1715), with his steam-powered pump of 1698, and his fellow Devon-born engineer Thomas Newcomen (1664–1729), with his atmospheric engine of 1714, who did so.

The more powerful steam-powered pumps and engines of Savery and Newcomen would soon begin to revolutionise mining (by the extraction of water from the workings), and other industries in Britain. However, with regard to water distribution the engines raised a new issue – could wooden pipes and their joints withstand the increased pressure to which they would now be subjected? One man certainly believed he had an answer to that problem. In 1725, William Edwards (*fl.*1720–30), of Monmouthshire, had taken out a patent on a type of pipe made from potter's clay.[7] Sceptics might see them as no more than a reinvention of the Roman clay pipes then being found at arch-aeological sites in London and elsewhere. Yet, as Englishman Stephen Switzer (1682–1745) details in Volume 1 of his *To a General System of Hydrostaticks and Hydraulicks* (1729), Edwards's design included a number of significant improvements. First the clay, on which 'great expense and trouble is bestowed', was subjected by workmen to 'pitting, washing, riddling, grinding it in a mill, treading it with men's feet, and working it with their hands into messes' before being formed into pipes with walls half an inch thick (Roman pipes often had walls some two inches thick). The Edwards pipes were next turned on a lathe to perfect the joints, which would be cemented or soldered when the pipes came to be laid. The finished clay pipes were next fired for five or six days and nights, making them as hard as flint and all for two shillings per yard for a 2-inch bore. Furthermore, to prevent 'all strugglings' that might occur when air as well as water was present within the pipes, Edwards 'contrived, and made brass valves, or clacks, placed within his said pipes, at a proper distance, which discharge the air'.[8]

Over three days in July 1727, tests were carried out in London on some 18 feet of the pipes under the supervision of the English nat-ural philosopher, clergyman, engineer and assistant to Isaac Newton, John Theophilus Desaguliers (1683–1744, FRS 1714, Copley Medal 1734, 1736, 1741). The high pressure necessary for the test was supplied by the new Newcomen engine installed in 1726 at the Thames-side water works of the York Buildings Company (they had previously had a Savery engine). The great tower that held their new engine stood at York Buildings, on the north bank of the Thames, and is clearly visible in a painting of 'The City of Westminster from near the York Water Gate, between 1746 and 1747', by Canaletto.[9] Desaguliers published his opinion of the test in the *London Evening Post* on 29 July 1727:

> With all the compressure of air and water that the engine was able
> to lay upon them, which amounted to a pillar of water above one

hundred foot, they sustain'd it without breaking any of the pipes; the so[l]der or cement stood the force as well as the pipes. The experiment having been try'd before me, and so well approv'd of, that I recommend them to be used in all buildings and aqueducts to convey water.[10]

Another mid-eighteenth-century London-based Welshman to take advantage of the developments in steam, though from a different innovative and inventive perspective, was the optician Francis Watkins (see Chapter 14). In 1750 he developed a model steam engine which, according to an announcement in the *London Evening Post* of 16 August, was 'very neatly finish'd, in brass, and containe'd within a frame of four feet by two, and six feet high, and raises near seven hogsheads an hour, with a small quantity of fire. It is purchased by a foreign prince, and will be sent abroad in a few days'. A single hogshead or barrel held 54 gallons, so, as Gee et al. (2014) noted, the model 'was more suited to pumping water through modest heights, not deep mining'. As such, it would have appealed not only to the mechanically curious, but the likes of gentry landowners engaged in draining their land.

In 1751, Watkins developed a second engine, which he displayed with an entry fee of half a crown. In 1752, he published *An Accurate Print of the STEAM ENGINE, with a mechanical and philosophical description of the parts and properties of that most curious and useful machine*. Sold at two shillings a copy, none appear to have survived; but it is suggested by Gee et al. that 'should a copy come to light, it may fill a significant gap in our knowledge of this phase in the development of the steam engine'. The 1750s were also the years in which the next major contributor to the development of the steam engine, James Watt, worked in London (1755–6) for the mathematical instrument maker John Morgan of Cornhill. Again according to Gee et al., Watt also spent time with John Neale, who was known to Watkins through their shared interest in electrical machines (see below): 'It is, therefore, not unreasonable to suppose that [Watt] also visited Watkins's shop … and, perhaps, while there witnessed the recently built model steam engine in action.'[11] Watt moved to Glasgow in 1757, and by 1769 he had begun building steam engines that improved on the earlier ones of Newcomen. By 1804, Cornish-born Richard Trevithick (1771–1833) had produced the first steam locomotive to pull a load on rails. It made its journey at the Penydarren ironworks in Merthyr Tydfil.

Also in 1757, and as James Watt moved to Glasgow, we find Lewis Morris in London making improvements to his microscope, while building another for his brother William back home in Anglesey (see Chapter 14). But mechanical things were preoccupying Lewis more generally, as he relates to his brother in a letter from London that year:

> My head this week turns upon wheels, that is, I am inserting in my Bockler's book of machines, all ye curiositys that I have pickd out of Monsr. Gallon's Machines of ye Royal Academy of Paris, 1735, and of Belidore's Hydraulics, 1748, Ramelli's Machines, 1588, the best of them all, though first wrote; and some improvements of my own upon all these.[12]

This preoccupation with things mechanical is evident in seventeen pages of notes in the surviving manuscripts of Lewis Morris. In them he is concerned with various kinds of mechanical motive power, including engines moved by the 'power of ye tide of ye air, of water and steam, and by weight – not so useful in large works as in small, such as clocks; for there must be another power to bring them to their first situation, after their natural motion is finish'd'. True to his practical nature, Lewis records that 'Having long considered the natural powers now known to drive machines or engines for ye use of man I have often thought that some improvements might be made that way additional to what hath been yet published'. True to his word, the notes contain his designs for 'a machine to bore stones' and, after duly noting the 'defects and difficulties in the machines or known engines drove by wind', plans for a wind engine and a horizontal windmill.[13]

Despite the developments in steam power, wind power would still be tried on the otherwise horse-drawn Oystermouth and Mumbles Railway as late as 1807. Running around Swansea Bay, the first rail passenger journey made anywhere in the world took place on its 'cast-iron tram plates attached by spikes to granite blocks' on 25 March 1807. Barely a month later, *The Cambrian* newspaper recorded how:

> An experiment of a novel kind was carried out on the Oystermouth tramroad yesterday to ascertain the practicability of a carriage proceeding to the Mumbles without horse, by the aid of wind alone. Some jolly sons of Neptune rigged a wagon with a long sail and the wind blowing strong and as fair as could be wished set off from our quay and after clearing the houses dropped anchor at the end of the tramroad in less than three quarters of an hour, having come a distance of about 4½ miles.[14]

The degree to which machines were becoming part of the landscape within Wales by the 1750s can be seen in the illustrated travel journals of a Swedish visitor: Reinhold Rücker Angerstein (1718–60). He travelled extensively in England and to a number of places in the Welsh borders, all the while keeping an illustrated travel journal. He visited the Wrexham area and the Bersham furnace, and he drew the blast furnaces at Abercarn, as well as the roasting ovens, water wheels, shearing mills and tinplate furnaces at Pontypool. At the same time, Angerstein notes the secrecy that often surrounded these places of day-to-day endeavour and potential innovation and invention. At Pontypool, he records, 'everything is kept very secret and … all strangers are forbidden to approach the works. Anybody who intends viewing them must be prepared to use every possible means of achieving his wish'. Having gained entry to works in Pontypool, with the help of a workman, Angerstein was caught and had to listen to 'a torrent of abuse in broken French' from the owner, John Hanbury.[15] Another instance of the lengths to which secrecy might be taken comes from the French traveller Gabriel Jars (b.1732). He did not visit Wales, but when *en route* from London to Wandsworth he entered 'a large industrial building enclosed by walls on every side'. During his four or five minutes inside what appears to have been a site for distilling and concentrating sulphuric acid, he noted a singular precaution taken by the owners against espionage: 'I only met there women from the Principality of Wales, of whom only a few knew more than a few words of English (this precaution is taken no doubt so that no one may divulge anything which happens in that laboratory).'[16] With a growing population and increasing industrial development, civil and industrial engineering in Wales developed significantly throughout the long eighteenth century but particularly so from the 1750s. It is to some examples of these later developments that we now turn.

Civil Engineering

Beyond industrial life, machines were part of many civil developments in Wales. In an entry for 26 May 1763 the diarist William Thomas noted 'Richard William, a carpenter and a shopkeeper in Llantrissent [sic] and a reputed Magician because by Mechanick in his time rose Llantrissent Great Bell etc.'[17] Later, in his *Tours in Wales 1804–1818*, the St David's born writer and lawyer Richard Fenton (1747–1821) suggested, as have many visitors since, that 'the masonry at Dolgelly [sic] merits particular notice' with its almost magical placement of truly massive and precisely cut stones. Fenton, though, simply records that 'From time immemorial they have built with very large stones, even to the top, lifting the Stones to the work from towards the middle course with an immense machine which takes above a day to erect, and worked by two men, every stone being of such a weight as to require a lever of that vast power'.[18]

The bridges built in Wales in the long eighteenth century might be said to match, at least in their beauty, later and more famous engineering works such as the Pontcysyllte aqueduct (1805) and the Menai and Conwy bridges (1826) of Thomas Telford, or the Conwy (1847) and Menai (1850) tubular rail bridges of Robert Stephenson. In other respects, though, the long-eighteenth-century versions are smaller in scale and of more traditional design and construction. However, two important exceptions to that design and construction proviso are the bridges at Pontypridd and Merthyr Tydfil built by artisan engineers William Edwards (1719–89) and Watkin George (*c*.1759–1822), respectively.

Born in Ty Canol farm at Groeswen, Caerphilly, William Edwards is another engineer and scientist of Wales with little formal education, but who learned from the skills of others: English from a blind baker, with whom he lodged, and the art of masonry from local stonemasons and the work of their medieval forebears at Caerphilly Castle. Once again, as with many engineers and scientists of Wales, religion played a significant role in Edwards's life – so much so, that in addition to being a stone mason, architect and engineer, he also became a Nonconformist minister and chapel builder. As a youth, Edwards was said to be 'obstinate, stubborn and self-willed', which was perhaps fortunate since his attempt to build a single-span stone bridge at Pontypridd between 1746 and 1756 reflects the phrase 'try, try and try again'. After experimenting with various configurations, the surviving bridge, with the inclusion of innovative weight-reducing holes within its structure, was successfully completed at the fourth attempt in 1756. Spanning some 42.7 metres it formed the largest extant single span in the world at the time and remained Britain's widest masonry span for seventy-two years. It is just one of a number of bridges built in south Wales by Edwards and, in later years, by three of his sons.[19]

Watkin George hailed from Trefethin, near Pontypool. He began his working life as a carpenter, but by 1792 he had become a partner with Richard Crawshay in the vast Cyfarthfa Ironworks at Merthyr. Although experienced in many types of metal working, it was his carpentry skills that must have proved most useful when he worked on the 185-metre Gwynne Aqueduct at Merthyr. Constructed between 1793 and 1796 as a timber trestle structure, it ran over the top of another significant Watkin George creation. This was the 1792–3 Pont y Cafnau cast-iron tramway bridge, designed to take limestone to the Cyfarthfa Ironworks. Not only did George use an A-frame construction at Pont Y Cafnau, instead of the arch used at earlier metal constructions such as Ironbridge, but the metal work of his structure replicated many carpentry features – even down to the use of dovetail and mortis and tenon joints. The structure remains in place and is now believed to be the world's oldest known cast-iron tramway bridge.[20]

Another engineer, Samuel Baldwyn Rogers (1778–1863), may have come to Pontypool from 'near Tintern', or possibly Ludlow, in 1808. As a metallurgical chemist and industrial engineer he spent a considerable part of his life in south Wales, dying, at the age of eighty-five, in Llanfoist, Monmouthshire. Much of his contribution was made outside our period, but as Williams (1959) records, Rogers, while employed at the Hydrogen Laboratory he established in Pontypool, 'made improvements in the manufacture of coke and sulphuric acid, and in 1810 invented hydro-pneumatic pumps for taking gas from coke-ovens, and also, at Risca in 1817, for supplying blast to smelting furnaces, refineries and smitheries'. After 1838, the pumps and extracted gas would be used to provide lighting 'in every department of the ironworks' at Nant-y-glo, Monmouthshire. Rogers also tried, between 1816 and 1818, to interest the ironmasters of south Wales in his innovative design for an iron bottom for their puddling furnaces. Puddling, whereby pig iron was converted into wrought iron through the application of heat and stirring inside a furnace or crucible, was an old technique but one that had been improved by Englishman Henry Cort. His early puddling furnaces utilised a sand base and they had significantly improved iron output per furnace. Iron-base, and air-cooled iron-base furnaces had also been tested, but they proved unsuccessful. By contrast, Rogers's iron base was water-cooled and it doubled the already improved output of the Cort furnace. Nevertheless, the ironmasters of south Wales remained unconvinced, even when Rogers tested it and offered it to them in 1819. Yet, by 1825, iron-based furnaces were being operated in some south Wales works with Rogers gaining nothing from his invention, other than the nickname 'Mr Iron Bottom'. His career post-1820 would see him develop many other innovative ideas, while writing scientific publications, including *An Elementary Treatise of Iron Metallurgy* (1857), and radical political tracts; the last reflecting his socialist and chartist credentials, as well as some early environmental concerns. Rogers would die and be buried in Llanfoist in 1863 having lived his last years in considerable poverty.[21]

In another fire and furnace innovation, William Harry (dates unknown), in the Swansea Valley, discovered a means of vitrifying the inside of a ceramic brick-built furnace. He patented the invention in 1817, but the process would be taken further by the Quaker William Weston Young (1776–1847) of Bristol and later Glamorgan. Employed

by Lewis Weston Dillwyn between 1803 and 1806 as a botanical illus-
trator on products of Dillwyn's Cambrian Pottery at Swansea, Young
also invested in the Nantgarw Pottery in Glamorgan. He also provided
illustrations for Dillwyn's innovative natural history study *The British
Confervae* (see Chapter 10). Combining other interests as a surveyor
and amateur geologist, Young utilised deposits of silica found at Craig
y Ddinas in the Neath Valley to develop the 'silica firebrick'. From these
a whole furnace could then be built. The 'Dinas brick', as it came to be
called, sold across America and Europe, and in Russia the word for a
silica refractory brick is said to be 'Dinas'.[22]

In a comprehensive study of Swansea entitled *Copperopolis*, Stephen
Hughes (2000) discusses a number of other innovative civil and industrial engineers. They include the architect and engineer William Jernigan,
who had 'a turn for mechanical inventions'. We first met Jernigan as
the architect of Francis Greville's technical college at Milford Haven
(see Chapter 6). As Wales's first professional architect, Jernigan worked
extensively on projects in Swansea, including the Mumbles Lighthouse
(1793). In 1798 he developed and adapted the light from being coal
fired to 'a single cast-iron and glass lantern, which he designed and had
made by the Neath Abbey Ironworks'. The Hopkins family of engineers,
Evan, the father and his sons David and Roger, came from Llangyfelach,
near Swansea, and worked on the Penydarren Tramroad (1800–2), the
Mumbles and Oystermouth railway (1804–7), and various other canal
and railway developments. Evan is also believed to have constructed
'one of the earliest powered railway inclined-planes, at the head of the
Vale of Neath' between 1803 and 1805. Also in Swansea, a 'Mr Powell'
constructed 'steam-powered (or more correctly, atmospheric-powered)
rotary engines which were among the first of this type to be built'.
Powell would later be accidentally scalded by steam. He died two days
later, 'Universally lamented'.[23]

In 1864, *The Engineer* journal published an obituary to another
departed Welsh engineer. It recorded the fact that: 'One of the most
gifted inventors and skilful mechanical engineers, who within the last
half century, have added so much not only to the wealth of England
but of the whole civilized world, has just passed away, at the ripe age

of seventy-five.'[24] The inventor and engineer in question was Richard Roberts (1789–1864), whose inventions, the journal went on to declare, 'rank in value with those of Arkwright'. Born in the Wales/England border village of Llanymynech, in modern Montgomeryshire, 'with the front door opening into one country and the back into another', Roberts chose the door into Wales for his nationality, and that into England for his career. In doing so, his life as an inventor and scientist of Wales followed the trajectory of a number of the gifted artisan-scientists already mentioned in this book: an early life trained by a local figure, in Roberts's case the Revd Griffith Howell (a minister and mechanic who is said to have 'invented a small seed drill for sowing any kind of grain or seed'); local recognition of his engineering and mechanical abilities, for Roberts the making of a spinning wheel for his mother; periods of local employment, on the Ellesmere canal at Llanymynech and in a local quarry; then, finally, the move to England, in Roberts's case to Manchester in 1816, at the age of twenty-seven. Once there, he became a member of the Manchester Literary and Philosophical Society, with his membership supported by the great chemist and scientist John Dalton (1766–1844, FRS 1822), and by one of the Society's founder members and later President of the Society, Thomas Henry (see below). In 1824 Roberts would play a prominent part in the launch of the Manchester Mechanics' Institution, which helped working men to gain a better education.

As with Samuel Baldwyn Rogers, the greatest of Roberts's inventions lie outside our period of interest. They include twenty-five or more patents taken out between 1822 and 1860; including many relating to the spinning and preparation of cotton, wool and other fibrous materials (hence the obituary comparison to Richard Arkwright). In 1847, there was also a metal punching machine, worked by Jacquard cards (an early form of punched computer card), which was used during the construction of Robert Stephenson's tubular rail bridge across the Menai Strait.[25]

No Roberts patents were assigned within the long eighteenth century itself, but his inventiveness as a young man was already apparent. Having arrived in Manchester in 1816 he invented a new form of gas meter and, in 1818, made 'a rifled brass cannon adapted to load at the breech'. This was an invention well ahead of its time, as Hills (2002) records: 'It was only during the Crimean War [1853–6] that Will Armstrong and Whitworth began serious experiments with various forms of rifling and Armstrong developed a breech-loading gun.' Roberts also produced a machine for accurately cutting the teeth on gearwheels. As a later advert declared, the machines were 'constructed as to be capable of producing ANY number of teeth required: they will cut BEVIL, SPUR, or worm gear, of any size and pitch not exceeding 30 inches diameter'. One of his machines is preserved in the Science Museum in London, along with his slide lathe and planing machine. All are dated to 1817.[26]

For other scientists of Wales, engineering involvement came from their own local and amateur endeavours, coupled with contact and friendship with professional engineers. Given the opening to a letter received in 1804 from the apothecary, inventor, electrical experimenter and scientist Timothy Lane (1734/5–1807, FRS 1770, APS 1772), John Lloyd of Wigfair, like Lewis Morris before him, appears to have invented and designed some form of rock boring machine. As Lane wrote:

> I am so much pleased with your contrivance for boring rocks that I shall wish to know whether you find it answer the purpose, if it does, with your permissions intend proposing to the Directors of the Scottish Mines Company, being myself a proprietor, to have one made by your ingenious workman, Galloway.[27]

In a letter of 1812, the engineer James Fox (1760–1835) of Derby provides Lloyd with details of the steam engines and 'water closets at our infirmary'.[28] These were presumably the innovative water closets designed by civil engineer and inventor William Strutt (1756–1830, FRS 1817). He played a significant role in the design and building of the Derby Infirmary, which opened in 1810. Based on earlier work by Joseph Bramah, the closets 'flushed automatically by the action of the door as a person left the room and simultaneously admitted fresh air into the space, though it is not fully clear how the waste was disposed of'.[29] Fox also notes in a subsequent letter to Lloyd that he expected 'to finish your washing machine this week'. Washing machines were installed at the Derbyshire Infirmary and so we might wonder whether Lloyd was planning a similar institution in north Wales, or perhaps he intended the washing machines, and the also discussed 'drying closets', for use in his mineral interests.[30]

Lloyd had a friendship too with the eminent Scottish engineer John Rennie Sr, who he sponsored for Fellowship of the Royal Society in 1797. In an undated letter Lloyd writes: 'My friend Mr Rennie, the celebrated engineer, spent a day and night here [Wigfair], in his way to order the new improvements that are to be made under his direction at Holyhead.'[31] In the course of his career John Rennie Sr had worked with Robert Stephenson on the Bell Rock Lighthouse in Scotland and he was responsible for the lighthouse at Holyhead, which was completed in the year of his demise (1821). It seems likely, though, that Lloyd's undated letter relates not to the Holyhead lighthouse but to Rennie's work on the construction of the Admiralty Pier in Holyhead, which began in 1810 and continued until 1824.

Another significant Rennie Sr design would be for a New London Bridge across the Thames in London to replace the still standing one constructed in 1176. Rennie's design links, albeit indirectly, to an 1801 mathematical correspondence between Wales-born mathematician Lewis Evans and the Savilian Professor of Geometry at Oxford, Abraham Robertson (see Chapter 14). Prior to acceptance of Rennie's design for the new London bridge, a Parliamentary Commission was established to enquire into the feasibility of a different design submitted by Thomas Telford. He proposed a single-arch iron bridge with a span of 600 feet. A number of eminent scientists, including the engineer James Watt, the Astronomer Royal Nevil Maskelyne, as well as Abraham Robertson were consulted by the commission. At some point, during these consultations, Robertson consulted Lewis Evans. Robertson had been made FRS in 1795 for his work on conic sections, and the mathematics of his 1801 letter to Evans related to a circle of 1,450 feet diameter, a segment of which corresponded to the arc of the proposed Telford arch. The letter also discussed the pressures likely to exist at the apex of the Telford arch, and along its length, one of the main concerns of the commission of enquiry.[32] In the event, Telford's proposal was dropped and the more traditional, multi-arch, Rennie Sr design went ahead. It would be built in London between 1824 and 1831 by John Rennie Jr and stood until sold to the United States in the 1960s. Today it stands at Lake Havasu City in Arizona.

Flight

Another inventive marvel of the later eighteenth century, which involved a degree of engineering skill in its design and development, was the balloon, and with it the advent of human flight. Before they actually appeared in Wales, balloons came to the notice of her scientists through their contact with scientific colleagues. The Montgolfier brothers had launched their first unmanned balloon in France in June 1783. By September, the earliest Welsh involvement in the subject came in correspondence between Benjamin Franklin and Richard Price. On 16 September, Franklin, acting as one of revolutionary America's representatives in France, wrote from his base at Passy, near Paris, informing Price that:

> All the conversation here at present turns upon the Balloons fill'd with light inflammable Air, and the means of managing them so to give Men the Advantage of Flying. One is to be let off on Friday next at Versailles, which it is said will be able to carry up a 1000 pounds weight.[33]

The flight in question was another Montgolfier ascent, but this time 'manned' by a sheep, a cock and a duck. In a postscript to his letter Franklin also notes: 'They make small balloons now of the same material with what is called gold-beaters leaf. Inclos'd I send one which being fill'd with inflammable air, by my Grandson, went up last night, to the ceiling in my chamber, and remained rolling about there for some time.' It's pleasant to imagine Richard Price inflating it himself before fulfilling Franklin's request to give it to Sir Joseph Banks.

The Montgolfier brothers had burned dampened wood and straw in order to fill their balloons, but on 1 December 1783 came the first manned flight in a balloon filled with hydrogen.[34] The ascent was made by Jacques Charles (1746–1823) and Nicholas-Louis Robert (1760–1820) and details of the flight quickly found their way to John Lloyd in north Wales, via a letter from Joseph Banks:

> My dear Lloyd,
> In answer to your Earthquake intelligence [Lloyd had written to Banks in November 1782 describing an earthquake that took place in north Wales on 2 October; the letter appeared in the *Philosophical Transactions* in 1783] I can inform you that two men flew on Monday the 2nd from Paris 27 miles in 2 hours by the assistance of a globe filled with inflammable air which they made to ascend and descend as they chose it should.
> After this flight one of them set sail upwards and sailed till he saw the mercury in his barometer fall from 28=4 lines to 18=10 lines at which he found the thermometer at 21 and he came down safe and well.[35]

Although the letter is undated, it clearly relates to the 36-kilometre/22-mile flight made by Charles and Robert. They had used a gas release valve, as well as sand ballast, to adjust their ascent and descent and their flight lasted 2 hours 5 minutes. It was also the first to provide meteorological readings taken from a barometer and thermometer above the earth's surface. Both men took part in the initial part of the flight, but Robert had then alighted, and, at dusk, Charles took off again on his own. As he rose, he re-entered into sunlight and saw his second sunset of the day. The conjunction of the above evidence, with the fact that Banks simply registers the date of the flight as being 'Monday the 2nd' (it was actually the 1st), without mentioning the month, allows the letter to be dated to the month in which the Charles/Robert flight took place – December 1783.

Writing to Franklin in April 1784, Richard Price made a prescient observation on the advent of ballooning: 'The discovery of air balloons seems to make the present time a new Epoch, and the last year will, I suppose, be always distinguish'd as the year in which mankind began to fly in France.'[36] Replying in August, Franklin agreed with Price's observation while noting at the same time: 'Some of the [balloon] artists have lately gone to England. It will be well for your Philosophes to obtain from them what they know, or you will be behind-hand, which in mechanic operations is unusual in Englishmen.'[37] For many years the first flight in Britain was believed to have been that of the Italian, Vincent Lunardi, which was made from Moorfields in London on 15 December 1784. However, in an earlier letter of October that year, addressed to Ezra Stiles, the President of Yale University, Richard Price already notes in a postscript that 'We have lately begun to fly here. Hitherto we have been behindhand in this instance. But an order begins now to prevail, which perhaps will cause us to get the start of France'.[38] In saying 'we have lately begun to fly here' Price is probably referring to the flight made by the Scottish apothecary James Tytler (1745–1804), who flew for ten minutes and for a distance of some 350 feet on 27 August 1784; the first successful flight in Scotland and Britain. Price may also have heard, perhaps from relatives in south Wales, of an even earlier attempt. This was made in Swansea on 14 August 1784, just two weeks before Tytler flew and four months before Lunardi. Brant (2011) details how two Frenchmen attempted to launch balloons in Swansea on that date.[39] Unfortunately, their first burnt up as it was ascending, and the second only reached chimney-height before collapsing. As a report of the event put it, in terms typical of the times, 'we have met with a double disappointment from this Frenchified scheme'.[40]

Price also records in his letter to Ezra Stiles how his friend, Joseph Priestley, had 'just discovered a method of filling the largest air balloon with the lightest inflammable air in a very short time and at a very little expense; which removes one of the great impediments to the improvement of the art of flying'.[41] Inflammable air (the term is often used at the time to describe hydrogen, and inflammable essentially has the same meaning as flammable at this time) for filling balloons was generally produced by pouring copious amounts of sulphuric acid onto scrap iron or iron filings, the resulting flammable hydrogen gas being then piped into the balloon. Priestley, who was clearly interested in the commercial possibilities of balloon filling, had developed an improved method of inflation using tubes of a wide calibre. Hydrogen made by the sulphuric acid method would be used in a further attempt at flight in Swansea in 1802. Braving a number of early failures, the breaking of a young boy's leg, and the uncertain fate of an aeronautical cat and dog, the flight was eventually undertaken by showman, Francis Barrett. He lifted off watched by a crowd of around 20,000. Having reached a height of some 60 feet and a distance of four fields he was stopped by an improvidently positioned tree.[42] Humorous though his efforts might now seem, Barrett, and all the early balloonists, wherever they flew, were true pioneer scientists and at the forefront of technological developments,[43] as were those scientists of Wales seeking to know about and to pass on details of the aeronauts' adventures.

The Longitudinarians

Despite humanity's entry into the airy realm by balloon, it would be a long time before the development of flight allowed for a safe crossing of the oceans. Making that crossing remained a sea-bound and often fraught business; a consequence not only of storm, mutiny and wartime dangers but the difficulty of accurately determining one's position at sea.

To find latitude at sea was relatively straightforward. Expressed as a position relative to two natural datum's – 0 degrees latitude at the equator, 90 degrees at the poles – it could be found by measuring the angular height of the sun or specific stars above the horizon. This was done using the instruments then available – the astrolabe; the fore-staff, whose design Joseph Harris attempted to improve (see Volume 2); the quadrant; and the octant, which utilised mirrors to allow more accurate measurements. The Octant was developed in the 1730s in America, by glazier Thomas Godfrey (1704–49), and in England by astronomer John Hadley (1682–1744). William Jones Sr would raise the resulting dispute over inventive priority at the Royal Society in 1734.

In an era before the Greenwich Prime Meridian was established in 1884, longitude, unlike latitude, could not be expressed in relation to any natural datum – nothing in the sky distinguishes longitude in the way that the equator and the poles do for determining latitude. Furthermore, since one hour of time equates to 15 degrees of longitude (from 360 degrees of the earth's circumference divided by twenty-four hours), and four minutes to 1 degree of longitude, it does not take long for errors in, or ignorance of, longitudinal position to result in geographic disasters. One miscalculation in October 1707 led to the loss of 2,000 naval and military souls, and four naval vessels, near the Scilly Isles.[44] In response, in 1714, the Westminster Government established the Longitude Board, and a prize of £20,000 for a practical method of determining longitude at sea. Many ideas, sometimes bizarre, ensued, but three principal methods were to the fore: magnetic variation; the lunar method; and accurate timekeeping.[45]

The variation of the compass needle away from geographic north as a result of magnetic variation (or declination) had long been known. In 1701, Edmond Halley produced a chart illustrating the pattern of this variation throughout the Atlantic, derived from measurements taken on his shipboard journeys in the ocean between 1698 and 1700. Sailors could then compare their readings with the map in the hope of establishing their longitude. However, with a limited number of readings and without constant updating, because of the changes in the degree of variation over time, the map's value to sailors remained small. Yet magnetic variation remained of constant interest as a means to determine longitude well into the eighteenth century. One scientist of Wales who held to its possibilities was the surgeon and physician Zachariah Williams (1668–1755), of Rosemarket, near Haverfordwest in Pembrokeshire. Having moved to London c.1726, he proposed his ideas on magnetic variation to the Board of Longitude, one of whose members was the President of the Royal Society, Isaac Newton, but Williams received no success with the Board. By 1729, Williams had entered the Charterhouse in London, which had been established in 1611 as a school for boys and a hospital for poor gentlemen. He would remain there for most, though not quite all, of the rest of his life (see below).

In 1733, Joseph Harris opened his *Philosophical Transactions* paper 'An Account of Some Magnetical Observations Made … in the

Atlantick or Western Ocean' with the observation that once properly established, 'the magnetic variation theory might be of great use for estimating the longitude in several parts of the world'.[46] Zachariah Williams too, continued to promote his ideas by writing letters in the hope of patronage and in 1740 published, apparently at his own expense:

> *The Mariner's Compass Completed: or, the complement of the art of navigation discover'd and propos'd: Being a dissertation concerning the magnetical variations of the mariner's compass, and the variation of that same variation, (according to a new hypothesis and system). As also directions for the use of an universal instrument, invented for ready knowing the variations of the mariner's compass, at all times and places.*

A second part appeared under a slightly different title in 1745.

In 1750, Williams became acquainted with Samuel Johnson who wrote several letters in 1751 supporting Williams and his ideas. In 1752, the ideas were brought to the attention of the Astronomer Royal, James Bradley. He noted that in some cases Williams's tables of variation agreed with the observations of others. However, in some cases they differed by 10 to 20 degrees (that is, 600 to 1,200 sea miles). A further issue with Williams's proposals, as with many other applicants to the Longitude Board, was a propensity to secrecy. As Kuhn (1984) relates, James Bradley recounted that:

> Williams had shown him a magnetical instrument by which the supposed tables [of variation] were constructed, but that Williams concealed from him the principle on which it was made and would not let him examine its internal construction. Bradley concluded that neither the tables nor the instrument could be relied on to ascertain the longitude at sea.[47]

In 1755, there appeared a final 'Williams' work: *An Account of an attempt to ascertain the Longitude at Sea, by an exact theory of the variation of the magnetic needle. With a table of variations at the most remarkable cities in Europe from the year 1660 to 1860.* In fact, the pamphlet was ghost-written by Samuel Johnson. It relates the whole magnetic variation saga, both in actual historical terms and in relation to Williams's part in the story. It records his meeting with the likes of Desaguliers, who introduced Williams to Edmond Halley, and reveals how Williams 'frequently showed upon a globe of brass, experiments by which [his] system was confirmed'. Zachariah Williams died in 1755, the year the pamphlet was published, but it reveals that on one point Williams scored an innovative goal with his tables, at least for a time.

When Williams first presented his tables to Isaac Newton and the other members of the Board of Longitude in *c*.1726 (see above), it was said that Newton refused to comment on them due to his advanced age (he died in 1727). However, as the pamphlet discloses, Newton had, in fact, been persuaded to look at them and saw there a prediction, by Williams, that the declination at London was still increasing, and that it would continue to do so until 1860. Newton thought the declination at London 'to be then stationary, and on the point of regression', and he duly declared Williams's system 'visionary', in the more derogatory sense of the word. At the time, however, the facts favoured Williams, as the declination did continue to increase for some decades. A retrograde motion, though, set in well before 1860.

The second method of determining longitude at sea – the lunar method – was known to work. It involved measuring the lunar distance – the angle between the moon and the sun or certain key stars – as well as the height of those objects above the horizon. This could be done using the instruments already noted above. After various corrections, such as those for parallax and atmospheric refraction, prepared tables or almanacs provided the time at which the same lunar distance had been seen at home-base. With knowledge of the time at the ship's current location and with the time at home-base known, longitude could be ascertained.

Nevil Maskelyne was a pioneer of these almanac publications and the data that they contained, which derived from, and relied on, the work and mathematical skill of calculators such as Thomas Simpson Evans at the Royal Observatory (see Chapter 14). There remained, though, difficulties with the lunar method because of: i) the complexity of the necessary corrections; ii) the time that it could take to undertake the necessary corrections and computations; and iii) from the weather, which could obscure the sighted objects or render the sea too rough for accurate measurement.

One way around the pitfalls of the lunar method was to use the timekeeper method. As we have noted above, one hour of time equates to 15 degrees of longitude, and four minutes equates to 1 degree of longitude, so knowing the precise time at a starting point of known longitudinal position, and the time at a ship's current position, would allow the ship's longitude to be calculated. However, this meant having two chronometers aboard the ship, one showing the current time aboard ship, the other permanently displaying the time at the home base. Unfortunately, maintaining accuracy in a clock subject to the vagaries of rough seas, salt air, temperature fluctuations and mishandling rendered the method unusable. The frustrations involved in attempts to supply a reliable chronometer are to be seen in letters of William Jones Sr held in the Royal Society archives. In a letter to James Hodgson, Master of the Royal Mathematical School at Christ's Hospital, Jones notes that 'The method I wrote of for finding ye time of day to a minute, (or indeed to absolute exactness) failed me and appears broke to pieces when I came to examples in which ye latitude and declination are contrary'.[48] He then increasingly turned his attention to clocks in relation to temperature, and to what he called 'logarithmic considerations', presumably relating to his interest in logarithmic tables. He presents the Society with lengthy proposals, such as 'A curious proposal and demonstrated theory for determining when, where, and how we may have ye time in ye place of observation to proper exactness', and 'The Curious Physical theory of the clocks with my last improvements'.[49]

Jones, who served as a Baptist minister at Wantage between 1713 and 1737, was reported to 'have given way to the study of the longitudes, in which study he made a considerable figure, but to the great damage of his ministerial abilities, the hurt of his temporal circumstances and the great damage of the church'.[50] The damage to his health may also be surmised from his letters. As one point in 1734 he asks the Royal

Society to 'excuse my instability I will as soon as heaven and health permit prepare and deliver (as I had done before if ready) my essays'. He then goes on to promise 'to be no f[u]rther troublesome, and to rid my head, heart and hand, as fast as I can of everything yet any ways pertaining to ye unfortunate name of longitude'.[51]

It would not be until 1759, ten years after Jones's death, that the problem of sufficiently accurate sea-borne chronometers to determine longitude would be resolved with the invention that year of H4, the last of a magnificent series of four chronometers created by the English carpenter and clockmaker John Harrison (1693–1776, Copley Medal 1749). Harrison would reap the rewards of the Longitude Prize (eventually) through a series of payments, with the last being made in 1773, when he was eighty, and fourteen years after the success of his last timepiece, H4.

The Timekeepers

In the previous chapter we noted the increasing number of clockmakers in Wales over the course of the long eighteenth century. There were also a few scientists of Wales who not only made innovative improvements to existing timekeepers but tried to invent new ones too.

Along with the steam engine already discussed above, Edward Somerset's *A Century of Inventions* (1663) contained two time-pieces. Invention 23 was a clock set in a castle, 'the water filling the trenches around it, shall shew by ebbing and flowing the hours, minutes and seconds, and all comprehensible motions of the Heavens and Counterlibration of the Earth, according to Copernicus'. The other, number 78, was for a continually going watch, though, as far as we know, neither invention left the drawing board.[52] In Chepstow, in the 1660s, a maker by name of Taynton is said to have constructed a clock 'that went with air, and another with water. And … a very small watch, most part of it being only of wood'.[53] A while later, when writing his *History of Oxfordshire* (1677), the one-time keeper of the Ashmolean Museum, Robert Plot, noted how another 'Aerotechnick' and Fellow of Jesus College Oxford had actually made a clock:

which moves by the air, equally compressed out of the bellows of a cylindrical form, falling into folds in its descent, much after the manner of Paper Lanterns. These, in place of drawing up the weights of other clocks, are only filled with air, admitted into them at a large orifice at the top, which is stop'd up again as soon as they are full with a hollow screw, in the head whereof there is set a small brassplate, about the bigness of a silver halfpenny, with a hole perforated scarce so big as the smallest pins head: through this little hole the air is equally expressed by weights laid on top of the bellows, which descending very slowly, draw a Clock-line, having a counterpoise at the other end, that turns a pulley-wheel, fastened to the arbor or axis of the hand that points to the hour: which device, though not brought to the intended perfection of the Inventor, that perhaps it may be by the help of a tumbrel or fusie, yet highly deserves mentioning, there being nothing of this nature that I can find amongst the writers of mechanicks.[54]

The 'ingenious' air technician in question was John Jones, who later became Chancellor of Llandaff cathedral (1691) and whose *Treatise on Opium* we noted in Chapter 12.[55]

One of the major innovations applied to timekeepers was the development of the pendulum by Christiaan Huygens in 1658. One of the subsequent concerns relating to the innovation was to free it, as far as possible, of any requirement to perform 'work' in relation to the clock – such as advancing the clock movement. The so-called 'free-pendulum', which simply regulated the clock without performing any 'work', was not introduced until 1898 by R. J. Rudd. However, the clockmaker John Tibbot (*c*.1757–1820) of Montgomeryshire has a claim, made in an 1817 letter to London-based Abraham Rees (see Chapter 16), to having invented a clock 'on an entirely new principle'. In a detailed horological 'P.S.' to his letter to Rees, Tibbot claimed the movements in his clock were 'detached from the pendulum 59½ seconds in every minute'. As Iorwerth Peate (1960) – founder and first curator at the National Folk Museum at St Fagan's – has written: 'I know of no clock with the pendulum swinging free for such a number of seconds, until the electrical clocks appeared much later.' Although Peate concludes that 'Tibbot's invention was of outstanding importance' a saga ensued with regard to its acceptance. In 1816, Tibbot submitted the clock for scrutiny by a committee of the London-based Royal Society of Arts, in the hope of it being recommended to the Government. With intimations that the assembled committee was comprised of 'common-clockmakers' (the lack of knowledge in the committee is suggested by some of their comments on the clock as recorded in the minutes of the Royal Society of Arts, see Peate (1950)), and the intrigues of others designed to oppose 'all strangers, and stragglers', Tibbot appealed to Rees to use his influence to have the clock removed from the Society 'to a proper place and fairly tried'. Unfortunately, that never seems to have happened, nor has the actual clock ever been subsequently found. Yet, as Peate (1950) suggests, if it could be found, 'this Montgomeryshire craftsman may well be ajudged the inventor of the free-pendulum – a claim which, on the evidence of his description [in his P.S. to Rees] may now be presumed'.[56]

The Electricians

Beyond clockwork it was, as we have already seen, wind and water, fire and steam that provided the motive powers of the long eighteenth century, but scientists of Wales were among those associated with another innovation destined, in later times, to become a power source beyond their wildest imaginings – electricity.

Static electricity, produced by rubbing the likes of amber or glass tubes or rods with cloth, had long been known, its actual existence being indicated by sparks, strange glows and the curious attraction of objects to the rubbed material. By the early-to-mid-eighteenth century a static charge had been made easier to produce, and its power significantly enhanced, by various mechanical devices, but it remained largely a source of entertainment rather than true scientific enquiry. The 1730s saw a more scientific approach develop and Stephen Gray (1666–1736, FRS 1732, Copley Medal 1731, 1732) become one of the most important experimenters. A dyer and astronomer from Canterbury, he fell on hard times and in 1720 entered the Charterhouse for poor

gentlemen in London. While there, Gray conducted numerous electrical experiments and made important electrical discoveries. In 1729,
when attempting to see whether a glass tube corked at both ends
could still hold a static charge by rubbing, he noted that the corks
became charged along with the tube. Furthermore, when he inserted
a wooden stick or rod into the cork the charge appeared at the end of
that, and was then taken up by an ivory ball when he added that on
to the stick. Having effectively discovered electrical conductivity, he
experimented to determine how far what he called this 'communicative
electricity' could be conducted. His experiment eventually extended to
some 765 feet and in the process the electricity remained unaffected
by gravity and bends in the conducting wire.[57] Gray also identified
conductors and non-conductors, and he made the famous 'Flying Boy'
experiment. Using a boy suspended by silken cords from the ceiling as
a scientific guinea pig, a static generator had been rubbed close to, but
not touching, the boy's feet. The child's body becoming so charged that
his hands and face attracted pieces of 'leaf-brass'. This was the first time
a human body had been electrically charged and it provided a powerful
illustration of electrical induction – an electrical charge created in a
suspended object without contact.

Witness to at least some of Gray's experiments were Zachariah
Williams (see above), who entered Charterhouse on 29 September 1729
on the recommendation of Sir Robert Walpole, and Zachariah's daughter, Anna Williams. She entered sometime around 1746 to live with
and look after her father; though this was against the men-only rules
of Charterhouse and led to their subsequent eviction in 1748. Anna
was an educated woman. Samuel Johnson noted that 'she understands
chemistry and many other arts with which ladies are seldom acquainted'.
She knew the Italian language and, in 1746, published a translation into
English of a French language work on the life of the Roman emperor
Julian. An article by 'B', in the *London Magazine* (1783), records that:

> her skill in geography was uncommon. She knew the relative
> situation of almost every place on the globe. Nor was she less
> acquainted with magnetism, and the powers of a lodestone. The

instruments which her father invented to ascertain his fancied discoveries remained in her possession till she died.[58]

Anna would be rendered blind by cataracts and an operation in 1752, undertaken at the instigation of Samuel Johnson and performed in his house, failed to improve matters. From that time on she lived in Samuel Johnson's London home.

Stephen Gray, according to Zachariah Williams, would repay information on magnetism from Williams with his own on electricity, and it is apparent that, well before she actually entered Charterhouse, Anna too, on visits to her father, aided Gray with his experiments. Much later she published a poem – *On the Death of Stephen Gray F.R.S.; The Author of the Present Doctrine of Electricity* – in her *Miscellanies of Prose and Verse* (1766). James Boswell wrote that, as with her father's work, Samuel Johnson substantially rewrote the piece, though by exactly how much is not known for certain. From a history of science perspective, however, the most interesting feature of the work is a codicil passage Anna added to the published poem: 'The Publisher of this Miscellany, as she was assisting Mr. Gray in his experiments, was the first that observed and notified the emission of the electrical spark from a human body.' Given the nature of Gray's experiments, and the fact that he was the first to electrically charge a human being, her claim is not in any way unlikely. Her observation, and notification of it, though, does not seem to have registered with Gray himself. In a paper in the *Philosophical Transactions*, of 1735/6, he gave the credit for the discovery of the electric shock to the French chemist, Charles Dufay (1698–1739), who had described it in a *Philosophical Transactions* paper of 1733/4. Gray died in 1736, aged seventy, so it is conceivable that he simply forgot about Anna's earlier observation and notification, and that she felt the need to reclaim her priority later. The truth of the matter is unlikely to be resolved now, but Anna clearly realised the importance of the/her discovery.[59]

A further electrical advance in the mid-1740s came with the invention of the Leyden jar. Allowing for the storage of a high voltage charge for the first time, it was improved on in 1747/8 by Benjamin Franklin. Having earlier conducted his famous key and kite experiment, to show lightning to be a form of electricity, Franklin found a method to connect together a series of Leyden jars, thus producing an early form of electrical battery. In fact, the year 1747 proved to be a busy one for electrical developments. In that same year the physician and scientist William Watson (1715–87, FRS 1741, Copley Medal 1745) took Gray's conductivity experiments further by investigating the speed of electrical conductivity over wires sometimes extending to 6,732 feet and 12,276 feet; with various eminent personages feeling the shock along the way. Watson published his results in the *Philosophical Transactions* and notes in one experiment carried out at Stoke Newington the presence of the presumably 'shocked' mathematician, William Jones Sr.[60]

Someone else who capitalised on the increasing mid-eighteenth-century interest in electricity was Francis Watkins. Just three months after setting up as an optician in London in 1747, the *General Advertiser* records that he had made to his own design:

> a most compleat Electrical Apparatus for performing, at any time of the day [the viewer to pay one shilling], all known Experiments, with several new ones, and a very singular one, wherein a Horizontal iron bar is forced several feet out of its Place, only by the Electrical Attraction of a Bit of Down weighing less than Half a Grain.

Anyone purchasing the machine could also obtain his 1747 handbook detailing the care and maintenance of his electrical machine, and the experiments 'shown in his course'. The book – *A Particular Account of the Electrical Experiments hitherto made publick* – could be bought at Homer's Head, Temple Bar, at the shop of Watkins's brother-in-law, the bookseller and publisher William Owen.[61]

Although Richard Price is said to have possessed a machine for generating static electricity in his study, he does not seem to have left any record of electrical experiments in his surviving oeuvre. He did, though, act as a conductor for the literary efforts of his electrician friends: Benjamin Franklin, Joseph Priestley and John Canton. For example, Joseph Priestley wrote a number of letters to Price between

1766 and 1768 while completing his monumental *The History and Present State of Electricity, with Original Experiments* (1767). The letters request Price to try and obtain difficult-to-find reference material, including works by the Professor of Physics at the Imperial Academy in St Petersburg, Franz Ulrich Theodosius Aepinus (1724–1802). In fact, Price would prepare a list of analogies between magnetism and electricity from the work of Apeinus, which Priestley included in his published work. Priestley also outlined in his letters the experiments he was currently conducting, together with his hope that Price would impart the information to Franklin and Canton:

> I take it for granted you [Price] have seen the letter I wrote, about a fortnight ago, to Dr Franklin. I desired he would show it to you, and Mr Canton. Writing upon a philosophical subject to any of you; I would have it considered as writing to you all. I mentioned to him an experiment I had made, and which I believe is a new one proving that there is a real current of electric matter from all electrified points, negative and positive … [a lengthy and detailed description follows here] … I beg you would communicate these experiments to Dr Franklin and Mr Canton, and let me know what they think of them. I shall be glad to reconcile them with the hypothesis of positive and negative electricity.[62]

Although unrecognised at the time, a major contribution to electrical science in the long eighteenth century came from the experimental work of Bridgend-born William Morgan, who we met earlier as the father of actuarial science (Chapter 12). In 1785, his uncle, Richard Price, who had witnessed the experiments along with a Mr Lane (probably the apothecary and inventor Timothy Lane (1734/5–1807, FRS 1770, APS 1772)), read at the Royal Society Morgan's paper on 'Electrical Experiments made in Order to Ascertain the Non-Conducting Power of a Perfect Vacuum'; it was later printed in the *Philosophical Transactions*.[63] As Morgan says in the paper 'The non-conducting power of a perfect vacuum is a fact in electricity which has been much controverted among philosophers'. He attempted to examine that non-conductivity using the comparatively simple equipment illustrated in Figure 31, and by a method, and with a result, best described by Morgan himself:

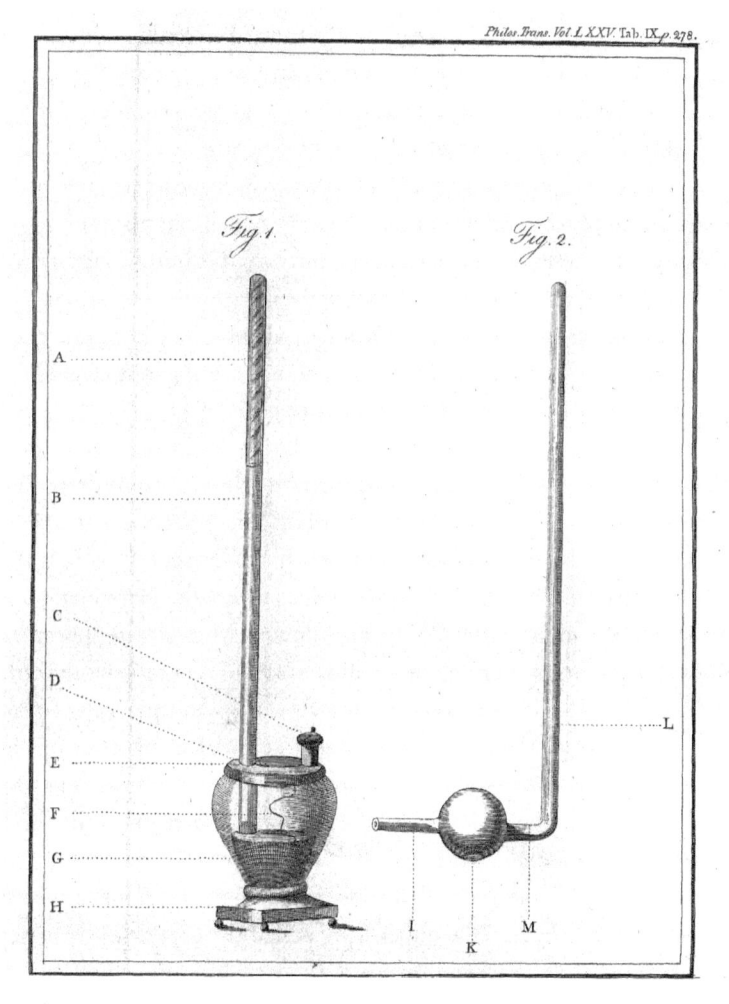

Philos. Trans. Vol. LXXV. Tab. IX. p. 278.

FIGURE 31 Equipment used by William Morgan to produce soft X-rays, as illustrated in his *Philosophical Transactions of the Royal Society* paper 'Electrical Experiments made in order to ascertain the non-conducting Power of a perfect Vacuum. &c.' (1785).

A mercurial gage B about 15 inches long, carefully and accurately boiled till every particle of air was expelled from the inside, was coated with tin-foil five inches down from its sealed end (A), and being inverted into mercury through a perforation (D) in the brass cap (E) which covered the mouth of the cistern (H), the whole was cemented together, and the air was exhausted from the inside of the cistern through a valve (C) in the brass cap (E) just mentioned, which producing a perfect vacuum in the gage (B) afforded an instrument peculiarly well adapted for experiments of this kind. Things being thus adjusted (a small wire (F) having been previously fixed on the inside of the cistern to form a communication between the brass cap (E) and the mercury (G) into which the gage was inverted) the coated end [of the gage] (A) was applied to the conductor of an electrical machine, and notwithstanding every effort, neither the smallest ray of light, nor the slightest charge, could ever be procured in this exhausted gage.[64]

With the equipment available to Morgan the conclusion he drew, that perfect vacuums do not conduct electricity, was 'correct' for the time. Morgan was able to produce a sufficiently good vacuum so that the voltage applied did not cause a flow of electrons. The introduction of a gas into the tube would allow such a flow, as Morgan later observes. Our modern understanding of conduction in a vacuum is more comprehensive. In today's 'perfect vacuums' – far in advance of those of Morgan's – very high potential differences can drive free electrons to flow. But the whole interplay between vacuum, source of electrons and electric field (potential difference) is complicated theoretically and, indeed, experimentally.[65]

During his experiment William Morgan had also noticed that 'if the mercury in the gage be imperfectly boiled the experiment will not succeed'. When this proved the case, it was because Morgan had created a partial rather than a perfect vacuum. In such cases, he also made an historically significant observation: 'the colour of the electric light which, in an air rarefied by an exhauster, is always violet or purple, appears in this case of a beautiful green.' As his great-great-great-granddaughter,

Nicola Bruton Bennetts, recently recounted in her 2022 biography of William Morgan, what he observed 'was the ionisation of a gas: the gas within his partial vacuum was split into ions which bombarded the glass tube and produced florescence – and soft X-rays'. Soft X-rays are rays of a lower energy and longer wavelength than hard X-rays. Morgan did not actually know these were what he had produced and observed but, as Nicola Bruton Bennetts also points out, his doing so proved 'the first link in a chain of investigation', one that would involve Humphry Davy and Michael Faraday and lead to the 1875 invention of the Crookes tube, by William Crookes, and from there to Röntgen's use of a Crookes tube in his discovery of X-rays in 1895.[66] William Morgan would conduct other electrical experiments including those described and illustrated in his unpublished manuscript 'On the law of electric attraction', which is in the archives of the Royal Society.[67]

Finally, there is William Morgan's brother, George Cadogan Morgan. We saw in an earlier chapter his involvement in a dispute over the use of pointed or blunt lightning conductors, but as the eighteenth century gave way to the nineteenth, he also made a contribution with the publication of his two volume *Lectures on Electricity* (1804). Although essentially a distillation of his work as a teacher, the volumes provide a valuable record of the state of electrical experimentation and equipment at the time of their publication (see Figure 32). They also reveal some of the innovative uses being suggested for the new science. George Morgan himself suggests that the electrical fluid might be used for the firing of explosives, and he goes on to suggest its potential for stimulating plant growth, though he recognises the evidence for this is not proven. Above all, though, there was its value to medicine.

The Medics

In an 'Introductory Lecture' to his *Lectures on Electricity*, George Morgan comments that 'Much has been attempted and professed by electrical empirics, in the art of medicine; but hitherto electricity has been chiefly employed as the instrument of quackery and imposture'. One person who tried to improve the use of electricity in medical settings

LECTURES

ON

ELECTRICITY.

BY

G. C. MORGAN.

VOL. II.

NORWICH:

PRINTED BY J. MARCH,

and sold by

J. JOHNSON, ST. PAUL'S CHURCH-YARD, LONDON.

1794.

Price 10s. 6d.

FIGURE 32 Title page, and electrical equipment illustrated in George Cadogan Morgan's two-volume *Lectures on Electricity* (Norwich, 1794).

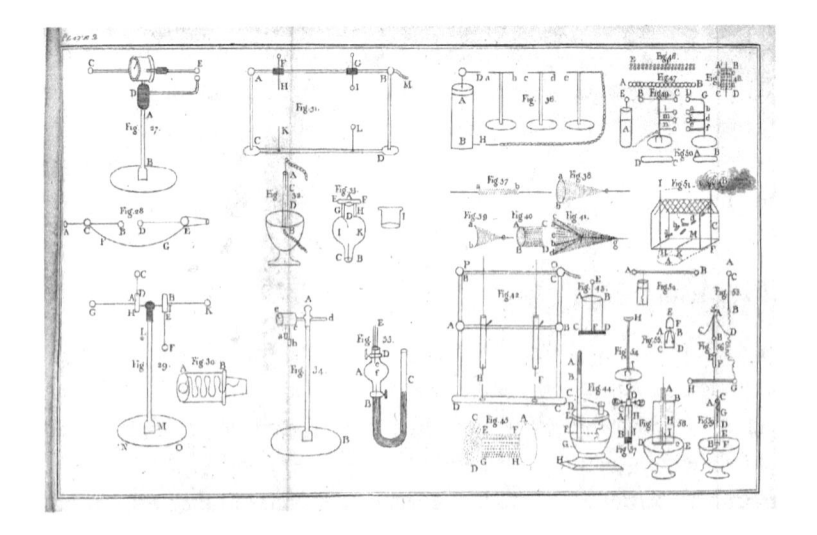

was Timothy Lane, who we noted above as one of the likely observers of William Morgan's electrical experiments. Lane invented what he called an Electrometer, its purpose being to monitor the amount of discharge from a Leiden jar or battery. He relates his invention in a letter to Benjamin Franklin that was subsequently published in the *Philosophical Transactions* of 1767. Much later, in 1804, John Lloyd of Wigfair appears to have consulted Lane concerning an injury that Lloyd had done to his hand. In a fascinating letter, Lane relates in considerable detail the use of electricity, including the giving of gentle shocks by the Electrometer, but always remembering that 'When the patient expresses a wish to have no more leave off for that time. Increase the shocks every day by degrees. Some days a person can bear larger shocks than at others. To fatigue the patient by giving too large or too many is improper.'[68]

We have already seen Francis Watkins's interest in electrical machines but, in association with the watch and globe maker John Neale, he also adopted what appears to be an innovative approach to medicine and electricity by desiring to collect the experiences of those who had been subjected to electrical treatments. Published in 1747, both Neale's *Directions for Gentlemen who have Electrical Machines, how to proceed in making their experiments* and Watkins's *A Particular Account of the Electrical Experiments hitherto made Publick* contained their request for such information (see Table 6).[69]

TABLE 6 Request for information about the experiences of those who had been subjected to electrical treatments

---

Whereas from several late influences of the success of electrical applications to persons afflicted with nervous disorders, especially of the paralytic kind, it appears highly probable, that great benefit may accrue from thence to the community; and that it will be well worth while to determine further by experience in what particular cases success might be expected, or despaired of; it is earnestly requested of all those who have made or undergone any such trials, that they would be pleased to transmit circumstantial and well attested accounts thereof, either to *Francis Watkins*, at Sir *Isaac Newton's Head, Charing-cross*, or to Mr *John Neale*, in *Leaden-hall-street*, to be inserted in a narrative in due time to be published, of the success of electrifying in the aforesaid disorders.

Watkins also developed a portable electrical machine 'scarce exceeding the size of an ordinary Tea-chest' that could be 'taken out [and] fixed up fit for use by one Person in a minute; and at the same time the person turns the machine with one hand, has the other at Liberty to do what is necessary'. It delivered the 'Electrical Virtue' and 'as soon as any Gentleman sees my machine, he will prefer it to any yet made, by reason of its portableness'.[70]

More specialised medical practitioners also combined medical science and technology in their invention of new tools of the trade, or their modification of old ones. In 1657 an acquaintance of John Aubrey published *Organon Salutis: an instrument to cleanse the stomach*. It was written by the Welsh judge and politician Walter Rumsey (1584–1660) of Llanover, in Monmouthshire. He later attended Gloucester Hall, Oxford and is said to have been taught by Francis Bacon and William Harvey.[71] In his work Rumsey detailed his invention of the Probang. This was invented by Rumsey to clear his throat of phlegm, with which, according to John Aubrey, 'he was much troubled ... sitting by the fire, spitting and spewing'. Rumsey took 'a fine tender sprig, and tied a rag at the end, and conceived he might put it down his throat, and fetch up the flegme'. It worked, with later versions being made of whale bone with a piece of sponge attached. Modern forms of the instrument are still in use, particularly in veterinary science where it is used to gently push obstructing bodies in the animal's oesophagus down into its stomach. Aside from also being a keen plantsman and composer, Rumsey also wrote on the medicinal uses of coffee, both as an electuary (a purgative) for use with the probang, and as a cure for the effects of drunkenness.

Another potential cure for the morning after the night before came from the inventive chemistry of Wrexham-born, Manchester-based, Thomas Henry (1734–1816). He invented a process for preparing magnesia alba, which became a remedy for heartburn and indigestion and proved useful as an antacid. He also utilised Joseph Priestley's 1767 discovery of carbonated water to begin the manufacture of artificial mineral waters. Henry built the very first factory for the purpose in 1781, while introducing some adjustments to Priestley's original apparatus for producing the carbonated water.

Among practising medics, the surgeon Griffith Rowlands, one of the pioneers of the surgery to join two ends of a broken bone (see Chapter 12), is said to have made a retractor and surgical saw of his own invention. Indeed, 'his mechanical ingenuity of devising ways and means to assist the inefficient resources of nature, was very considerable, and in numerous instances proved extremely satisfactory in its application'.[72]

Others were inspired to innovation and invention by their own particular circumstances. Surgeon David Daniel Davis, who we met previously as the first professor of midwifery at University College London, developed five different forms of forceps for different birth scenarios. His concerns were to reduce 'injury to the baby as well as minimising maternal trauma', concerns probably relating to the death of his own child from injuries suffered at birth from the existing technologies of the day.

Notes

1. With thanks to Stefano Barazza, Davide Crivelli and Dimitra Fimi, see *welshcrucible.org.uk/recovering-a-forgotten-welsh-inventor-philip-vaughan/* (accessed 26 May 2024).

2. See Chris Evans and Louise Miskell, *Swansea Copper: A Global History* (Baltimore, 2020).

3. John Davies, *A History of Wales* (London, 1993), p. 325.

4. See J. R. Harris, *The Copper King: Thomas Williams of Llanidan* (1964; Ashbourne, 2003).

5. See, for example, Reginald Nichols, *Pontypool and Usk Japan Ware* (Pontypool, 1981); also, Harris, *The Copper King*; Evans and Miskell, *Swansea Copper*; Stephen Hughes, *Copperoplis: Landscapes of the Early Industrial Period in Swansea* (Aberystwyth, 2000); Ronald Rees, *King Copper: South Wales and the Copper Trade 1584–1895* (2000; Cardiff, 2012); Keith Strange, *Merthyr Tydfil: Iron Metropolis* (Stroud, 2005).

6. A. Wolf, *A History of Science, Technology and Philosophy in the 16th & 17th Centuries*, 2 vols (1935; London, 1962), II, pp. 545, 547.

7. 'P[ublic] R[ecord] O[ffice] SP (Dom.) 35/76/21. Patent of William Edwards, February 26, 1724/1725': see Larry Stewart, *The Rise of Public Science, Rhetoric, Technology, and Natural Philosophy in Newtonian Britain, 1660–1750* (Cambridge, 1992), p. 338 n. 3, and all of Chapter 11. Also, in Bennet Woodcroft, *Subjects Matter Index (Made from titles only) of Patents and Inventions 1672–1815, Part II, N–W* (London, 1854), p. 589; another, perhaps additional, patent by William Edwards is noted as: 'Patent No. 480, 28 June 1725, Making pipes of clay or earth for conveying water underground from place to place'.

8. Stephen Switzer, *An Introduction to a General System of Hydrostaticks and Hydraulicks Philosophical and Practical*, 2 vols (London, 1729), I, pp. 116–17. See also, in the same volume, 'Notes upon Book I', (n.p.) and the entry 'Chap VII. Pg. 97. Line 18'.

9. See *https://collections.britishart.yale.edu/catalog/tms:14267* (accessed 31 October 2022).

10. Audrey T. Carpenter, *John Theophilus Desaguliers: A Natural Philosopher, Engineer and Freemason in Newtonian England* (London and New York, 2011), p. 135.

11. Brian Gee (ed.) with Anita McConnell and A. D. Morrison-Low, *Francis Watkins and the Dolland Telescope Patent Controversy* (Farnham, 2014), pp. 60–4.

12. *ML-*1, p. 451.

13. See Hugh Owen, *The Life and Works of Lewis Morris (Llewelyn Ddu o Fôn) 1701–1765* (Anglesey, 1951), pp. 189–93, citing NLW MS 67A f. 169–86.

14. See Rob Gittins, *Rock & Roll to Paradise: A History of the Mumbles Railway* (Llandysul, 1982), p. 8.

15. Torsten Berg and Peter Berg (trans.), *R R Angerstein's Illustrated Travel Diary, 1753–1755* (London, 2001), pp. 147–56, 159–66, 321–7.

16. J. R. Harris, *Industrial Espionage and Technology Transfer: Britain and France in the Eighteenth Century* (1998; Abingdon and New York, 2017), p. 232.

17. R. T. W. Denning (ed.), *The Diary of William Thomas 1762–1795* (Cardiff, 1995), p. 73.

18. John Fisher (ed.), *Tours in Wales (1804–1813) by Richard Fenton* (London, 1917), p. 94.

19. See Hughes, *Copperopolis*, pp. 132–41.

20. See *www.merthyr-history.com* (accessed 28 May 2024).

21. For all details, see E. I. Williams, 'Samuel Baldwyn Rogers', *NLWJ*, 11 (1959), 99–102.

22. See *https://coflein.gov.uk/en/site/405050?term=dinas%20brick* (accessed 2 November 2022).

23. See 'Copper-men, Colliers & Mechanicks', in Hughes, *Copperopolis*, pp. 131–54.

24. Quoted in Revd Dr Richard L. Hills, *Life and Inventions of Richard Roberts 1789–1864* (Ashbourne, 2002), p. 2.

25. The patent application for the punching machine is dated to 5 March 1847. See Hills, *Life and Inventions of Richard Roberts*, pp. 105–10.

26. For all details see Hills, *Life and Inventions of Richard Roberts*.

27. NLW MS 12420D, f. 1.

28. NLW MS 12418D, f. 58.

29. Paul Elliott, 'The Derbyshire General Infirmary and the Derby Philosophers: The Application of Industrial Architecture and Technology to Medical Institutions in Early-Nineteenth-Century England', *Medical History*, 46 (2000), 65–92 (at 80).

30. NLW MS 12418D, f. 58, 59. A plan of the washing machines at Derby can be found in Elliott, 'Derbyshire General Infirmary', 79.

31. NLW MS 12424C, f. 44.

32. OHSM archives, Lewis Evans papers, MS 18 Part 3 Letters: Abraham Robertson to Lewis Evans, 27 September 1801. For full details of the Commission and its

enquiries, see Edward Cresy, *An Encyclopædia of Civil Engineering: Historical, Theoretical, and Practical* (London, 1847), pp. 475–93.

33. W. Bernard Peach and D.O. Thomas (eds), *The Correspondence of Richard Price*, 3 vols (Durham NC and Cardiff, 1983–94), II (1991), p. 193.

34. A small, unmanned, hydrogen filled balloon had been launched on 27 August 1783 from the Champ-de-Mars by Jacques Charles and the Robert brothers, with Benjamin Franklin among the onlookers.

35. NLW MS 12415C, f. 60.

36. *Correspondence of Richard Price*, II, p. 214.

37. *Correspondence of Richard Price*, II, p. 224.

38. *Correspondence of Richard Price*, II, p. 237.

39. The two Frenchmen are not named; however, it is of note that one Frenchman who arrived in Britain in August 1784 was Jean Blanchard (1753–1809). He travelled within Britain but is not known to have flown until 16 October 1784 in London. Later, in 1785, he made the first ever balloon crossing of the English Channel with John Jeffries. He also made the first balloon flight in America in 1793.

40. Clare Brant, *Balloon Madness: Flights of Imagination in Britain 1783–1786* (Woodbridge, 2017), p. 94.

41. *Correspondence of Richard Price*, II, p. 237 and n. 9.

42. Brant, *Balloon Madness*, pp. 268–70; Louise Miskell, *'Intelligent Town': An Urban History of Swansea, 1780–1855* (2006; Cardiff, 2012), pp. 2–5.

43. It is of interest that the 'father of aeronautics' Sir George Cayley (1773–1857) was very much influenced by the mathematics and physical sciences taught to him by Price's nephew, George Cadogan Morgan, who ran a school at Southgate, in Middlesex, which Cayley attended from 1792. That same year, 1792, Cayley had his first thoughts on heavier-than-air flight. See J. A. D. Ackroyd, 'Sir George Cayley, the Father of Aeronautics, Part 1: The Invention of the Aeroplane', *Notes and Records: The Royal Society Journal of the History of Science*, 56 (2002), 167–81.

44. See Dava Sobel, *Longitude* (London, 1996), pp. 11–13; N. A. M. Rodger, *The Command of the Ocean*, (2004; London, 2005), p. 172.

45. For a detailed discussion of the search for longitude, see William J. H. Andrews (ed.), *The Quest for Longitude* (Cambridge MA, 1996).

46. J. Harris and George Graham, 'An Account of some Magnetical Observations Made in the Months of May, June and July 1732 in the Atlantick or Western Ocean; Also the Description of a Waterspout by Mr Joseph Harris'. Communicated by Mr George Graham F.R.S., *Philosophical Transactions of the Royal Society*, 38 (1733), 75–9.

47. Albert J. Kuhn, 'Dr. Johnson, Zachariah Williams, and the Eighteenth-Century Search for Longitude', *Modern Philology*, 82 (1984), 40–52 (at 47).

48. RSC MS EL/i1/185.

49. RSC MS LBO/21/13.

50. Brian Bowers, 'A Baptist Man of Science', *The Baptist Quarterly*, 38 (2000), 305–7. See also Chapter 4, note 18 of this volume.

51. RSC MS LBO/21/13.

52. See William Linnard, *Wales Clocks and Clockmakers* (Ashbourne, 2003), p. 49.

53. Linnard, *Wales Clocks and Clockmakers*, p. 47.

54. Robert Plot, *The Natural History of Oxfordshire* (Oxford, 1677), p. 230.

55. The modern Atmos clock was invented by Jean-Léon Reutter and developed by Jaeger-LeCoultre watchmakers in 1928. It winds automatically and uses temperature and atmospheric pressure changes to obtain its energy. Friction needs to be as limited as possible and the clocks use a torsion pendulum, which is more efficient than an ordinary pendulum. Clocks using atmospheric pressure and temperature changes were first invented by Dutchman Cornelius Drebbel (1572–1633). He made such clocks for James I.

56. For all details, see Iorwerth Peate, 'John Tibbot, Clock and Watch Maker', *Montgomeryshire Collections*, 48 (1944), 176–85; Iorwerth Peate, *Clock and Watch Makers in Wales* (1945; Cardiff, 1960), pp. 22–4; Iorwerth Peate, 'John Tibbot's Inventions', *Montgomeryshire Collections*, 51 (1950), 38–50.

57. Date and details given in A. Wolf, *A History of Science, Technology, and Philosophy in the Eighteenth Century* (1938; London, 1952), pp. 215–6.

58. 'Memoirs of Mrs Anne Williams', *London Magazine*, 52 (1783), 517–21.

59. A full discussion of Anna's discovery and the arguments over priority can be found in Bodil Hoist and N. Zimmer, 'The Discovery of the Electric Shock', *Science*, 298/5602 (20 December 2002), pp. 2327–8.

60. William Watson, 'A Collection of the Electrical Experiments Communicated to the Royal Society by Wm. Watson F.R.S.', *Philosophical Transactions of the Royal Society*, 45 (1748), 49–120.

61. For all details and quote, see Gee et al., *Francis Watkins*, pp. 50–2.

62. *Correspondence of Richard Price*, I (1983), p. 38. For details of Price's contributions to, and editing of, Priestley's works, see Paul Frame, *Liberty's Apostle, Richard Price his Life and Times* (Cardiff, 2015), pp. 60–2.

63. William Morgan, 'Electrical Experiments made in Order to Ascertain the Non-Conducting Power of a Perfect Vacuum', *Philosophical Transactions of the Royal Society*, 75 (1785), 272–8.

64. Morgan, 'Electrical Experiments', 272–3.

65. I am extremely grateful to John Tucker for providing this explanation.

66. See all of Chapter 6 in Nicola Bruton Bennetts, *William Morgan: Eighteenth-Century Actuary, Mathematician and Radical* (Cardiff, 2020).

67. RSC L&P/10/105, available online.

68. NLW MS 12420D f. 2, letters to John Lloyd.

69. Table 6's content is presented as in Gee et al., *Francis Watkins*, p. 53, and originally transcribed from Watkins' *Electrical Experiments*.

70. See Gee et al., *Francis Watkins*, p. 54.

71. Bennett Alan Weinberg and Bonnie K. Bealer, *The World of Caffeine: The Science and Culture of the World's most Popular Drug* (London and New York, 2001), p. 97.

72. Joseph Hemingway, *A History of the City of Chester*, 2 vols (Chester, 1831), I, pp. 363–6 (at p. 365).

16

ENCYCLOPAEDIC KNOWLEDGE

I have retired into a Little Villa of my own, where my Garden,
Orchard & Farm, & some small mineworks (take a good part of my
time) and a little knowledge in Physic & Surgery, which brings me
the visits of the poor; Botany having been my favourite Study, is now
of use to *them*. Natural Philosophy and Mathematics have taken
up much of my attention from childhood, and I have a tolerable
Collection of Fossils, Shells, &c. from most parts of the World, &
a valuable Collection of Instruments and apparatus's on that head.
Models of engines also hath taken up a part of my thoughts and
in this branch of Mech[s] I have made some Improvements, beyond
what has been published on that Subject in Britain or France.[1]

Lewis Morris writing to Samuel Pegge, 11 February 1761

In reflecting on his life in this way, in a letter written just four years
before his demise, Lewis Morris effectively outlines the course of
science presented in the preceding chapters of this book, from the collection and classification of fossils, minerals, shells and botanical and
zoological specimens, through to a realisation of their usefulness and the
potential to develop that usefulness through technological and mechanical advance. Morris also presents himself as something of a polymath,
a person of learning in numerous subjects with an ability to relate those
bodies of knowledge to the solving of problems. The polymath was
a guise that became increasingly difficult to maintain after the mid-
eighteenth century, given the volume and ever widening nature of the
scientific knowledge by then presenting itself to the individual and to
society at large. The contribution of scientists of Wales to this expanding body of knowledge has been the subject of the preceding chapters.

In this final chapter we consider their contribution to making that body of knowledge available to a long-eighteenth-century audience.

Compendia of Knowledge

Whether they appear as a paper in a scientific journal, a volume on a specialised topic, or a work embracing a variety of different subjects, all science publications are essentially compendia of knowledge. Many scientists of Wales produced such compendia in the course of the long eighteenth century (see Table 7). They include single papers, such as William Morgan's pioneering work on conductivity in a vacuum, as well as volumes on a wide range of topics including fossils and zoology, navigation, surveying, the mathematics of life insurance and annuities, and the wonders of electricity.

These compendia, though, did not really give the reader a sense of the wider knowledge base available at any particular time in the long eighteenth century. That more comprehensive purpose fell to the encyclopaedists.

Encyclopaedias of Knowledge

The word 'encyclopaedia' derives from the Greek *enkyklios paidea*, which is usually translated as either 'the circle of knowledge' or 'general education'. In contrast to a dictionary, which deals exclusively with the meaning of individual words, encyclopaedias tackle the understanding of things, with the earliest surviving example being the *Historia Naturalis* ('Natural History') by the Roman writer Pliny the Elder (*c.*23–79). Completed about 77, the *Historia* details the knowledge of its time in subjects such as astronomy, meteorology, botany, zoology, mineralogy and geography.[2] Later, in Christian medieval Europe, the *Sepher Rezial Hemelach* ('Book of the Angel Rezial') appeared. The angel Rezial, acting as a messenger from God, was charged with giving to the first man – Adam – the secret knowledge of the ways of nature, the Adamic knowledge that many later scientists felt they were endeavouring to recover, as we saw in Chapter 1.[3] Today, surviving copies of the book

TABLE 7 Selection of Compendia of Knowledge
produced by Scientists of Wales

A Selection of Compendia of Knowledge produced by Scientists of Wales
(1650–1820)

1655 (published 1663), *A Century of the Names and Scantlings of Inventions by me already practised*, by Edward Somerset

1660, *Humane Industry: or, a History of most Manual Arts, Deducing the Original, Progress, and Improvement of them. Furnished with a variety of Instances and Examples, shewing forth the excellency of Humane Wit*, by Thomas Powell

1665, *Another Collection of Philosophical Conferences of the French Virtuosi, upon questions of all sorts; for the improving of Natural Knowledge*. Translated from the French by George Havers and John Davies of Kidwelly

1699, *Lithophylacii Britannici Ichnographia*, by Edward Lhwyd

1702, *A New Compendium of the Whole Art of Practical Navigation*, by William Jones Sr

1706, *Synopsis Palmariorum Matheseos* ('A New Introduction to Mathematics'), by William Jones Sr

1718, *Hanes y Byd a'r Amseroedd* ('History of the World and Times'), by Simon Thomas

1725, *Golwg ar y Byd* ('A View of the World'), by Dafydd Lewis

1729, *The Description and Use of the Globes and the Orrery*, by Joseph Harris

1730, *A Treatise of Navigation*, by Joseph Harris

1758, *An Essay on Money and Coins*, by Joseph Harris

1768, *British Zoology*, by Thomas Pennant

1771, *Observations on Reversionary Payments*, by Richard Price

1779, *The Doctrine of Annuities and Assurances on Lives and Survivorships*, by William Morgan

1785, *Electrical Experiments Made in Order to Ascertain the Non-Conducting Power of a Perfect Vacuum*, by William Morgan

1793, *Pharmacopoeia*, by Nathaniel Williams

1794, *Lectures on Electricity*, by George Cadogan Morgan

1798, *A Complete Treatise of Land Surveying*, by William Davis

1813, *Welsh Botanology*, by Hugh Davies

1816, *Daearyddiaeth* ('Geography'), by Robert Roberts

---

are believed to be derived from a medieval encyclopaedic compilation of various Cabbalistic writings.[4] Also within the medieval period, encyclopaedists with a 'passion for producing meticulous compilations and analyses extending to all nature and to human life as well' were active in what Starr (2013) calls the *Lost Enlightenment* of Central Asia (800–1200).[5] In later centuries, encyclopaedic compilations from Moorish Spain, with a few Christian examples from elsewhere, prefigure those of long-eighteenth-century Europe.

Among the encyclopaedias of that long century in Europe was the *Lexicon Technicum: Or, An Universal English Dictionary of Arts and Sciences: Explaining not only the Terms of Art, but the Arts Themselves.* Compiled and published in London as two volumes (1704 and 1710) by the English clergyman and mathematician John Harris, the second volume contains a list of some 1,300 subscribers.[6] Among them are a number of Welshmen, including 'Henry Rowlands of Anglesey' – this is probably the author of *Idea Agriculturae* (written 1704, published 1764) and the short work on the origin of fossils discussed in Chapter 9. Other Anglesey subscribers include Robert Humphreys, who is described as a 'clerk', Pierce Lloyd Esq. of 'Laniden', and a John Wynn Esq. of 'Wainwenn'. Then, along with Thomas Wilkins of 'Lemmington in Flintshire', Thomas Wynn Esq. of Dyffrin-Aled in Denbighshire, and Nicholas Williams of Rhydowen in Carmarthenshire, we have a reference to 'The Reverend Mr William Jones'. The last mentioned is likely to be the mathematician and creator of the symbol for pi, π, William Jones Sr, who became a minister at a Baptist church in Wantage, near Reading, by at least 1713, and probably as early as 1709.[7] When compiling his *Lexicon*, John Harris declared that he made use of 'the best Original Authors' he could find. As related in Chapter 11, it is likely that Jones Sr was one of those original authors and that he contributed articles to the *Lexicon* on navigation.[8] This conclusion is circumstantially supported by the fact that Jones not only dedicated his own *A New Compendium of the Whole Art of Practical Navigation* (1702) to Harris but also declared it to have been composed while he was living under Harris's roof.

Another important eighteenth-century encyclopaedic development in Britain came with the publication of Ephraim Chambers' two-volume

Cyclopaedia: or, an Universal Dictionary of Arts and Sciences; containing the Definitions of the Terms, and Accounts of the Things signify'd thereby, in the Several Arts, both Liberal and Mechanical, and the Several Sciences, Human and Divine. Born in Westmoreland, Chambers (*c.*1680–1740) published the *Cyclopaedia* in 1728, with later editions, by various authors, appearing up to the 1780s. In 1744, a proposal to translate Chambers' work into French led to it becoming the inspiration and basis for the monumental *Encyclopédie* compiled in France between 1751 and 1777 by Denis Diderot and Jean le Rond d'Alembert.[9]

A significant Welsh contribution to the story of the Chambers *Cyclopaedia* came in 1778 when a revised single-volume edition appeared. This was followed between 1781 and 1786 by an expanded four-volume edition. These revised and expanded editions were undertaken by the scientifically minded Wales-born, London-based minister Abraham Rees (1743–1825). Born in Llanbrynmair, in Montgomeryshire, Rees was educated in a school at Llanfyllin before moving to a Grammar School in Carmarthen. Later, he became an assistant tutor in Coward's Academy in London while also studying there for the ministry. The subjects that he tutored at the academy give ample evidence of the breadth of his scientific interests; they include: 'mathematics, statics, hydrostatics, optics, spherical geometry, and the use of applied mathematics in navigation, geography, and astronomy'. Rees went on to tutor in mathematics and Hebrew between 1786 and 1796 at the New College in Hackney where, between 1786 and 1787, his friend and colleague Richard Price taught Newton's fluxions and the mathematics of annuities and life assurance. Rees was also pastor (1783–1809) of the Old Jewry meeting House in the City of London when, on 5 November 1789, Price welcomed the opening events of the French Revolution in his controversial *Discourse on the Love of our Country*. Price died in April 1791 and Rees, along with Joseph Priestley and others, would be pallbearers at Price's internment in Bunhill Fields cemetery, in the City of London; Rees would be interred at Bunhill following his own death in 1825.[10]

Rees's contribution to the revised Chambers *Cyclopaedia* is, by his own admission, an extensive one:

It will be sufficient to observe, in general, from a regular account which the Editor [Rees] has preserved, that the number of NEW ARTICLES amounts to more than FOUR THOUSAND FOUR HUNDRED; several of which do not occur in any DICTIONARY OF SCIENCE, which he has had an opportunity of consulting.[11]

While some of these new articles benefited from material published in other encyclopaedic volumes – including the second edition (1777–83) of the Edinburgh-based *Encyclopaedia Britannica* – they were all additional to the original articles contained in the Chambers edition of 1738, many of which were revised and condensed by Rees. In the 'Preface' to his edition of Chambers' work, Rees also gratefully acknowledges the new material that he had received from others but, like Chambers, he omits naming any individual contributors – bar one. The exception being 'the Rev. Dr [Richard] Price', 'to whom the public are indebted for several articles, that will be thought to enrich this edition, on the subjects of ANNUITIES, ASSURRANCE, FUNDS, LIFE-*Annuities*, *Bills of* MORTALITY, and SURVIVORSHIP.'[12] For his efforts on the new Chambers edition, Rees would be elected a Fellow of the Royal Society in 1786, with Richard Price as one of his sponsors. He would also be elected to the Linnean Society (1811), the American Philosophical Society in Philadelphia, and the Royal Society of Literature.[13] As Werner (1994) notes:

Viewed in the context of many other encyclopaedic efforts, including Harris's *Lexicon technicum*, the *Encyclopédie*, or indeed the *Encyclopaedia metropolitana* (a great pictorial dictionary issued during the first half of the nineteenth century), Rees's eighteenth-century revision of Chambers is a significant and original text in the history of encyclopaedias.[14]

Having been involved in one encyclopaedic enterprise Rees decided, sometime in the 1770s, to embark on another, but this time one of his own making. The result would be truly monumental.

Rees's Cyclopaedia

Rees's *Cyclopedia; or, Universal Dictionary of Arts, Sciences, and Literature* appeared in seventy-nine parts between January 1802 and July 1820.[15] Two text parts generally form a volume, so the complete edition ran to thirty-nine text volumes, five of plates, and an atlas – forty-five volumes in total. The estimates of its content are staggering – 39 million words with some 500 articles of at least 10,000 words. Printed by a consortium of printers, including the King's printer, Andrew Strahan, the complete work could be bought for £85 in 1820 (see Figure 33).

In contrast to the Chambers *Cyclopaedia*, and Rees's own edition of that work, there is no royal, aristocratic or other form of dedication in Rees's later production. Instead, a relatively brief six-page 'Preface' is followed by the alphabetically arranged encyclopaedia of topics. However,

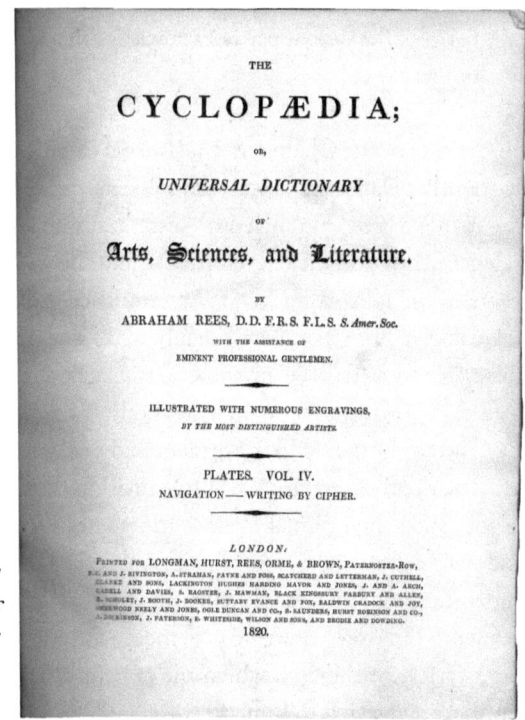

FIGURE 33 Title page to the fourth volume of plates from Abraham Rees's forty-five-volume *Cyclopaedia; or Universal Dictionary of Arts, Sciences, and Literature* (London, 1802–20).

as in the earlier works, page numbers seem to have been considered unnecessary – presumably in light of the alphabetic arrangement of the text.

Although most encyclopaedias had adopted alphabetical listings by the eighteenth century, some compilers still worried that the method hid the overall complexity of individual subjects from the reader. For example, a general alphabetic entry on the subject of medicine might obscure links to associated topics such as alchemy, chemistry, vision and optics, bone structure, specific diseases, and the many other branches that make up the subject of medicine. In such cases the purpose of an encyclopaedia as a 'circle of knowledge' would be lost. Consequently, as Yeo (1991) notes:

> from the eighteenth century the prefaces of encyclopaedias were devoted to explanations of the ways in which the particular publication did in fact exemplify logical or systematic relationships between the various parts of knowledge that lay scattered throughout its pages.[16]

In the case of the Chambers *Cyclopaedia* of 1738, an attempt was made to identify relationships between subjects and associated topics by allying text to a 'tree of knowledge' – essentially an illustration in which the science subjects in the encyclopaedia, and the various branches or topics thereof, were displayed in a form reminiscent of the family tree of the genealogist. For Chambers, the allying of this tree to an associated text necessitated a 'Preface' of some fifty pages. He also employed cross-referencing within the main text, and he introduced a nine-page index designed to further help the reader see the relationship between subjects and their component topics. When revising the Chambers edition Rees republished the original Preface in its entirety. He also expanded the index to eighty pages comprising some 57,000 entries grouped under broad headings such as 'Science' or 'Industry'. His aim: to enable the reader to discover 'where he may meet with miscellaneous information on a variety of subjects, which the general titles of such articles could not have suggested to him'.

Rees had abandoned such concerns by the time his own alphabetically arranged *Cyclopaedia* appeared in the early years of the nineteenth century. In a short, six-page 'Preface' he accepts that the sciences are 'mutilated and mangled' by an alphabetical arrangement, but he felt the problem could be ameliorated by the use of references to other works, and by the cross-referencing of topics within the alphabetically listed subjects. These, he believed:

> when judiciously distributed and arranged, will serve, like the index of a book, but much more effectually, to conduct the reader from one subject to another; they will enable him to perceive their relation to each other; and they will direct him how to collect and combine the dispersed parts of any science into one entire and regular system.[17]

The scale of Rees's new encyclopaedic enterprise meant that he was once again indebted to numerous contributors. In contrast to the Chambers *Cyclopaedia* (and Rees's own edition of that work), he names some of these contributors in his 'Preface'. According to Woolrich (2004), most were Nonconformists, and they came from a wide variety of occupations – journalists writing for technical journals, teachers and lecturers at various institutions, and several who 'were active in radical politics; one who was goaled for sedition and another indicted for treason'.[18] Many were also leaders in their particular fields of endeavour. In the arts they included the likes of the sculptor John Flaxman (1755–1826); the music historian Charles Burney (1726–1814); the landscape engraver John Landseer (1762/3–1852), father of Edwin Landseer whose lions still grace Trafalgar Square; and the Cornish artist John Opie (1761–1807), who painted the portrait of Abraham Rees that was later engraved as the frontispiece to the *Cyclopaedia*. Opie is also said to have painted a now lost portrait of the Welsh scientist and electrical experimenter George Cadogan Morgan.[19]

Fellowship of the Royal Society, as well personal acquaintance, must have aided Rees in enlisting the help of at least some of his eminent scientific contacts. They included the founding father of actuarial science, William Morgan, writing on annuities; Humphry Davy (1778–1829) and John Dalton (1766–1884) on chemistry; the Italian-born, London-based Tiberius Cavallo (1749–1809) on electricity, magnetism, mechanics and machinery; Charles Koenig (1774–1851), the German naturalist and keeper at the British Museum, wrote on geology with Robert Bakewell (1767–1843), the author of an *Introduction to Geology* (1813) and *Introduction to Mineralogy* (1819); the surgeon William Blair (1766–1822) wrote on ciphers, and in doing so produced what has been described as 'one of the finest articles in English on cryptology'; the mathematician and developer of 'Ivory's Theorem', James Ivory (1765–1842), wrote on conic sections and curvilinear geometry; and the founder of the Linnean Society, James Edward Smith, contributed on botany. European interest in Rees's publication is made evident in a letter to Smith from the Austrian botanist Franz von Jacquin (1766–1839) who notes the *Cyclopaedia* as being available in Vienna in 1818.[20]

Rees also acknowledges the contributions of a 'Mr S. Bevan'. This is most likely Silvanus Bevan III (1743–1830), the son of Swansea-born Timothy Bevan who, with his brother Silvanus Bevan II, founded the Plough Court apothecary in London (see Chapter 12). Silvanus Bevan III had initially joined the Plough Court pharmacy but, in 1767, two years after his father's death, he left to become a partner in Barclay's Bank. Also mentioned by Rees is the Unitarian minister and historical writer Dr Thomas Rees (1777–1864); no relation to Abraham Rees, he is recorded as a contributor of material on botany who also 'paid particular attention to the arrangement of the Plates'.

The use of illustrations in encyclopaedias was not new. Both Harris's *Lexicon Technicum* and the Chambers *Cyclopaedia* had included them, but their content is dwarfed by the 1,107 engraved plates and sixty-one folded maps found in Rees's *Cyclopaedia*. The excellence of their engraving, as illustrated, for example, by Figure 34, has rendered them a firm favourite among today's antiquarian print sellers.

During the course of the long eighteenth century, Wales produced a number of engravers and illustrators who deserve to be recognised as artisan scientists. Even though his precise origins are unknown

FIGURE 34 Examples of engraved plates
from Abraham Rees's *Cyclopaedia*.

(perhaps Dutch, perhaps Welsh)[21] they include the London-based Robert Vaughan (*c*.1600–*c*.63), who engraved illustrations for Elias Ashmole's *Theatrum Chemicum Britannicum* (1652) and had working connections to Wales. For example, he engraved the title page for Thomas Powell's *Elementa Opticæ* of 1651 (see Figure 4), and three monumental brasses for the Wynn (aka Wynne) family in the Gwydir Chapel of Llanrwst Church, Denbighshire; including one for Sir Owen Wynne whose interest in alchemy, chemistry and metallurgy was noted in Chapter 13.[22] Caernarfonshire-born Moses Griffith (1747–1819) provided illustrations of landscapes, buildings and natural history specimens for Thomas Pennant's works, and Robert Baugh (1748?–1832) 'of Llandysilio' engraved the 1795 map of north Wales by John Evans (see Figure 12). However, the known engravers for Rees's *Cyclopaedia* were Englishmen. They include the likes of poet and artist William Blake (1757–1827), who engraved some of the sculpture plates; Thomas Milton (1743–1827), who produced well over 200 plates on natural history; the biographer and engineer Aaron Arrowsmith (1750–1823), who produced the maps; and Wilson Lowry (1762–1824), who created magnificent engravings illustrating architecture, scientific instruments and machinery. The one Welsh representative known to have worked on the plates was the botanical and animal draughtsman Sydenham Edwards (1768–1819). He was born in Usk to a Lloyd Pittell Edwards and Mary Reece who had been married at Llantilio Crossenny Church, where Edwards would be baptised. Having later moved to London, Edwards produced almost 100 signed natural-history plates for the Rees *Cyclopaedia*, as well as a number of unsigned ones. He would also provide botanical illustrations for numerous magazines and journals including, *Flora Londinensis*, *The New Flora Britannica*, *The Botanical Register*, *Collectanea Botanica*, *The Complete Dictionary of Practical Gardening* and *The Temple of Flora*. Between 1787 and 1815 Edwards is said to have been responsible for all but seventy-five of the 1,721 water colour drawings for *The Botanical Magazine*. He also produced zoological and ornithological illustrations, and a series depicting dog breeds for inclusion in his own *Cynographia Britannica* of 1799–1805.[23]

Rees records that the number of plates had expanded far beyond the original intention as new subjects were introduced into the work as it progressed. He also concluded that:

> as these plates constitute a character of excellence peculiar to this Cyclopaedia, it is thought that the circumstances of their being additional embellishments of the work, besides that of their being indispensable as explanatory of the articles to which they refer, will be a sufficient apology for the increase of their number.

In a review of Rees's earlier work on the Chambers *Cyclopaedia*, Werner (1994) notes how Rees's plates stand out since they are not mere decorations. Instead, they act as:

> complementarities between written texts and illustrations. They underscore the use of drawings as an imaginative as well as utilitarian part of an encyclopaedia. Ultimately indeed these plates point to the presence of a novel kind of encyclopaedia not generally thought of as consonant with eighteenth-century English encyclopaedic practices. The mode is that of an encyclopaedia conceived of less as a reference book with illustrations (a mechanical or limited version of the genre) than as the more unified, even organic, *pictorial* encyclopaedia.[24]

As Werner also notes, this trend is developed fully in Rees's later *Cyclopaedia* where in many instances, and particularly in relation to machinery, the plates are fully integrated with the associated text. As can be seen in Figure 35, each part of Mr Watt's steam engine is labelled with a number or letter. Each of these labelled parts is then described as to its nature and function in the associated text on 'steam engines'. In this way a complete and full understanding is gained, not only of the construction, function and operation of the machine as a whole, but of its individual constituent parts as well.

Rees's work on his multi-volume *Cyclopaedia*, which represents a comprehensive illustration of the 'world of new ideas' that we have

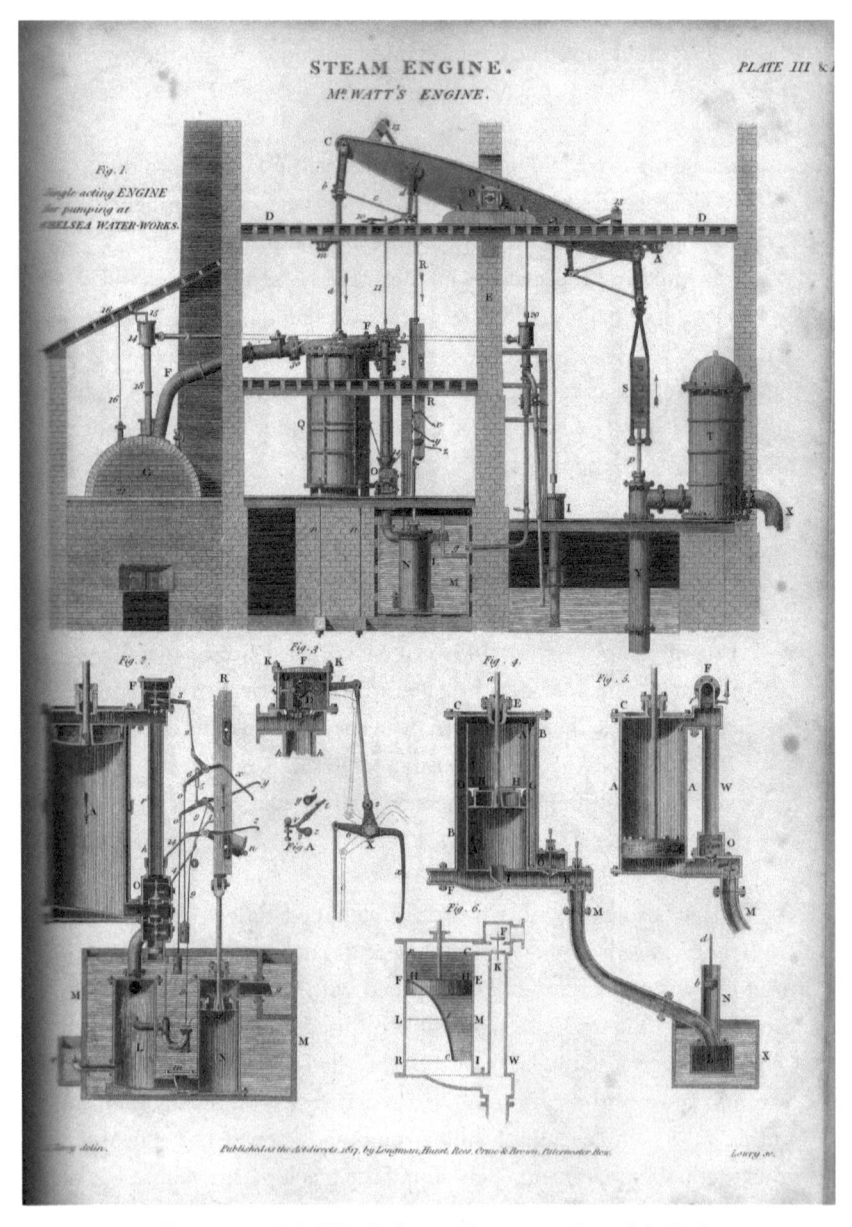

FIGURE 35 Mr Watt's Steam Engine with its labelled
parts, from Abraham Rees's *Cyclopaedia*.

been exploring, had been the incessant labour of twelve or fourteen-hour-days for almost twenty years, and it had been completed 'very much to the relief of the Editor's mind'. He also held out high hopes for both his earlier work on the Chambers edition, and for his own *Cyclopaedia*. When dedicating the former to George III, Rees noted how it contained discoveries and improvements unknown at the time of the king's accession and these 'by extending the boundaries of Science and enlarging the empire of the human mind, diffuse a glory over this country, unattainable by conquest or dominion.' He went even further in the Preface that followed the dedication with a comment equally as applicable to his later multi-volume work. With due modesty, we might also see it as an encomium on the aim and purpose of this current volume:

> In enumerating the advantages recommending this work, we must not forget to mention, that it records and transmits to future ages many inventions and improvements, which might otherwise sink into oblivion; that it forms a compendious history of science; that it furnishes the outlines of its gradual progress and advancement; and that, by preserving a summary of what has already been done and discovered, it lays a foundation for further discoveries and improvements. In this latter view of its importance and use, it may not be improperly compared to a map of science, in which the line that terminates the *terra cognita* is distinctly marked out for the direction of those, whose ingenuity and industry are employed in extending the boundaries of knowledge, and in exploring those regions that are still unknown.

* * *

The narrative in this first volume of *A World of New Ideas* has taken us from the proto-scientific, quasi-magical world of alchemy, at the start of the long eighteenth century, to the world of electricity and steam power at its close. We have seen and examined the contribution made by scientists of Wales to the new ideas of that changing world, both within Wales itself and in the other nations of these Isles. In Volume 2, the narrative centres on the same period but in an extended *terra cognita* – 'The Wider World'. Only then, in the final chapter of that second volume, will an attempt be made to draw together what the lives and works of our Scientists of Wales say to us about their long eighteenth century, and what value that knowledge might have for our own times.

Notes

1. AML-II, p. 511.
2. Daisy Dunn, *In the Shadow of Vesuvius: A Life of Pliny* (2019; London, 2020) discusses the *Natural History* in relation to Pliny the elder and his nephew Pliny the younger. Pliny the Elder, as a result of his inquisitiveness regarding natural phenomena, as well as humanitarian concerns, would die in the eruption of Vesuvius in AD 79.
3. In his *Magia Adamica* of 1650 Thomas Vaughan also 'asserted that Raziel taught Adam "the *Mysteries* of both *Worlds, Aeternall* and *Temporall*"', noted in Thomas Willard, *Thomas Vaughan and the Rosicrucian Revival in Britain 1648–1666* (Leiden, 2022), p. 118.
4. Peter Harrison, *The Fall of Man and the Foundations of Science* (2007; Cambridge, 2009), see Chapter 1: 'Adam's Encyclopaedia'.
5. S. Frederick Starr, *Lost Enlightenment: Central Asia's Golden Age from the Arab Conquest to Tamerlane* (Princeton, 2013), p. 7.
6. John Harris, *Lexicon Technicum: Or, An Universal English Dictionary of Arts and Sciences: Explaining not only the Terms of Art, but the Arts Themselves*, 2 vols (London, 1704 and 1710), available online.
7. See Brian Bowers, 'A Baptist Man of Science', *The Baptist Quarterly* (2000), 305–7. Bowers suggests Jones's ministry extended from 1713 to 1737; however, it could also have been underway since 1709 following the death of the previous incumbent in that year. For important details of Jones Wantage see Chapter 4 n. 18 of this volume.
8. See Paul Quarrie, 'The Scientific Library of the Earls of Macclesfield', *Notes and Records: The Royal Society Journal of the History of Science*, 20 (2006), 5–24 (at 14).
9. See Philipp Blom, *Encyclopédie: The Triumph of Reason in an Unreasonable Age* (London, 2004), pp. 35–40, 45.
10. All biographical details are from the *ODNB* and *DWB* entries with the quote on subjects taught by Rees taken from the *ODNB* entry.
11. Abraham Rees (ed.), Ephraim Chambers, *Cyclopædia: or, An Universal Dictionary of Arts and Sciences* (London, 1778), 'Preface to the New Edition', p. iii.
12. Rees (ed.), Chambers *Cyclopaedia*, 'Preface to the New Edition', p. iv.
13. Details from Abraham Rees *ODNB* entry.

14. Stephen Werner, 'Abraham Rees's Eighteenth-Century Cyclopaedia', in Frank A. Kafker (ed.), *Notable Encyclopaedias of the Late Eighteenth Century: Eleven Successors of the Encyclopédie* (Oxford, 1994), pp. 183–99.

15. Abraham Rees, *Cyclopedia; or, Universal Dictionary of Arts, Sciences, and Literature* (London, 1802–20). Available online; at the very end of the Wikipedia entry on 'Rees's *Cyclopaedia*' is digital access to every volume of the original British edition of the *Cyclopaedia*. Access is also provided to the later American edition. There is also a separate Wikipedia entry titled 'List of Rees's *Cyclopaedia* articles'; for the entry on contributors see note 18 below.

16. See Richard Yeo, 'Reading Encyclopedias: Science and the Organization of Knowledge in British Dictionaries of Arts and Sciences', *Isis*, 82 (1991), 24–49 (at 24).

17. Rees, *Cyclopaedia*, 'Preface' (n.p.).

18. *ODNB*. For the listing of contributors, see Rees, *Cyclopaedia*, 'Preface' (n.p.); also, Wikipedia entry 'List of contributors to Rees's Cyclopaedia'. The latter gives some details of the subjects written about, where they are known.

19. Following Morgan's death in 1798, his wife and immediate family immigrated into the United States. It would seem likely they took any portrait of their father with them. If so, it may well have been destroyed in a fire at the family's residence in Stockbridge, Massachusetts. George Cadogan Morgan's eldest son, also called George Cadogan Morgan, died in the blaze. See Mary-Ann Constantine and Paul Frame (eds), *Travels in Revolutionary France and a Journey Across America* (Cardiff, 2012), pp. 189, 203 n. 38.

20. Linnean Society archives, James Edward Smith Correspondence, GB-110/JES/COR/5/103, available online.

21. See T. A. Glenn, 'Robert Vaughan of Hergest and Robert Vaughan Engraver', *Archaeologia Cambrensis*, (1934), 291–301. Glenn suggests possible Dutch origins, but see end note by Isaac J. Williams on possible Welsh origins (at 300–1).

22. For the brasses, see J. M. Lewis, *Welsh Monumental Brasses: A Guide* (Cardiff, 1974), pp. 24, 78–82.

23. All details from Kevin L. Davies, 'The Life and Work of Sydenham Edwards FLS, Welshman, Botanical and Animal Draughtsman, 1768–1819', *Minerva: The Journal of Swansea History*, 9 (2001), 30–58 (a list of Edwards works and illustrations can be found at 53–7).

24. Werner, 'Abraham Rees's eighteenth-century Cyclopaedia', pp. 196–7.

SELECT BIBLIOGRAPHY

In this first volume of '*A World of New Ideas*', all modern and historical books and manuscripts mentioned in the text are fully referenced in the endnotes accompanying each chapter. This 'Select Bibliography' concentrates primarily on works written by our scientists of Wales during the long eighteenth century (1650–1820). It is hoped that this will provide a useful resource for future studies. Modern texts included here are restricted to those containing significant biographical content and/or extensive correspondence relevant to the scientists in this study.

It is realised that the exclusion here of so much of the reference material included in the endnotes may be frustrating, but it is hoped to include a full bibliography to both volumes of '*A World of New Ideas*' in Volume 2.

* * *

Adams, John, 'Descriptions of *Actinia crassicornis*, and some British Shells', *Transactions of the Linnean Society*, 3 (1797), 252–4.

— 'The Specific Characters of some Minute Shells discovered on the Coast of Pembrokeshire, with an account of a new marine animal', *Transactions of the Linnean Society*, 3 (1797), 64–9.

— 'Description of Some Marine Animals found on the Coast of Wales', *Transactions of the Linnean Society*, 5 (1800), 7–13.

— 'Descriptions of Some Minute British Shells', *Transactions of the Linnean Society*, 5 (1800), 1–6.

Aikin, Arthur, *Journal of a Tour through Wales and part of Shropshire: with Observations in Mineralogy and other Branches of Natural History* (London, 1797).

[Anon.], 'Enquiries touching agriculture for arable and meadows', *Philosophical Transactions of the Royal Society*, 1 (1665), 91–4.

[Anon.], 'An Account of a storm of rain that fell at Denbigh in Wales: communicated to Dr Hans Sloane, R. S. Secr.', *Philosophical Transactions of the Royal Society*, 25 (1706), 2342–4.

[Anon.], *Proposals for Enriching the Principality of Wales: Humbly submitted to the Consideration of his Countrymen; by Giraldus Cambrensis* (London, 1755).

[Anon.], 'Memoirs of Mrs Anne Williams', *London Magazine*, 52 (1783), 517–21.

Augustus, William, *Erra Pater, neu Ddarogynydd yr Amserau/The Husbandman's Perpetual Prognostication* (Carmarthen, 1794).

Beer G. R. de (ed.), *Tour on the Continent, 1765, by Thomas Pennant, Esq.* (London, 1948).

Bennetts, Nicola Bruton, *William Morgan: Eighteenth-century Actuary, Mathematician and Radical* (Cardiff, 2020).

Bentley, Richard, *The Folly and Unreasonableness of Atheism Demonstrated from the Advantage and Pleasure of a Religious Life, the Faculties of Human Souls, the Structure of Animate Bodies, and the Origin and Frame of the World* (London, 1699).

Berg, Torsten, and Peter Berg (trans.), *R. R. Angerstein's Illustrated Travel Diary, 1753–1755: Industry in England and Wales from a Swedish Perspective* (London, 2001).

Barrington, Daines, 'Letter from Hon. Daines Barrington FRS to William Heberden, MD, FRS, giving an Account of some experiments made in North Wales, to ascertain the different quantities of rain, which fell in the same time, at different heights', *Philosophical Transactions of the Royal Society*, 35 (1771), 294–7.

Bayes, Thomas and Richard Price, 'An Essay toward Solving a Problem in the Doctrine of Chances', *Philosophical Transactions of the Royal Society*, 53 (1763), 370–418.

Bevan, William, *Llyfr Meddyginiaeth, ir anafys ar chlwfus* (n.p., 1733).

Beynon, Tom (Trans. and ed.), *Howell Harris: Reformer and Soldier (1714–1773)* (Caernarvon, 1958).

— *Howell Harris's Visits to London* (Aberystwyth, 1960).

Bowers, Brian, 'A Baptist Man of Science', *Baptist Quarterly* (2000), 305–7.

Bridges, Jake E., 'A Science Fit for the Chapel: Astronomy in Nineteenth Century Wales' (unpublished MA thesis, University of Alberta, 2018).

Burnet, Thomas, *The Theory of the Earth; Containing an Account of the Original of the Earth, and all Changes which it hath already undergone, or is to undergo till the consummation of all Things* (London, 1691).

Butler, Joseph, *Analogy of Religion, Natural and Revealed* (London, 1736).

[Cadogan, William] *An Essay upon Nursing, and the Management of Children, from their birth to three years of age* (London, 1748).

— *A Dissertation on the Gout, and all Chronic Diseases, jointly considered, as proceeding from the Same Causes; what those causes are; and a Rational and Natural Method of Care Proposed. Addressed to all Invalids* (Dublin, 1771).

Calendar of Wynn (of Gwydir) Papers (Aberystwyth, 1926).

Cannon, Garland (ed.), *The Letters of Sir William Jones*, 2 vols (Oxford 1970).

Carpenter, Audrey T., *John Theophilus Desaguliers: A Natural Philosopher, Engineer and Freemason in Newtonian England* (London and New York, 2011).

Carter, P. W., 'History of Botanical Exploration in Monmouthshire', *Botanical Society of the British Isles – Welsh Bulletin*, 66 (1999), 18–30; the paper (available online) ends with a source listing for all of Carter's county-by-county botanical history volumes relating to Wales.

Chapman-Huston, Desmond, and Ernest C. Cripps, *Through a City Archway: The Story of Allen and Hanburys 1715–1954* (London, 1954).

Charles, B. G., *George Owen of Henllys* (Aberystwyth, 1973).

Claromontii, Caroli (= Dr Charles Clermont), *De Aere, Locis, & Aquis Terrae Angliæ; Deque Morbis Angulorum Vernaculis* (London, 1672); contains 'Observationes Medicae Cambro-Britannicae', pp. 77–180.

Cohen, I. Bernard, and Anne Whitman (trans.), *Isaac Newton: The Principia, Mathematical Principles of Natural Philosophy, A New Translation* (Los Angeles CA and London, 1999).

Cone, Carl, *Torchbearer of Freedom: The Influence of Richard Price on 18th Century Thought* (Lexington KY, 1952).

Constantine, Mary-Ann, and Paul Frame (eds), *Travels in Revolutionary France and a Journey Across America* (Cardiff, 2012).

Cook, Alan, *Edmond Halley: Charting the Heavens and the Seas* (Oxford, 1998).

Coward, Adam N. (ed.), S. Lucilius Verus (Edmund Jones), *A Spiritual Botanology* (Newport, 2017).

Cresy, Edward, *An Encyclopædia of Civil Engineering: Historical, Theoretical, and Practical* (London, 1847).

Cripps, Ernest C., *Plough Court: The Story of a Notable Pharmacy 1715–1927* (London, 1927).

Cule, John, *Wales and Medicine; a source-list for printed books and papers showing the history of medicine in relation to Wales and Welshmen* (Aberystwyth, 1980).

Davies, Evan, 'An Account of what happened from Thunder in Carmarthenshire … communicated to the Royal Society, By John Eames, FRS as he received it in a Letter from Mr Evan Davies', *Philosophical Transactions of the Royal Society*, 36 (1729–30), 444–8.

Davies, Hugh, *Welsh Botanology: A Systematic Catalogue of the Native Plants of Anglesey in Latin, English, and Welsh* (London, 1813).

Davies, John, (trans.), *Reflections upon Monsieur Des Cartes's Discourse of a Method for the Well-guiding of Reason, and Discovery of Truth in the Sciences* (London, 1654). See also G. Havers.

Davies, John H. (ed.), *The Letters of Lewis, Richard, William and John Morris of Anglesey (Morrisiaid Mon) 1728–1765*, 2 vols (Aberystwyth, 1907, 1909).

Davies, Kevin L., 'The Life and Work of Sydenham Edwards FLS, Welshman, Botanical and Animal Draughtsman, 1768–1819', *Minerva, the Journal of Swansea History*, 9 (2001), 30–58.

Davis, David Daniel, *A Treatise on Insanity, by P. Pinel, M. D. Translated from the French by D. D. Davis, M. D.* (Sheffield, 1806).

Davis, William, *A Complete Treatise of Land Surveying, by the Chain, Cross, and Offset Staffs Only* (London, 1798).

Delbourgo, James, *Collecting the World, the Life and Curiosity of Hans Sloane* (2017; London 2018).

Denning, R. T. W. (ed.), with J. B. Davies and G. H. Rhys, *The Diary of William Thomas 1762–1795* (Cardiff, 1995).

Derham, William, *Physico-Theology: Or, a Demonstration of the Being and Attributes of God, from his Works of Creation* (London, 1713).

— *Astro-Theology Or, a Demonstration of the Being and Attributes of God, from a Survey of the Heavens* (1715; London, 1721).

Descartes, René, *A Discourse on the Method of Rightly Directing One's Reason and of Seeking Truth in the Sciences* (Leiden, 1637).

— *The Meditations* (Paris, 1641).

— *Principia Philosophiae* (Amsterdam, 1644).

Dickson, Donald R. (ed.), *Thomas and Rebecca Vaughan's Aqua Vitæ: Non Vitis*, Medieval and Renaissance Texts and Studies, vol. 217 (Temple AZ, 2001).

Dillwyn, Lewis Weston, 'Letter from L. W. Dillwyn, Esq. to W. H. Pepys jun. Esq. respecting the effects of the oxymuriatic acid on the growth of plants', *Philosophical Magazine*, 11/42 (1801), 158–9.

— 'Catalogue of the more rare Plants found in the Environs of Dover, with Occasional Remarks', *Transactions of the Linnean Society*, 6 (1802), 177–84.

— *British Confervae* (London, 1809).

— *A Descriptive Catalogue of Recent Shells* (London, 1817).

Dixon, Peter S., 'Notes on Important algal herbaria, IV, The Herbarium of Lewis Weston Dillwyn (1778–1855)', *British Phycological Bulletin*, 3/1 (1966), 19–22.

Dodson, James, *The Anti-Logarithmic Canon* (London, 1742).

Dowson, Duncan, and Bernard J. Hamrock, *History of Ball Bearings*, NASA Technical Memorandum 81689 (February 1981).

Drake, Stillman (ed.), *Mathematical Collections and Translations in Two Tomes by Thomas Salusbury … in Facsimile* (London and Los Angeles CA, 1967).

Duncan, Andrew, *A Letter to Dr Robert Jones of Carmarthenshire* (Edinburgh, 1782).

Durston, Thomas, *Cyfarwyddiadau i Fesurwyr* (c.1700).

Dybikowski, J., *On Burning Ground: An Examination of the Ideas, Projects and Life of David Williams* (Oxford, 1993).

Edwards, Haydn, E., *Griffith Davies: Arloeswr a Chymwynaswr* (Cardiff, 2023).

Edwards, Nancy, and Brynley F. Roberts, *Edward Lhuyd 1660–1709* (Aberystwyth, 2010).

Elderton, William, 'William Morgan F.R.S. 1750–1833', *Transactions of the Faculty of Actuaries*, 14 (1931–4).

Elstob, Ivy (ed.), with E. S. Rohde, *The Garden Book of Sir Thomas Hanmer* (1659; London, 1933).

Emery, F. V., '"The Best Naturalist Now in Europe": Edward Lhuyd F.R.S. (1660–1709)', *THSC*, 1 (1969/70), 54–69.

— *Edward Lhwyd FRS 1660–1709* (Cardiff, 1971).

— 'Edward Lhuyd and a Natural History of Wales', *Studia Celtica*, 12–13 (1977–8), 247–58.

— 'Edward Lhuyd and Snowdonia', *Nature in Wales*, new series, 4/1–2 (1985), 3–11.

'Espinasse, Margaret, *Robert Hooke* (London, 1956).

Evans, Chris (ed.), with G. G. L. Hayes, *The Letterbook of Richard Crawshay 1788–1797* (Cardiff, 1990).

Evans, Dewi W., and Brynley F. Roberts (eds), *Edward Lhwyd 1660–1709, Llyfryddiaeth a Chyfarwyddiadur/A Bibliography and Readers' Guide* (Aberystwyth, 2009); contains a full listing of all Lhwyd's publications.

— *Edward Lhwyd* Archæologia Britannica *Texts and Translations* (Aberystwyth, 2009).

Evans, Eifion, *Bread of Heaven: The Life and Work of William Williams, Pantycelyn* (Bridgend, 2010).

Evans, John, *Cyfrifydd Parod* (n.d.).

Evans, Lewis, 'Observations of a Polaris, for determining the north Polar distance of that star at the beginning of the year 1813', *The Philosophical Magazine*, Series 1, 43/194 (1814), 446–7.

— 'Mr Evan's on The Woolwich Observatory', *The Monthly Magazine*, 41 (1 July 1816), 483–4.

— 'An Improved Demonstration of Sir Isaac Newton's Binomial Theorem, on Fluxional Principles, more especially calculated for the Young Student in Mathematics', *The Philosophical Magazine*, Series 1, 64/318 (1824), 270–2.

Evans, R. Paul, 'The Life and Work of Thomas Pennant (1726–1798)' (unpublished PhD thesis, University College of Wales, Swansea, 1993).

Evans, W. A., 'Three Denbighshire Scientists', *Transactions Denbighshire Historical Society*, 17 (1968), 183–90.

Fahrenheit, Daniel Gabriel, 'Barometri Novi Descriptio', *Philosophical Transactions of the Royal Society*, 33 (1724–5), 179–80.

Fenton, Edward, *The Diaries of John Dee* (1998; Charlbury, 2000).

Fisher, John (ed.), *Tours in Wales (1804–1813) by Richard Fenton* (London, 1917).

Frame, Paul, 'A Further Seven Uncollected Letters of Richard Price', *Enlightenment and Dissent*, 27 (2011), 143–60.

— *Liberty's Apostle, Richard Price his Life and Times* (Cardiff, 2015).

Franklin, Michael J., *Orientalist Jones: Sir William Jones, Poet, Lawyer, and Linguist, 1746–1794* (Oxford, 2011).

Gardiner, William, *Tables of Logarithms for All Numbers from 1 to 102100* (London, 1742).

Gee, Brian, with Anita McConnell and A. D. Morrison-Low (eds), *Francis Watkins and the Dolland Telescope Patent Controversy* (Farnham, 2014).

General Board of Health, *Papers Relating to the History and Practice of Vaccination* (London, 1857).

Gilbert, Lord Bishop of Sarum, *A Sermon Preached at the Funeral of the Honourable Robert Boyle; at St Martin's in the Fields Jan 7 1691/2* (London, 1692).

Glenn, T. A., 'Robert Vaughan of Hergest and Robert Vaughan Engraver', *Archaeologia Cambrensis*, (1934), 291–301.

Greenlaw, Joanna, *Swansea Clocks, Watch and Clockmakers of Swansea and District (Llanelli, Llandeilo, Neath)* (Swansea, 1997).

Griffith, Alexander, *Mercurius Cambro-Britannicus, or, News from Wales* (London, 1652).

Griffith, George, *A Welsh Narrative, corrected, and taught to speak true English, and some Latine ... Containing a narration of the disputation between Dr Griffith and Mr Vavasor Powell, neer New Chappell in Montgomery-shire, July 23, 1652* (London, 1652).

Griffiths, Martin, 'Joseph Harris of Trevecka, Scientist, Artisan, Servant of the Crown', *The Antiquarian Astronomer*, 6 (2012), 19–33.

Gruffydd, Geraint, 'Yny Lhyvyr Hwnn (1546): The Earliest Welsh Printed Book', *BBCS*, 22/2 (1969), 105–16.

Gunther, R. T., *Early Science in Oxford, vol. XIV: Life and Letters of Edward Lhwyd* (Oxford, 1945).

Hall, Rupert A., *Isaac Newton: Adventurer in Thought* (Oxford, 1992).

Halley, Edmond, 'A Letter from Mr Halley of June the 7th 97 concerning the Torricellian Experiment tryed on the top of Snowdon-hill and the success of it', *Philosophical Transactions of the Royal Society*, 19 (1697), 582–4.

Harris, J. R., *The Copper King: Thomas Williams of Llanidan* (1964; Ashbourne, 2003).

Harris, John, *Lexicon Technicum: Or, an Universal English Dictionary of Arts and Sciences: Explaining not only the Terms of Art, but the Arts Themselves*, 2 vols (London, 1704 and 1710).

Harris, John, *Vox Stellarum & Planetarum* (Caerfyrddin, 1797).

Harris, Joseph, *A Treatise of Navigation* (London, 1730).

— *The Description and Use of the Globes and the Orrery* (London, 1731).

— 'Proposals for printing by Subscription a Treatise upon Microscopes', *Bibliothèque britannique*, 19/1 (1742), 211–3.

— *An Essay on Money and Coins, Part I* (London, 1757).

— *An Essay on Money and Coins, Part II* (London, 1758).

— *A Treatise of Optics: Containing elements of the science; in two books* (London, 1775).

Harris, Joseph, and George Graham, 'An Account of some Magnetical Observations Made in the Months of May, June and July 1732 in the Atlantick or Western Ocean; Also the Description of a Waterspout by Mr Joseph Harris', Communicated by Mr George Graham F.R.S., *Philosophical Transactions of the Royal Society*, 38 (1733), 75–9.

Harvey, Thomas, *John Owen's Latine Epigrams Englished by Tho. Harvey, Gent.* (London, 1677).

Havers, G., and John Davies (trans.), *Another Collection of Philosophical Conferences of the French Virtuosi, upon … Natural Knowledge* (London, 1665).

Hemmingway, Joseph, *History of the City of Chester*, 2 vols (Chester, 1831).

Henry, Thomas (trans.), *Essays Physical and Chemical by M. Lavoisier* (London, 1776).

Herbert, Edward, *De Veritate, prout distinguitur a revelatione, a verisimili, a possibili, et a falso* (Paris, 1624).

Hills, Revd Dr Richard L., *Life and Inventions of Richard Roberts 1789–1864* (Derbyshire, 2002).

Hoist, Bodil, and N. Zimmer, 'The Discovery of the Electric Shock (Letters)', *Science*, 298/5602 (2002), 2327.

Hook, David, 'John Davies of Kidwelly, a Neglected Literary Figure of the Seventeenth Century', *The Antiquarian*, 11 (1975).

Hume, David, *An Enquiry Concerning Human Understanding* (London, 1748).

Hunter, Michael, Antonio Clericuzio and Lawrence M. Principe (eds), *The Correspondence of Robert Boyle*, 6 vols (London, 2001).

Hutchinson, F. E., *Henry Vaughan: A Life and Interpretation* (Oxford, 1947).

Jacobs, J. (ed.), *Epistolae Ho-elianae: The Familiar Letters of James Howell* (London, 1892).

Jardine, Lisa, and Alan Stewart, *Hostage to Fortune: The Troubled Life of Francis Bacon 1561–1626* (London, 1998).

Jeffers, R. H., 'Edward Morgan and the Westminster Physic Garden', *Proceedings of the Linnean Society of London*, 164/2 (1953), 102–33.

Jenkins, Geraint H., '"The Sweating Astrologer": Thomas Jones the Almanacer', in R. R. Davies, Ralph A. Griffiths, Ieuan Gwynedd Jones and Kenneth O. Morgan (eds), *Welsh Society and Nationhood: Historical Essays Presented to Glanmor Williams* (Cardiff, 1984), pp. 161–77.

Jenkins, Geraint H., Ffion Mair Jones and David Ceri Jones (eds), *The Correspondence of Iolo Morganwg*, 3 vols (Cardiff, 2007).

Johnes, Thomas, *The Rules and Premiums of the Society for the Encouragement of Agriculture and Industry in the County of Cardigan* (Hafod, 1804).

— *A Cardiganshire Landlord's Advice to His Tenants* (Hafod, n.d. *c*.1804).

Jones, Bassett, *Lapis Chymicus Philosophorum Subjectus* (Oxford, 1648).

— *Herm'aelogium; or an Essay at the Rationality of the Art of Speaking: as a Supplement to Lillie's Grammar* (London, 1659).

— 'Lithochymicus, or A Discourse of a Chymic Stone', in Robert M. Schuler (ed.), *Alchemical Poetry 1575–1700: From Previously Unpublished Manuscripts* (1995; Abingdon, 2014).

Jones, David, *A Compleat History of Europe* (1698; London, 1699).

Jones, Dewi, *The Botanists and Guides of Snowdonia* (Llanrwst, 1996).

Jones, Edmund, *A Geographical, Historical, and Religious Account of the Parish of Aberystruth* (Trevecka, 1779).

— *A Relation of Apparitions of Spirits in the County of Monmouth and the Principality of Wales* (Trevecca, 1780).

Jones, John, *Novarum dissertationum de morbis abstrusioribus tractatus primus* (London, 1683).

— *De morbis Hibernorum: speciatim vero de dysentaria Hibernica* (London, 1698).

— *The Mysteries of Opium Revealed* (London, 1700).

Jones, Robert, *An Inquiry into the State of Medicine* (Edinburgh, 1781).

Jones, Thomas, *Newyddion oddi wrth y Sêr* (1680).

Jones, William Sr, *A New Compendium of the Whole Art of Practical Navigation* (London, 1702).

— *Synopsis Palmariorum Matheseos, or, A New Introduction to the Mathematics* (London, 1707).

— *Analysis per quantitatum series, fluxiones ac differentias* (London, 1711).

— 'Of Logarithms, by the late William Jones, Esq; F.R.S. Communicated by John Robertson, Lib. R. S.', *Philosophical Transactions of the Royal Society*, 61 (1771), 455–61.

— 'Properties of Conic Sections; deduced by a Compendious Method. Being a Work of the late William Jones, Esq; F.R.S. which he formerly communicated to Mr John Robertson, Libr. R. S. who now addresses it to the Reverend Nevil Maskelyne, F.R.S. Astronomer Royal', *Philosophical Transactions of the Royal Society*, 63 (1773), 340–60.

Jones, William Garel, 'The Life and Works of Henry Rowlands' (unpublished MA thesis, University of Wales, Bangor, 1936).

Jurin, James, 'Invitatio ad Observationes Meteorologicas communi consilio instituendas', *Philosophical Transactions of the Royal Society*, 32 (1723), 422–7.

Keys, John, *The Practical Bee-master* (London, 1781).

— *Keys on Bees: The Ancient Bee-master's Farewell* (Dublin, 1796).

King-Hele, Desmond, *Doctor of Revolution: The Life and Genius of Erasmus Darwin* (London, 1977).

Knight, Thomas, *An Essay on the Transmutation of Blood containing the aetiology or an account of the immediate cause of putrid fevers* (London 1725).

— *A Vindication of a Late Essay on the Transmutation of Blood*, also containing 'A Dissertation Concerning the Manner of Operation of Chalybeat Medicines' (London, 1731).

— *A Critical Dissertation upon the Manner of the Preparation of Mercurial Medicines, and their Operation on Human Bodies* (1734).

— 'A Letter from Mr. T. Knight to Sir Hans Sloane … concerning Hair Voided by Urine', *Philosophical Transactions of the Royal Society*, 41 (1739–41), 705–7.

— *Reflections upon Catholicons or Universal Remedies, with some remarks on the natural heat in animals, and the luminous emanations from human bodies [also with] Sundry Experiments and Observations made upon the Human Calculus* (London, 1740, 1749).

Lewys, Dafydd, *Golwg ar y Byd* (Caerfyrddin, 1725).

Lhwyd, Edward, 'An Account of a sort of paper made from Linum Asbestinum found in Wales', *Philosophical Transactions of the Royal Society*, 14 (1684), 823–4.

— 'Part of a letter from Mr. Edward Floyd, Cim. Ashm. Oxon. To Dr M. Lister, giving an account of locusts lately observed in Wales', *Philosophical Transactions of the Royal Society*, 18 (1694), 45–7.

— 'A note concerning an extraordinary hail in Monmouthshire, extracted out of a letter sent from Mr Edward Lhwyd to Dr Tancred Robinson', *Philosophical Transactions of the Royal Society*, 19 (1697), 579–80.

— 'Part of a Letter from Mr. Edw. Lhwyd to Dr. Martin Lister, Fell. of the Coll. of Phys. and R. S. Concerning several regularly figured stones lately found by him', *Philosophical Transactions of the Royal Society*, 20 (1698), 279–80.

— *Lithophylacii Britannici Ichnographia*, (London, 1699).

Linden, Diederick Wessel, *A Treatise on the three medicinal mineral waters at Llandrindod, in Radnorshire, with some remarks on mineral and fossil mixtures in their native veins and beds; at least as far as respects their influence on water* (London, 1756).

— 'An Account of a mineral water at Llangyba', *Gentleman's Magazine*, 36 (1766), 328.

— *An experimental and practical enquiry into the ophthalmic, antiscrophulous, and nervous properties of the mineral waters of Llangybi, in Caernarvonshire* (London, 1767).

Linnaeus, Carl, *Systema Naturae* (Holmiae/Stockholm, 1758).

Linnard, William, 'The First Treatise on Forest Trees in Wales', *Journal of the Welsh Bibliographical Society*, 11 (1975–6), 247–50.

— *Wales, Clocks and Clockmakers* (Ashbourne, 2003).

Linnard, William, and Robin Gwyndaf, 'William Morris of Anglesey: A Unique Gardening Book and a New Manuscript of Horticultural Interest', *THSC*, (1979), 7–30.

Lloyd, Evan, *The Methodist: A Poem* (London, 1766).

Lloyd, Nesta, 'Meredith Lloyd', *Journal of the Welsh Bibliographical Society*, 11 (1975–6), 133–92.

Loudon, John Claudius, *An Encyclopaedia of Gardening* (London, 1822).

Luft, Diana, *Medieval Welsh Medical Texts, Volume 1: The Recipes* (Cardiff, 2020).

MacDonald, Neil, Cerys A. Jones, Sarah J. Davies and Cathryn Charnell-White, 'Historical weather accounts from Wales: an assessment of their potential for reconstructing climate', *Weather*, 65/3 (2010), 72–7.

Martin, Joanna (ed.), *The Penrice Letters 1768–1795* (Swansea and Cardiff, 1993).

McConnell, Anita, *Jesse Ramsden (1735–1800), London's Leading Scientific Instrument Maker* (2007; Abingdon, 2016).

McConnell, Anita, and Alison Brech, 'Nathaniel and Edward Pigott, Itinerant Astronomers', *Notes and Records: The Royal Society Journal of the History of Science*, 53/3 (September 1999), 305–18.

Miskell, Louise (ed.), *Robert Morris and the First Swansea Copper Works, c.1727–1730* (Newport, 2010).

Moivre, Abraham de, 'A Letter from Mr Abraham De Moivre F.R.S. to William Jones, Esquire, F.R.S. concerning the easiest method for calculating the value of annuities upon lives, from tables of observations', *Philosophical Transactions of the Royal Society*, 43 (1744), 65–78.

Moody, Jennifer Stanesby, 'The Trefeca Meridian and Transit of Venus', *Brycheiniog*, 41 (2010), 51–64.

Moore-Colyer, Richard J., 'Thomas Johnes of Hafod (1748–1816): Translator and Bibliophile', *WHR*, 15/1 (1990), 399–415.

— (ed.), *A Land of Pure Delight: Selections from the letters of Thomas Johnes of Hafod, Cardiganshire (1748–1816)* (Llandysul, 1992).

More, Henry [Alazonomastix Philalethes], *Observations upon Anthroposophia Theomagica, and Anima Magica Abscondita* (London, 1650).

— *The Second Lash of Alazonomastix; Conteining a Solid and Serious Reply to a Very Uncivill Answer to Certain Observations upon Anthroposophia Theomagica, and Anima Magica Abscondita* (Cambridge, 1652).

Morgan, A. de, 'On a Point Connected with the Dispute between Keil and Leibnitz about the Invention of Fluxions', *Philosophical Transactions of the Royal Society*, 136 (1846), 107–9.

Morgan, George Cadogan, 'Observations and Experiments: on the Light of Bodies in a State of Combustion. By the Rev. Mr. Morgan; communicated by the Rev. Richard Price, LL.D. F.R.S.'; *Philosophical Transactions of the Royal Society*, 75 (1785), 190–212.

— *Lectures on Electricity*, 2 vols (Norwich, 1794).

— *Directions for the Use of a Scientific Table in the Collection and Application of Knowledge* (London, 1796).

Morgan, William, *The Doctrine of Annuities and Assurances on Lives and Survivorships, stated and explained* (London, 1779).

— 'Electrical Experiments made in Order to Ascertain the Non-Conducting Power of a Perfect Vacuum', *Philosophical Transactions of the Royal Society*, 75 (1785), 272–8.

— 'On the Probabilities of Survivorship between Two Persons of Any Given Ages, and the Method of Determining the Value of Reversions Depending on Those Survivorships', *Philosophical Transactions of the Royal Society*, 78 (1788), 331–49.

— 'On the Method of Determining, from the Real Probabilities of Life, the Value of Contingent Reversions in Which Three Lives are Involved in the Survivorship', *Philosophical Transactions of the Royal Society*, 79 (1789), 40–54.

— *A Review of Dr Price's Writings, on the Subject of the Finances of the Kingdom: to which are Added the Three Plans communicated by him to Mr Pitt in the year 1786, for Redeeming the National Debt* (London, 1792).

— *Additional Facts addressed to the Serious Attention of the People of Great Britain respecting the Expence [sic] of the War, and the State of the National Debt* (London, 1796).

— *Facts addressed to the Serious Attention of the People of Great Britain respecting the Expence [sic] of the War, and the State of the National Debt* (London, 1796).

— *Appeal to the People of Great Britain, on the Present Alarming State of the Public Finances, and of Public Credit* (London, 1797).

— *Memoirs of the Life of The Rev. Richard Price*, (London, 1815).

— *Sermons on Various Subjects, by the late Dr. Richard Price, D.D. F.R.S.* (London, 1816).

— *The Principles and Doctrine of Assurances on Lives, Annuities on Lives, and Contingent Reversions, stated and explained* (London, 1821).

Morris, Lewis, *Plans of Harbours, Bars, Bays and Roads in St George's Channel* (London, 1748).

Morris, Owen, *The 'Chymick Bookes' of Sir Owen Wynne of Gwydir: An Annotated Catalogue* (Cambridge, 1997).

Muller, Mark, 'The Inventor's Talented Daughter', *Pembrokeshire Life* (January 2006), 18–9.

Musgrave, Toby, *The Multifarious Mr Banks: From Botany Bay to Kew, the Natural Historian Who Shaped the World* (2020; New Haven CT and London, 2021).

Needham, Joseph, *Science and Civilisation in China* (Cambridge, 1954).

Newman, William R., *Newton the Alchemist: Science, Enigma, and the Quest for Nature's 'Secret Fire'* (Princeton NJ, 2019).

Newton, Isaac, *Of Nature's Obvious Laws & Processes in Vegetation* (*c.*1670–*c.*5), unpublished manuscript at Smithsonian Institution, Dibner 1031B.

— *Philosophiae Naturalis Principia Mathematica* (1687), see Bernard I. Cohen and Anne Whitman.

Nuttall, Geoffrey, 'The Correspondence of John Lewis, Glasrug, with Richard Baxter and with Dr. John Ellis, Dolgelley' *Merionethshire Historical and Records Society*, 2/2 (1954), 120–34.

— *Howel Harris 1714–1773: The Last Enthusiast* (Cardiff, 1965).

Ogborn, M. E., 'The Theory of Simple and Compound Interest an Eighteenth-century Manuscript', *Journal of the Institute of Actuaries*, 75/1 (1949), 73–5.

— *Equitable Assurances* (London, 1962).

Oliver, P. Graham, 'John Adams FLS of Pembroke (1769–1798): a forgotten Welsh naturalist and conchologist', *Archives of Natural History*, 46/2 (2019), 183–202.

Oliver, P. Graham, K. Talbot, B. Fredriksson, V. Tomlinson, M. Lewis and D. Fraser, 'William Lyons of Tenby (1776–1849) and his conchology collection in the Tenby Museum & Art Gallery with recognition of type material', *Colligo*, 3/1 (2020).

Oliver, R. C. B., 'Diederick Wessel Linden', *NLWJ*, 18/3 (1974), 241–67.

Owen, Henry, *The Intent and Propriety of the Scripture Miracles*, 2 vols (London, 1773).

Owen, Hugh (ed.), 'Additional Letters of the Morrises of Anglesey (1735–1786)', *Y Cymmrodor*, 49/1, 2 (London, 1947, 1949).

— *The Life and Works of Lewis Morris (Llewelyn Ddu o Fôn) 1701–1765* (Anglesey, 1951).

Owen, T. J., 'Hugh Davies: The Anglesey Botanist', *Transactions of Anglesey Antiquarian and Field Club* (1961), 39–52.

Parry, John H., *The Cambrian Plutarch* (London, 1834).

Pasley, William Charles, *Observations on the Expediency and Practicability of Simplifying and Improving the Measures, Weights and Money, used in this Country, without materially altering the present standards* (London, 1834).

Peach, W. Bernard, and D. O. Thomas (eds), *The Correspondence of Richard Price*, 3 vols (Durham NC and Cardiff, 1983–94).

Peate, Iorwerth, 'John Tibbot, Clock and Watch Maker', *Montgomeryshire Collections*, 48 (1944), 176–85.

— 'John Tibbot's Inventions', *Montgomeryshire Collections*, 51 (1950), 38–40.

— *Clock and Watch Makers in Wales* (Cardiff, 1960).

Pennant, Thomas, *British Zoology* (1761–6; London, 1768).

— 'Account of the different species of the birds, called Pinguins', *Philosophical Transactions of the Royal Society*, 58 (1768), 91–9.

— *A Tour in Scotland* (Chester, 1769).

— *Indian Zoology* (London, 1769).

— 'An Account of two new tortoises; in a letter to Matthew Maty, M.D. Sec. R.S.' (1771), *Philosophical Transactions of the Royal Society*, 61 (1771), 266–73.

— *The Synopsis of Quadrupeds* (Chester, 1771).

— *A Tour in Scotland and Voyage to the Hebrides* (Chester, 1772).

— *A Tour in Wales* (London, 1778).

— 'An account of the Turkey', *Philosophical Transactions of the Royal Society*, 71 (1781), 67–81.

— *Genera of Birds* (London, 1781), published 1773 in Edinburgh without plates.

— *The History of Quadrupeds* (London, 1781).

— *Arctic Zoology* (London, 1784).

Pierce, James, *The Life and Work of William Salesbury: A Rare Scholar* (Talybont, 2016).

Pigott, Edward, 'Account of nebula in Coma Berenices', *Philosophical Transactions of the Royal Society*, 71 (1781), 82–3.

— 'Determinations of the Longitudes and Latitudes of some remarkable Places near the Severn', *Philosophical Transactions of the Royal Society*, 80 (1790), 385–90.

Pigott, Nathaniel, 'Double stars discovered in 1779 at Frampton House, Glamorganshire', *Philosophical Transactions of the Royal Society*, 71 (1781), 84–6.

Pike, Samuel, *Philosophia Sacra: Or, the Principles of Natural Philosophy. Extracted from Divine Revelation* (London, 1753).

Plot, Robert, *The Natural History of Oxfordshire* (London, 1677).

Powell, Thomas, *Elementa Opticæ* (London, 1651).

— *Humane Industry: or, a History of most Manual Arts, Deducing the Original, Progress, and Improvement of them* (London, 1661).

Price, Richard, *A Review of the Principal Difficulties and Questions in Morals* (London, 1758), see D. D. Raphael.

— 'A Demonstration of the Second Rule in the Essay toward the Solution of a Problem in the Doctrine of Chances', *Philosophical Transactions of the Royal Society*, 54 (1764, appeared 1765), 296–325.

— *Four Dissertations* (1767; London, 1772).

— 'A Letter from Richard Price, D.D. F.R.S. to Benjamin Franklin, LL.D F.R.S. on the Effect of the Aberration of Light on the Time of a Transit of Venus over the Sun', *Philosophical Transactions of the Royal Society*, 60 (1770), 536–40.

— *An Appeal to the Public on the Subject of the National Debt* (London, 1772).

— *The State of the Public Debts and Finances at the signing of the Preliminary Articles of Peace in January 1783. With a Plan for Raising Money by Public Loans, and for Redeeming the Public Debts* (London, 1783).

— *Observations on the American Revolution and the Means of Making it a Benefit to the World* (London, 1784), in D. O. Thomas (ed.), *Price, Political Writings* (Cambridge, 1991).

— 'The Evidence for a Future Period of Improvement in the State of Mankind' (1787), in D. O. Thomas (ed.), *Price, Political Writings* (Cambridge, 1991).

— *Sermons on the Security and Happiness of a Virtuous Course* (1787; Philadelphia, 1788).

See below at Thomas, Stephens and Jones (Aldershot, 1993) for a complete listing of Price's works and their various editions.

Pryce, W. T. R., and T. Alun Davies, *Samuel Roberts Clockmaker: An Eighteenth-Century Craftsman in a Welsh Rural Community* (Cardiff, 1985).

Rail, Tony, 'A Previously Unpublished Letter from Thomas Belsham to Samuel Fawcett, 21 April 1791', *Enlightenment and Dissent*, 30 (2015), 136–49.

Raphael, D. D. (ed.), *A Review of the Principal Questions in Morals by Richard Price* (Oxford, 1974).

Rees, Abraham (ed.), *Ephraim Chambers Cyclopædia: or, An Universal Dictionary of Arts and Sciences* (London, 1778).

— *Cyclopedia; or, Universal Dictionary of Arts, Sciences, and Literature* (London, 1802–20).

Rees, George, *A Treatise on the Primary Symptoms of Lues Venerea, with a Concise Account of the English Writers on that Subject* (London, 1802).

— *A Treatise on Diseases of the Uterus* (London, 1805).

— *Observations on Spasms of the Stomach* (London, 1810).

— *A Treatise on Hemoptysis or Spitting of Blood* (London, 1813).

Rendle-Short, John, and Morwenna Rendle-Short, *The Father of Child Care: Life of William Cadogan (1711–1797)* (Bristol, 1966).

Risse, Guenter B., 'Explaining Brunonianism: A Biography of Edinburgh's Master of Conviviality', *www.researchgate.net/publication/344483633_ EXPLAINING_BRUNONIANISM_A_BIOGRAPHY_OF_ EDINBURGH'S_MASTER_OF_CONVIVIALITY* (accessed 27 May 2024).

Roberts, Brynley F., *Edward Lhwyd c.1660–1709: Naturalist, Antiquary, Philologist* (Cardiff, 2022).

Roberts, Gareth, and Fenny Smith (eds), *Robert Recorde: The Life and Times of a Tudor Mathematician* (Cardiff, 2012).

Roberts, G. T., 'Humphrey Edwards (1730–1788)', *Transactions of the Caernarvonshire Historical Society*, 25 (1964), 13–22.

Roberts, Gordon, *Robert Recorde: Tudor Scholar and Mathematician* (Cardiff, 2016).

Roberts, John, *Arithmetic: mewn Trefn Hawdd ac Eglur* (1768).

Roberts, Robert, *Daearyddiaeth* (Caerleon, 1816).

Robinson, Nicholas, *A Compleat Treatise of the Gravel and Stone with all their Causes, Symptoms and Cures, accounted for* (London, 1721).

— *A New Theory of Physick and Diseases Founded on the Newtonian Philosophy* (London, 1725).

— *A New Method of Treating Consumptions* (London, 1727).

— *A Discourse on the Nature and Causes of Sudden Death* (London, 1729).

— *A New System of the Spleen and Vapours and Hypochondriack Melancholy* (London, 1729).

— *A General Scheme for a Course of Medical Lectures Intended for the Improvement of Young Physicians, and Gentlemen* (n.p., n.d.).

Ross, J. E., *Radical Adventurer; the Diaries of Robert Morris, 1772–1774* (Bath, 1971).

Rowlands, Griffith, 'Observations on the Hydrocele', *The Medical and Physical Journal*, 8/41 (1802), 34–5.

— 'A Case of an Un-united fracture of the Thigh, cured by sawing off the ends of the bone', *Medico-Chirurgical Transactions*, 2 (1811), 47–51.

Rowlands, Henry, *An Account of the Origin and Formation of Fossil-Shells* (London, 1705); published anonymously.

— *Idea Agriculturae* (Dublin, 1764).

Roy, William, 'Experiments and Observations made in Britain, in order to obtain a Rule for Measuring Heights with the Barometer', *Philosophical Transactions of the Royal Society*, 67 (1777), 653–787.

Rudrum, Alan (ed.), with Jennifer Drake-Brockman, *The Works of Thomas Vaughan* (Oxford, 1984).

Rumsey, Walter, *Organon Salutis: an instrument to cleanse the stomach: as also divers experiments of the virtue of tobacco and coffee, how much they conduce to preserve humane health* (London, 1657).

Rusnock, Andrea, *The Correspondence of James Jurin (1684–1750)* (Amsterdam, 1996).

Rutty, John, 'On the Vitriolic Waters of Amlwch in the Isle of Anglesey', *Philosophical Transactions of the Royal Society*, 51 (1759), 470–7.

Salusbury, Thomas, *Mathematical Collections and Translations in Two Tomes* (London, 1661 and 1665). See also at Stillman Drake (ed.).

Schuyler, Robert Livingston, *Josiah Tucker: A Selection from his Economic and Political Writings* (Carmel IN, 2021), pp. 555–8. See 'Bibliography' for a full listing of Tucker's many writings.

Scur, Ruth, *John Aubrey: My Own Life* (2015; London 2016).

Shuckburgh-Evelyn, George, 'On the Variation of the Temperature of Boiling Water', *Philosophical Transactions of the Royal Society*, 69 (1779), 362–75.

Smith, Adam, *Inquiry into the Nature and Causes of the Wealth of Nations* (London, 1776).

Snyder, Henry, 'David Jones, Augustan Historian and Pioneer English Annalist', *Huntington Library Quarterly*, 44/1 (winter 1980), 11–26.

Somerset, Edward, *A Century of the Names and Scantlings of Inventions by me already practised* (London, 1663).

Steinicke, Wolfgang, 'William Herschel, Flamsteed Numbers and Harris's Star Maps', *Journal of the History of Astronomy*, 45/3 (2014), 287–303.

Stevenson, David (ed.), *Letters of Sir Robert Moray to the Earl of Kincardine* (Aldershot, 2007).

Switzer, Stephen, *An Introduction to a General System of Hydrostaticks and Hydraulicks Philosophical and Practical*, 2 vols (London, 1729).

Teignmouth, Lord, *Memoirs of the Life, Writings, and Correspondence of Sir William Jones* (London, 1804).

Thomas, Alban, 'Advertisement', *Philosophical Transactions of the Royal Society*, 26 (1708), 77–80.

— *List of the Fellows of the Royal Society* (1718).

Thomas, Ben Bowen, *The Old Order based on the Diary of Elizabeth Baker (Dolgelley 1778–1786)* (Cardiff, 1945).

Thomas, Beryl, and D. O. Thomas, 'Richard Price's Journal for the Period 25 March 1787 to 6 February 1791. Deciphered by Beryl Thomas with an Introduction by D. O. Thomas', *NLWJ*, 21/4 (1980), 366–413.

Thomas, D. O., John Stephens and P. A. L. Jones, *A Bibliography of the Works of Richard Price* (Aldershot, 1993).

Thomas, H. L., 'An anatomical description of a male rhinoceros', *Philosophical Transactions of the Royal Society*, 91 (1801), 145–52.

Thomas, Jenkin (a.k.a. Jenkin Thomas Philipps), *Historia Atheismi, breviter delineata à Jenkino Thomasio Cambro-Britanno, V.D.M. & M.D* (Basel, 1709).

Thomas, John, *Annerch i Ieuengtyd Cymru* (1795).

Thomas, Roland, *Richard Price: Philosopher and Apostle of Liberty* (London, 1924).

Thomas, Simon, *Hanes y Byd a'r Amseroedd* (1718).

Thomas, T. H., 'Sydenham Edwards of Usk', *Reports and Transactions, Cardiff Naturalists' Society*, 43 (1910), 15–19.

Toorians, Lauran, 'Jenkin Thomas Philipps, Every Inch a Welshman and a Poet Moreover', *Studia Celtica*, 56 (2022), 107–22.

Torrens, Hugh, '"Mineral Engineer" John Williams of Kerry (1732–95)', *Montgomeryshire Collections*, 84 (1996), 67–102.

Tucker, John V., 'Richard Price and the History of Science', *THSC*, 23 (2017), 69–86.

Tucker, Joseph E., 'John Davies of Kidwelly (1627?–1693), Translator from the French: with an Annotated Bibliography of his Translations', *The Papers of the Bibliographical Society of America*, 44/2 (1950), 119–52.

Tucker, Josiah, *The Elements of Commerce and Theory of Taxes* (privately printed, 1755).

— *Instructions to Travellers* (Dublin, 1758).

— *A Treatise concerning Civil Government* (London, 1781).

Turner, Dawson, and Lewis Weston Dillwyn, *The Botanists Guide Through England and Wales* (London, 1805).

Turton, William, *A General System of Nature*, 7 vols (Swansea, 1800–6).

— *Treatise on cold and hot baths: with directions for their application to various diseases. To which is added a letter to … the President and Managers of the Jennerian Society on the introduction and success of the cow pock in the principality of Wales* (Swansea, 1803).

— *British Fauna: Containing a Compendium of the Zoology of the British Islands: arranged according to the Linnaean system* (Swansea, 1807).

Vaughan, Henry, *Thalia Rediviva* (London, 1678).

Vaughan, Thomas, see Alan Rudrum (ed.) for the complete works.

Wakin, Peter, and Joan Day, 'Joseph Harris in Bristol, 1748', *Bristol Industrial Archaeological Society Journal*, 15 (1982), 8–12.

Walker, Raymond, 'The Dillwyns as Naturalists: Lewis Weston Dillwyn', *Minerva: The Journal of Swansea History*, 11 (2003), 20–42.

Walters, Gwyn, 'Richard Price and the Carmarthen Academy', *The Price-Priestley Newsletter*, 4 (1980), 69.

Walters, Gwynfryn, 'Bibliotheca Llwydiana: Notes on the Sale Catalogue (1816) of John Lloyd's Library', *NLWJ*, 10/2 (1957), 185–204.

Watkins, Francis, *A Particular Account of the Electrical Experiments hitherto made publick, with [a] Variety of new ones; and full Instructions for performing them: To which is annex'd The Description of a compleat Electrical Machine, and its Apparatus, with the Way of using it* (London, 1747).

— *L'Exercise du Microscope* (London, 1754).

Watkins, William, *A Treatise on Forest-Trees* (London, 1753).

Watson, William, 'A Collection of the Electrical Experiments Communicated to the Royal Society by Wm. Watson F.R.S.', *Philosophical Transactions of the Royal Society*, 45 (1748), 49–120.

Watt, Robert, *Bibliotheca Britannica; or a General Index to British & Foreign Literature*, 4 vols (Edinburgh, 1824).

Werner, Stephen, 'Abraham Rees's eighteenth-century *Cyclopaedia*', in *Notable Encyclopaedias of the Late Eighteenth Century: Eleven Successors of the Encyclopaédie* (Oxford, 1994).

Whiston, William, *A New Theory of the Earth: From its Original, to the Consummation of all Things* (London, 1696).

Willard, Thomas, *Thomas Vaughan and the Rosicrucian Revival in Britain 1648–1666* (Leiden, 2022).

Williams, Anna, *Miscellanies of Prose and Verse* (London, 1766).

Williams, Caroline, *A Welsh Family from the Beginning of the 18th Century* (London, 1893).

Williams, David, *Essays on Public Worship, Patriotism, and Projects of Reformation* (London, 1773).

— *A Treatise on Education. In which the general method pursued in the public institutions of Europe; and particularly those of England; that of Milton, Locke, Rousseau, and Helvetius are considered; and a more practical and useful one proposed* (London, 1774).

— *Lectures on Education. Read to a Society for Promoting Reasonable and Humane Improvements in the Discipline and Instruction of Youth*, 3 vols (London, 1789).

— *The History of Monmouthshire*, 2 vols (London 1796).

Williams, E. I., 'Samuel Baldwyn Rogers', *NLWJ*, 11/1 (1959), 99–102.

Williams, Edward, William Mudge and Isaac Dalby, 'An Account of the Trigonometrical Survey carried on in the Years 1791, 1792, 1793, and 1794, by Order of His Grace the Duke of Richmond, late Master-General of the Ordnance', *Philosophical Transactions of the Royal Society*, 85 (1795), 414–591.

Williams, Gryffith, *Truth Vindicated Against Sacrilege, Atheism, and Prophaneness* (London, 1666).

Williams, Jack, *Robert Recorde: Tudor Polymath, Expositor and Practitioner of Computation* (London, 2011).

Williams, John, *The Natural History of the Mineral Kingdom in Three Parts*, 2 vols (Edinburgh, 1789).

Williams, Mathew, *Speculum Terrarum & Coelorum: neu Ddrych Y Ddaear A'r Ffurfaven* (Aberhonddu, 1804).

Williams, Nathaniel, *Pharmacopoeia, Or Medical Admonitions, English and Welsh* (Trevecka, 1793).

Williams, Perrot, 'Observations upon dissecting the body of a person troubled with the stone' *Philosophical Transactions of the Royal Society*, 32 (1723), 326.

— 'Part of Two Letters concerning a method of procuring the small pox, used in South Wales. From Perrot Williams MD Physician at Haverford West, to Dr Samuel Brady, Physician to the Garrison at Portsmouth', *Philosophical Transactions of the Royal Society*, 32 (1723), 262–4.

— *Some Remarks upon Dr. Wagstaff's Letter Against Inoculating the Smallpox ...* (London, 1725).

Williams, Richard, *An Analysis of the Medicinal Waters of Llandrindod* (London, 1817).

Williams, T. P. T., 'The lost geology book of the Reverend Henry Rowlands of Llanidan', *Transactions of the Anglesey Antiquarian Society & Field Club*, (2007), 11–24.

Williams, Walter, *Discourse Preached at St. Paul's Covent Garden, on St. David's Day, March 1, 1731*, (n.p., 1731).

Williams, William, *Bywyd a Marwolaeth Theomemphus o'i Enedigaeth i'w Fedd/ The Life and Death of Theomemphus, from his Birth to his Grave* (London, 1764).

Williams, Zachariah, *The Mariner's Compass Completed: or, the complement of the art of navigation discover'd and propos'd* (London, 1740).

— *An Account of the attempt to ascertain the Longitude at Sea, by an exact theory of the variation of the needle. With a table of variations at the most remarkable cities in Europe, from 1660 to 1860* (London, 1755).

Winthrop, R. C. (ed.), *Correspondence of Hartlib, Haak, Oldenburg and Others of the Founders of the Royal Society, with Governor Winthrop of Connecticut, 1661–1672* (Boston MA, 1878).

Withey, Alun, *Physick and the Family: Health, Medicine and Care in Wales, 1600–1750* (Manchester, 2011).

Woodcroft, Bennet, *Subjects Matter Index (Made from titles only) of Patents and Inventions 1672–1815, Part II, N–W* (London, 1854).

Woodward, John, *An Essay Toward a Natural History of the Earth* (London, 1695).

Woolston, Thomas, *Six Discourses on Miracles* (London, 1727–9).

Wright, Richard, 'A Letter on the Same Subject, from Mr. Richard Wright, Surgeon at Haverford West, to Mr. Sylvanus Bevan, Apothecary in London', *Philosophical Transactions of the Royal Society*, 32 (1772/3), 267–9.

Wynne, Ellis, *Gweledigaetheu y Bardd Cwsc* (London, 1703).

Principal websites

Archives Wales, *https://archives.wales*

British Library, *www.bl.uk*

Biodiversity Library, *https://biodiversitylibrary.org*

Curious Travellers, *https://curioustravellers.ac.uk*

Dissenter Academies, *https://qmul.ac.uk/sed/religionandliterature/dissenting-academies*

Dictionary of Welsh Biography, *https://biography.wales*

Early Modern Letters Online, *http://emlo-portal.bodleian.ox.ac.uk*

Early Tourists in Wales, *https://sublimewales.wordpress.com*

Eighteenth Century Collections Online, via NLW

Google

Hartlib Papers at Sheffield University, *http://dhi.ac.uk/hartlib*

Hathi Trust, *www.hathitrust.org*

Internet Archive, *https://archive.org*

JSTOR, via NLW

Medical Heritage library, *www.medicalheritage.org*

History of Science Museum, University of Oxford, *https://hsm.ox.ac.uk*

National Library of Ireland, *www.nli.ie*

National Library of Scotland, *www.nls.uk*

National Library of Wales, *www.library.wales*

National Museum of Wales, *https://museum.wales*

Natural History Museum Collections, *https://nhm.ac.uk/our-science/collections.html*

Newton Project, *www.newtonproject.ox.ac.uk*

Oxford Dictionary of National Biography, *www.oxforddnb.com*

Royal College of Physicians, *https://history.rcplondon.ac.uk/inspiring-physicians*

Royal College of Surgeons, *https://rcseng.ac.uk/museums-and-archives/archives*

Royal Society Library, *https://royalsociety.org*

Science Museum, *https://sciencemuseum.org.uk*

Stanford Encyclopedia of Philosophy, *https://plato.stanford.edu*

Wellcome Library, *https://wellcomecollection.org*

Welsh Journals Online, *https://journals.library.wales*

Welsh Newspapers Online, *https://newspapers.library.wales*

Wikipedia

INDEX